PELICAN
A 420

THE ANCIENT EXPLORERS
M. CARY AND E. H. WARMINGTON

THE ANCIENT EXPLORERS

*

M. CARY

E. H. WARMINGTON

WITH FIFTEEN MAPS

PENGUIN BOOKS

Penguin Books Ltd, Harmondsworth, Middlesex
U.S.A.: Penguin Books Inc., 3300 Clipper Mill Road, Baltimore 11, Md
AUSTRALIA: Penguin Books Pty Ltd, 762 Whitehorse Road,
Mitcham, Victoria

—

First published by Methuen 1929
This revised edition published in Pelican Books 1963

—

Copyright © M. Cary and E. H. Warmington, 1929

—

Made and printed in Great Britain
by Cox and Wyman Ltd
London, Reading, and Fakenham
Set in Monotype Bembo

This book is sold subject to the condition
that it shall not, by way of trade, be lent,
re-sold, hired out, or otherwise disposed
of without the publisher's consent,
in any form of binding or cover
other than that in which
it is published

CONTENTS

LIST OF MAPS	7
PREFACE	9
1 INTRODUCTORY	11
2 THE MEDITERRANEAN AND THE BLACK SEA	21
3 THE ATLANTIC	43
4 INDIAN WATERS	73
5 THE CIRCUMNAVIGATION OF AFRICA	110
6 EUROPE	132
7 ASIA	157
8 AFRICA	202
9 RESULTS OF ANCIENT EXPLORATIONS	222
10 IMAGINARY DISCOVERIES	233
NOTES	247
INDEX	295

LIST OF MAPS

1	Europe and the Mediterranean	40–1
2	Coast of North-West Africa	65
3	Eastern Waters	84–5
4	Indian Waters	108–9
5	Africa	113
6	The Balkans and Lower Danube Lands	141
7	Central Europe	149
8	Eastern France and the Western Alps	154
9	Caucasus, Caspian and South-West Siberia	164–5
10	The Levant	170–1
11	The Near East	182–3
12	Asia	200–1
13	The Nile Basin	204
14	The Sahara	217
15	The World	223

PREFACE

THE authors have described in this book the opening up of the Old World by the discoverers of ancient times, and have given a view of ancient geography in the making. The book is intended as an introduction and a companion to, but not as a substitute for, the standard works of H. Berger (*Geschichte der wissenschaftlichen Erdkunde der Griechen*, 2nd ed.), H. F. Tozer (*A History of Ancient Geography*, 2nd ed.), J. O. Thomson (*History of Ancient Geography*), and E. H. Warmington (*Greek Geography*; in Dent's Library of Greek Thought). These works deal more particularly with the results of ancient exploration, and especially with ancient geographical theories; but they provide no full or connected account of the actual explorations which produced those results and gave birth to those theories. In the present work the authors have concerned themselves with the activities of ancient travellers rather than with the speculations of ancient scholars; but they have also traced the growth of geographic knowledge which sprang from these activities, and have explained their importance as a prelude to the discoveries of the fifteenth and sixteenth centuries.

Chapters 1, 2, 3, 6, and the second part of 8, are the work of Mr Cary, Chapters 4, 5, 7, and the first part of 8, of Mr Warmington; Chapters 9 and 10 are the result of their joint effort. The history of ancient sea-travels is shown in Chapters 2 to 5, while Chapters 6 to 8 tell of land-explorations continent by continent. Each author prepared the maps which illustrate his chapters and sections.

A general description of the seas and lands in which the activities of the ancient explorers took place will be found in M. Cary, *The Geographic Background of Greek and Roman History*.

The authors wish to thank all those connected with the printing and publishing of this book; and also Miss M. Price and Mr H. Jackson, of Mill Hill, for reading through the proof

PREFACE

sheets. Mr Warmington, moreover, thanks the Cambridge University Press for permission to make use of material in the text and the map of his book entitled *The Commerce between the Roman Empire and India*, published in 1928.

                                    M. CARY
                                    E. H. WARMINGTON

August 1929

CHAPTER I

INTRODUCTORY

FOREWORD

The history of ancient exploration must be written on somewhat different lines from that of medieval and modern discovery. Ancient pioneers had not altogether the same aims as those of later ages; their equipment varied in many respects from that of explorers in the Great Age of Discovery during the fifteenth and sixteenth centuries, and can hardly be compared with that of present-day adventurers in unknown regions. Furthermore, the materials available for writing the story of ancient discovery are very diverse from those out of which books on modern travel are composed.

1. OBJECTS OF ANCIENT EXPLORATION

A broad distinction may be drawn between the motives of ancient and of later discoverers. Those of later ages have been in a large measure idealists, those of antiquity had almost exclusively worldly aims. This of course is not to deny that the hope of commercial gain has run like a red thread through the history of exploration in all periods, nor that some notable voyages of discovery in modern times have been undertaken for a military or political purpose. Yet it remains a characteristic of medieval and modern exploration that it has largely been inspired by zeal for religion or for science. Among the pioneers of medieval wayfaring in unknown lands the typical figure is that of the religious missionary. The opening up of the routes to the Far East, which was the chief achievement of medieval Europe in the field of exploration, was largely the work of three friars, John of Carpini, William of Rubrouck, and Odoric of Pordenone; and the greatest of all medieval travellers, Marco Polo, albeit a layman and a merchant, set out with a commission from the Pope to explain the Christian doctrine to the emperor of Cathay. In the Great Age of Discovery the religious motive was not lost out of sight: to mention but a few leading names, Prince Henry of Portugal, Columbus, and Champlain mingled religious objects with secular. In the seventeenth and eighteenth centuries exploration was largely in the hands of missionaries, who ranged over the Roof of the World in Asia, and in America put California and the region of the Great Lakes on the map. In Central Africa Livingstone and Stanley alike were concerned to root out negro slavery in its last strongholds. Scientific observation was no doubt an incident rather than an avowed object of medieval exploration. On the other hand it has entered largely into the calculations of modern travellers since William Dampier went on a British man-of-war to study the geography of Terra Australis (1699); and ever since Captain Cook was sent to Tahiti (1768) to observe a transit of Venus it has become the predominant object of discovery.

In ancient times, conversely, missionary enterprise contributed nothing to the opening up of the globe. Religious propaganda played a far smaller part among ancient polytheistic societies than among Jews, Moslems, and Christians. Moreover, when ancient missionaries set out on their travels, they made their way, not to the outermost fringes, but to the central portions of the known world. Jews and Christians, votaries of Isis and Mithra and Cybele, all these sought their converts in the very centres of ancient civilization; and St Paul's famous journey past Derbe and Lystra to Iconium and Antioch was made along one of the most frequented high roads of the Roman Empire. Scientific interest, it is true, was not wholly lacking in ancient explorers. Among the Greeks, who created geography as a science (Chap. 9), sheer curiosity was doubtless a strong impelling force. One of the first travellers who visited Egypt for the sake of its sights was Solon of Athens. In a later age Greek pioneers in Arabia made a point of seeking out its archaeological remains (Chap. 7). Of Alexander the Great it is specially recorded that he reckoned new lore of plants and animals among the prizes of victory. Megasthenes studied the Hindus as Marco Polo observed the Cathayans. The stories of Pytheas and Eudoxus show that these voyagers at any rate had a scientific turn of mind (Chaps. 3 and 5). Among the Roman men of arms both Julius Caesar and Septimius Severus took advantage of their campaigns in Britain to measure the length of a midsummer day in high latitudes.[1] Yet pure love of science counted for little among

INTRODUCTORY

the Romans, and for nothing among the Phoenicians, whom nobody yet has accused of disinterested motives. Even among the Greeks, it must be confessed, the scientific traveller was an exception.

We shall therefore do ancient explorers no injustice in saying that their motives were pre-eminently mundane. The greater number of them had no object other than the pursuit of riches.[2] They were lured on alike by true reports of great fortunes acquired by actual visitors to foreign lands, and by fictitious tales of Eldorados beyond the ordinary range of travel.[3] This absorption in the quest of wealth need not be put down to greater acquisitiveness on the part of the ancients as compared with the moderns. It rests rather on the fact that two of the chief exploring nations of antiquity, the Phoenicians and the Greeks, were driven by hard circumstance to a life of adventure. Both these peoples had their homes in mountainous countries which hardly gave room for human habitation save on a ragged fringe of narrow coastal plains. On these small pockets of territory the limits of cultivation were soon reached, and surplus populations were forced, in the words of Professor Myres, 'to go either overseas or underground'. Under this pressure the Phoenicians and Greeks set out in search of wider expanses of cultivable land, of metalliferous areas, or of markets for international trade. Among the pioneers of modern world-travel they may most fitly be compared with the Portuguese and the Dutch, who experienced the same pressure of population on subsistence.

In the early modern era some notable discoveries have been made by bands of soldiers seeking new worlds to conquer: the names of Cortés and Pizarro will occur to every reader. In more recent times several regions of Central Africa have been explored by military officers with or without armies: Abyssinia by Lord Napier of Magdala, Nigeria by Sir Garnet Wolseley, and the Central Sudan by Captain Marchand. In South America the armies of Bolivar and San Martin have made journeys comparable with those of Alexander and Hannibal. Yet in the history of exploration in later ages there is nothing to compare with the systematic opening up of the European continent by the soldiers of Rome. Among the Greeks the list of military explorers is not so long as among the Romans, but it includes the name of Alexander, who upset the balance of an old world by calling in a new, and contributed more than any other ancient personage to the widening of the Mediterranean peoples' horizon. In fine, ancient explorers, with hardly an exception, were intent on some form of worldly success.

2. EQUIPMENT

Turning from ends to means, we shall find that the equipment of ancient travellers was considerably more primitive than that of the early modern discoverers, and not to be compared with that of present-day explorers. If we are to appreciate correctly the achievements of the ancient pioneers, we shall need to consider the deficiencies in their apparatus.

(a) *Land Travel.* In the means of locomotion on land the disparity between ancient conditions and those of the fifteenth and sixteenth centuries was comparatively slight. Previous to the invention of the motor-car and aeroplane, explorers on land have merely had the option of walking or of riding on pack-animals. In pedestrian prowess the men of antiquity were no whit inferior to those of subsequent ages. The normal march of a heavily laden Roman soldier was estimated at twenty Roman or eighteen English miles, and trained messengers were expected to cover fifty Roman or forty-four English miles day in day out.[4] Neither was the stock of animals which accompanied ancient explorers appreciably poorer. The ordinary breeds of Mediterranean horses were of small stature, in fact they might almost be called ponies, but ponies when used by modern explorers have proved well suited to long-distance travel under trying conditions of climate and food-supply.[5]

(b) *Travel by Sea.* For sea voyages the ancients used sailing ships which seldom were of greater size than 250 tons burthen. As a rule these craft carried only one low mast with one broad sail of square rig, and a square foresail on the bowsprit.[6] Their steering gear consisted of a paddle or of a linked pair of paddles. Given the squatness of their lines, and the deficiencies of their canvas, they could attain but a moderate speed: an average of 500 stades, say 55 nautical miles, in a summer day was all that was expected of them. To us, who think in terms of Mauretanias, or maybe of Cutty Sarks, the ancient vessels may appear absurdly small and slow; yet they will bear comparison in most respects with the ships of the early modern period. In point of speed the ships of the Great Age of Discovery had far less advantage over those of the Phoenicians and Greeks than the clippers of the eighteenth and nineteenth centuries possessed over the craft of the fifteenth and sixteenth.[7] The chief defect of the ancient vessels lay in the unhandiness of their steering-gear and rigging, which would not allow them to sail close to the wind or to tack across it.[8] A good example of this lack of manoeuvring power is furnished by the well-known story of St Paul's voyage to Rome. Caught by a gale on his beam, the skipper endeavoured to turn into the wind, but his craft fell away from it and drifted dangerously with her broadside exposed to the seas.[9] The ships of Columbus's day, which carried a mizzen-mast and lugsails thereon, could no doubt travel closer hauled than those of antiquity. On the other hand, ancient ships, with their broad and shallow hulls, would ride comfortably on a moderate swell, and could be beached without much difficulty.

(c) *Measuring Instruments.* A far more serious handicap to ancient explorers lay in their lack of adequate appliances for measuring distance or time, and of instruments for the taking of bearings. In the computation of mileage on land early travellers used the primitive method of counting steps: hence the name of 'bematistae' or 'steppers' for Alexander's road-surveyors.[10] This system, it is true, could be made to yield reasonably accurate results. On the other hand ancient navigators, not having discovered the use of the log, could only measure their progress by dead-reckoning from the supposed mean daily speed of their craft, and by making allowance for the varying conditions of wind and current. This rough-and-ready method gave rise to some grossly erroneous estimates of naval distances. Herodotus overstated the length of the Black Sea by fully two-thirds. Pliny nearly doubled the true length of the Sea of Azov. Nearchus exaggerated the mileage of his Indian Ocean cruise, and Pytheas the circumference of Britain, in a similar proportion. It is true that in this respect the ancient sailors were no worse off than those of early modern times: Columbus judged, or misjudged, his rate of progress by eye alone.[11]

For the measurement of time the sundials[12] of the ancients were about as useful as clocks that stop at nightfall and whenever the sun goes in. Their 'clepsydrae', or water-clocks, could at least be kept going continuously, but they were rather cumbrous and required frequent refilling. Julius Caesar used such an instrument for measuring the interval between dusk and dawn in an English midsummer. But we do not know whether such timepieces were habitually carried by travellers. The clocks of the later Middle Ages, with all their friction and unevenness of motion, were a great improvement upon the best of ancient instruments.

For the taking of latitudes the ancients used a simple pointer which cast its shadow on a disc or a hemispherical bowl. This instrument, if accurately suspended in a vertical plane, must have given readings as good as those of the early modern astrolabe, though of course it was less precise than the present-day sextant. In the hands of men of science like Eratosthenes and Hipparchus it yielded important contributions to geographic knowledge; but it does not appear to have been extensively used by practical travellers. Hence the great majority of figures of latitude in Ptolemy's Geography could not be derived from direct measurement of the sun's inclination, but had to be inferred from the observed length of a midsummer night or from the data of maps and itineraries. The fixing of longitudes remained throughout antiquity a mere matter of dead-reckoning. The astronomer Hipparchus, it is true, made the brilliant suggestion that the relative longitude of two places could be ascertained if observers at either place were to note the moment of first obscuration in a lunar eclipse moving from west to east;[13] but this method was never applied in practice. The comparison of the local noon-time with that of a standard meridian, which has recently been perfected by the use of wireless signals,[14] was not even thought of by ancient travellers. The suggestion put forward by Columbus, that longitude could be determined by observing variations in the declination of the magnetic from the true north, was simply inconceivable to the ancients, who knew nothing of a magnetic as distinct from an astronomic pole.

The ignorance of the ancients in this last respect illustrates one of their worst deficiencies in the matter of equipment, their lack of the compass.[15] The magnetic properties of iron had not indeed escaped their attention,[16] but they were not put to any practical use. For the

INTRODUCTORY

purpose of finding their bearings ancient travellers had no recourse but to sun, stars,[17] and winds. In the Mediterranean regions these are not, to be sure, such treacherous guides as in our British latitudes. The yearly ration of sunshine in the Mediterranean seldom falls below 2,000 hours; in summer more particularly one almost cloudless day follows another, and the night skies are correspondingly clear. In summer the Mediterranean winds shift according to simple and regular laws, thus facilitating the use of wind-points as means of orientation. In Homer's time the Greeks made use of four wind-points, in the days of Aristotle they used no less than twelve.[18] But at best the ancients possessed no real substitute for the compass, no device that was always available and nearly always accurate. The difficulties into which they accordingly fell may be plentifully illustrated from faulty orientations in ancient geographers. The cases in which these are 45 degrees out of their reckoning are so numerous as not to be worth collecting; in some instances the error amounts to a full right angle. Thus Herodotus imagined the pass of Thermopylae as running from north to south,[19] and Strabo described the Pyrenees as extending in the same direction.[20] Under such conditions ancient travellers proceeding without some familiar landmark to guide them stood in danger of losing their way irretrievably. A good illustration of this risk occurs in the late Roman historian Priscus, who relates how a party of Roman envoys travelling by night through an unknown region to the camp of Attila found themselves face to face with the rising sun when they thought they were proceeding westward.[21] A similar mistake was made by Xenophon and the Ten Thousand when they marched eastward along the river Phasis in an attempt to reach the Black Sea.[22]

(*d*) *Food and Drink*. The question of victuals, always a difficult one for explorers, was more formidable in antiquity than at the present day, though perhaps no worse than in early modern times. The ancient Mediterranean peoples, being accustomed to a more or less vegetarian diet, could cheerfully endure long successions of meatless days, and were hardly the worse for their lack of concentrated foods. If we may judge by the rations of Roman soldiers, which consisted of two pounds (about twenty-six ounces on the average) of wheaten flour and of very little else, explorers could have kept up their strength on

17

a diet composed mainly of cereals, i.e., of food that is not specially difficult to preserve in good condition. The principal adjunct to the bread ration would no doubt be olive oil, which keeps better than most edible fats. The lack of anti-scorbutic articles such as potatoes and fresh fruit which caused havoc among modern ships' crews before the days of Captain Cook, may be reckoned as the most serious deficiency in the menu of ancient travellers. But for these the problem of drink was probably more troublesome than that of food. The thick earthenware jugs in which they commonly conveyed their wine or water, being unglazed and slightly porous, no doubt allowed of sufficient evaporation to keep their contents cool, but their deadweight was very great as compared with that of the modern iron or aluminium containers. On the other hand their wooden pitch-lined chests,[23] though light, would not keep the water palatable for long. The lack of adequate provision of water on shipboard was a serious handicap to naval explorers as late as the eighteenth century; on the smaller craft of the ancients the deficiency must have been even more troublesome.[24] Lastly, the pharmacopoeia of the ancients, though fairly elaborate,[25] was not equal to the manifold needs of exploring parties. Considering how often a few doses of quinine have saved the lives of modern pioneers, we may be sure that the lack of this specific in particular brought many an ancient traveller to grief.[26]

(*e*) *Weapons*. But of all disadvantages under which ancient explorers laboured, the greatest was their deficiency in the matter of weapons. The margin of strength which the more civilized peoples of antiquity possessed over the more primitive ones in regard to lethal instruments was so narrow that exploring parties must often have been more or less at the mercy of the folk whose countries they discovered. Even with the possession of firearms the earlier modern settlers in North America had some difficulty in holding their own against the more bellicose of the Indians. The Norsemen who settled in those parts about A.D. 1000 were definitely unable to make good their footing wth such means of defence as they possessed, and had to abandon their otherwise promising colonies. Had some ancient mariner anticipated Columbus in the discovery of America, he would have run a big risk of being eaten for his pains by Caribbean cannibals. A lesser but still appreciable handicap for the ancient adventurers was their inability to distil fifty-per-cent 'fire-water'. Greek slave-traders knew full well the value of new wine as a means of persuading the backward tribes on the 'Guinea Coasts' of antiquity, but they had no such quick-acting weapons as the 'anchors of brandy', with which modern kidnappers used to work their will on their negro victims. When ancient travellers encountered the hostility of native populations, they had no such easy play as Cortés and Pizarro among the ill-armed inhabitants of Central and South America, or the British among Australian Blackfellows; indeed, they might find themselves facing impossibly long odds.

From the above considerations it may be gathered that many an ancient voyage which at first sight appears a simple undertaking will show up on further thought as a quite respectable achievement.

3. HISTORICAL MATERIALS

The story of ancient explorations cannot be compiled out of the journals and reports of the actual travellers, which have only survived in the rarest instances. It has to be pieced together out of scattered allusions and descriptions in the general literature of Greece and Rome. This task is rendered none the easier by the fact that our Greek and Roman sources were sometimes lacking in the requisite care and understanding, so that their accounts are not always clear and trustworthy. Still more difficult is the interpretation of the various archaeological materials (coins, pottery, bronze-ware, etc.) with which our scanty literary records have to be supplemented; the value of these indeed can only be assessed by experts. Under these conditions it is impossible to offer to the reader such interesting and arresting quotations as may be culled, e.g., from the *Travel Book* of Marco Polo, Nansen's *Farthest North*, and the Journals of Stanley or of Captain Scott: the heroism and the resourcefulness of ancient discoverers can only be appreciated by an effort of imagination. It is equally impossible to conceal from the reader all the difficulties of interpreting the ancient evidence: the smooth flow of the narrative must therefore be interrupted by occasional passages of a controversial nature.

But even so, the general outlines of the history of ancient exploration stand out clearly enough, and the tale which they set forth needs no commendation. The contribution of the ancients to the opening up of the world was by no means negligible. Therefore all those who attach importance to geographic discovery as one of the chief factors in world history will desire that the record of ancient exploration should be preserved.

CHAPTER 2

THE MEDITERRANEAN AND THE BLACK SEA

1. GENERAL FEATURES OF THE MEDITERRANEAN SEA

THE Mediterranean Sea was one of the earliest schools of navigation, and the chief training-ground for naval exploration among ancient peoples. Though not uniformly favourable to the primitive mariner, it was on the whole well designed to lure him out and to give him confidence. The most forbidding feature of the Mediterranean is the frequency of its winter gales, which change direction rapidly and create incalculable cross-seas in confined waters.[1] In ancient times these gales effectually closed the seas for all ordinary sailings from October to April, so that the season for navigation was confined to summer. Another discouragement to early mariners lay in the fact that Mediterranean rivers, unlike those of the Atlantic seaboard, do not provide a suitable preparatory school of seamanship. With the exception of the Nile, these waterways are ill-suited for travel, and their mud-choked estuaries are a positive danger to shipping. Nevertheless, the Mediterranean entices more than it deters. In summer a brisk breeze from a northerly quarter blows with the regularity of a trade-wind. Inshore navigation is facilitated by the general absence of tideways, shoals, and breakers, by the rarity of strong currents, and by the alternation of land and sea breezes with which a coaster can help himself along when the open-sea wind fails him. Harbourage is plentiful under the lee of numerous headlands and islands. Best of all, these headlands and islands serve admirably as direction-posts. Unaided by compass or sextant, a seaman may sail the whole length of the Mediterranean without losing his bearings.[2] He can pick his way by an unbroken series of landmarks, which in the general absence of summer cloud or fog loom up clear at a distance of fifty miles or more. A Mediterranean explorer, therefore, is like a wayfarer on land who follows out a well-defined path. Under such conditions it is not surprising that the 'wet lanes' (as Homer aptly called them) of the

Mediterranean should have been traced out at an early stage of human history.

2. EARLY EGYPTIAN VOYAGES

A Greek tradition recorded that 'Danaus the Egyptian had invented fifty-oared ships', and some modern scholars maintain that sea-going ships of all kinds are derived from Egyptian prototypes.[3] However this may be, the Egyptians certainly mastered the art of shipbuilding, and plied on the Nile by oar and sail, at a very remote date. Their earliest recorded voyage by sea took place under a Pharaoh named Snefru, about 3200 B.C. This ruler commemorated the 'bringing of 40 ships of 100 cubits with cedar-wood' from Byblus in Phoenicia.[4] Henceforth the Egyptians probably maintained regular intercourse with the Lebanon coast, whose forests furnished better ship-timber than the Nile valley. Perhaps they also initiated that naval commerce with Crete whose first beginnings can be traced back into the fourth millennium.[5] But it is far from certain that the Egyptians were the active partners in this traffic (see p. 24). In any case, their interest in seafaring gradually declined, and in the course of the second millennium B.C. they became an essentially home-staying people, who retained but a vague knowledge of Crete as 'an isle of the sea', and described Asia Minor as a 'cluster of islands'.[6] Though the Egyptians were competent ship-masters, they had little need and less desire for maritime adventure. Their slender appetite for foreign conquest was appeased by the land campaigns of the eighteenth-dynasty Pharaohs in Syria (c. 1500 B.C., see p. 159). Their need for overseas products was not extensive, and as soon as the foreigner undertook to deliver these at their doors they left the Levantine carrying-trade in his hands. It is true that about 600 B.C. an Egyptian king planned the boldest of all ancient voyages of exploration; but significantly enough he entrusted its execution to alien seamen (see pp. 111–112.).

3. SARGON OF AKKAD

In a history of Mediterranean exploration one does not expect to find names of kings from the Euphrates lands. Yet one Mesopotamian ruler, Sargon of Akkad (*c.* 2800 B.C.), has left a record of naval adventure which requires a passing notice at this point. Sargon apparently claimed that he 'crossed the sea in the west' and 'thrice conquered Anaku-ki (Tin-land?), Kaptara-ki (Crete?), and the lands beyond the upper sea.[7] The allusion to Tin-land (if such it was) suggests that Sargon extended his conquests to the gates of the Atlantic. Nay more, it has been maintained, on the strength of supposed Babylonian place-names in Spain, that this country was colonized from Mesopotamia.[8] But the precise meaning and whereabouts of Kaptara-ki and Anaku-ki are uncertain; the Babylonian derivations of Spanish names are extremely hazardous; and the antecedent improbability of Spanish conquests by Mesopotamian rulers is enormous. The most that can be asserted on Sargon's behalf is that he may have made naval raids from Phoenicia or Cilicia with Cyprus as his probable objective.[9] On the other hand it is possible that his exploits do not belong to the Mediterranean at all, but to the Persian Gulf.[10] In any case Sargon made no serious contribution to Mediterranean exploration.

4. THE MINOANS

The typical advantages of the Mediterranean Sea as a fairway of traffic are nowhere more clearly exhibited than in the Aegean area. Steady summer winds blowing across a clear sky, a lavish development of coast-line, and a profusion of islands that almost reduce navigation to a simple ferry-service, make the call of the Aegean well-nigh irresistible. And in no other region of the Mediterranean do growing populations feel more acutely the law of diminishing returns, or realize sooner the need of new homes or new markets overseas. In the Aegean area sea-travel was probably as old as the sea-going ship.

Of prehistoric navigation in their seas the Greeks preserved no more than a dim tradition. Yet they remembered one Minos, lord of Cnossus in Crete, who had exercised a 'thalassocracy' or dominion of the seas before the Trojan War. The clue thus provided by ancient legend has been followed up by modern archaeologists, who have verified and amplified Greek tradition. Excavations by Sir Arthur Evans at Cnossus, and by other scholars at various Aegean sites, have disclosed a pre-Hellenic or 'Minoan' civilization which had come into being by 3000 B.C. and reached its height in the second millennium. This civilization was plainly based on maritime intercourse, which was largely, if not wholly, in the hands of Aegean seamen. The ships of the Minoan sailors are represented on Cretan seal-stones. Unlike the Egyptian boats, whose lines always recalled those of the river-punts out of which they developed, Aegean vessels were built with a keel and ribs and carried a high forecastle.[11] This type of ship was better suited to open-sea navigation than the magnified rivercraft of the Egyptians.

The Minoans have left numerous traces of their maritime ventures in foreign waters. Intercourse between Crete and Egypt, as we have seen (p. 22), was established in the fourth millennium. The question whether the Egyptians first found Crete or the Minoans first explored Egypt cannot be settled on present evidence. But the increasing abundance of Cretan pottery on Egyptian sites of the twelfth and of the eighteenth dynasty suggests that by 2000 B.C. at any rate the Minoans had become the active partners.[12] The voyages of the Minoans to Egypt are also attested by Egyptian frescoes of c. 1500–1450 B.C., in which foreigners of a distinctively Cretan type are represented as bringing typically Cretan objects of merchandise.[13] It has even been suggested that Minoan navigators instigated the construction of a jetty of 700–800 feet, and of a breakwater of $1\frac{1}{4}$ mile, of which traces have recently been found near Alexandria.[14] More probably these were the original (but soon discarded) harbour-works of Ptolemaic Alexandria (c. 300 B.C.).[15] On the other hand we may recognize Minoan craft in the 'Keftiu ships' which served the conquering Pharaoh Thothmes III on his expeditions to Syria (c. 1475 B.C.),[16] for the Keftiu are probably to be identified with the Minoans.[17] Presumably Cretan ships first made their way to Egypt by heading straight for the African coast and following it to the Nile delta; and it is not unlikely that their usual starting-point for this trip was a harbour on the south coast, which was connected with Cnossus by a well-constructed road.[18] There is no evidence of travel by the alternative route past Cyprus and Phoenicia before 2000 B.C. But from this time Minoan objects begin to make their appearance on these intermediate stages, and from 1500 B.C. Aegean pottery became so common on Cyprus as to suggest colonization.[19]

The Minoans therefore may be credited with the first extensive exploration of the Levant. Did they fulfil a similar function in the western Mediterranean? Greek legend associated king Minos with Sicily, and the modern archaeologist has on this point also confirmed the ancient tradition. Some of the evidence brought forward to prove Minoan influence in the western Mediterranean is indeed open to dispute; but enough remains to establish a strong *prima facie* case.[20] The blocks of liparite in the palace of Cnossus were undoubtedly quarried in the Aeolian Islands.[21] Plates of raw copper, of a type which served the Minoans as money, have been discovered in Sicily and Sardinia.[22] The standard forms of copper daggers and bronze swords which spread from Italy into Central Europe were probably of Minoan origin,[23] and the bronze double-axes of the Balearic Islands reproduce a distinct Minoan shape.[24] In Sicily and South Italy Minoan pottery of the period 1400–1100 B.C. has been unearthed,[25] while in southern and eastern Spain native imitations of this type are not uncommon.[26] There is as yet no sure proof of Minoan voyages into the Gulf of Lions or into the Adriatic, though it is not unlikely that amber was brought from Venetia or Illyria on Cretan vessels.[27] In any case it is clear that in the second millennium at the latest Aegean sailors passed the Straits of Messina; it is not unlikely that they paid visits to Spain, and they may even have travelled beyond the Straits of Gibraltar to the tin and silver emporium of Tartessus, which lay at or near the mouth of the Guadalquivir.[28]

In the course of the second millennium the greater part of the Mediterranean lay open to the Minoan navigators. But by 1000 B.C. the seafaring activity of the Aegean peoples had been suspended. This interruption was due to the general unsettlement of the Aegean area by a succession of immigrant peoples, of whom the Achaeans and the Dorians were the most notable. Of these new-comers the Achaeans adopted many of the outward marks of Aegean civilization; they followed the Minoans westward and eastward, and established colonies in Asia Minor, Cyprus, and South Italy.[29] According to some scholars they imitated or even supplanted the Cretans in their overseas trade.[30] But the Achaeans were rovers and corsairs rather than merchants. In the poems of Homer their predatory habits are but half concealed; in the annals of the Egyptians and the Hittites they are shown up without reserve.[31] Moreover, if the Achaeans were no worse than the Normans or the Crusaders, the Dorians were no better than the Germans who broke up the settled conditions of the Roman Empire. Under such circumstances the Aegean area relapsed into a Dark Age, and the results of Minoan exploration were temporarily lost.

5. THE PHOENICIANS

At the beginning of the first millennium B.C., while maritime enterprise lay dormant in the Aegean area, Mediterranean exploration was carried on by the Phoenicians. Among the pioneers of ancient civilization the Phoenicians are the mystery people *par excellence*. Though they played an important part in the development of the alphabet, they have left singularly few records of themselves; and in the numerous lands which they visited their material leavings are surprisingly scanty. Hence the date and the scope of their explorations have been the subject of much discussion.[32]

In the literature of other ancient peoples the Phoenicians are made to appear almost ubiquitous. In the Old Testament their voyages on 'Tarshish ships' (ships that ply to Tartessus) find frequent mention.[33] Besides, Greek and Roman writers assign to them a preponderant part in the opening up of the Mediterranean. Homer and Herodotus attest their presence in Greek waters at the dawn of Greek history. Thucydides implies that they had colonized the Aegean area at an even earlier date than king Minos. A well-known legend represented Cadmus, the founder of Thebes, as a Phoenician. Thucydides further remarks that the Phoenicians preceded the Greeks in the colonization of Sicily; various Greek and Latin authors date their settlements in North Africa and Spain back to the second millennium; and Strabo declares his belief that the Odyssey was based on Phoenician records of travel.[34] To these statements modern scholars have added the testimony of Semitic place-names, which they claim to have discovered in almost every corner of the Mediterranean. The ablest and most forcible exposition of the Phoenician case will be found in M. Bérard's learned and ingenious book, *Les Phéniciens et l'Odyssée*,[35] where Strabo's theory of a Phoenician source for the Odyssey is developed and fortified with a mass of etymology. According to M.

Bérard the exploration of the Mediterranean was begun as well as completed by the Phoenicians; the 'Keftiu' of Egyptian records came from Phoenicia, not from Crete; the Cretans were a stay-at-home people whose wine and oil, the foundations of their prosperity, were traded on their behalf by the Phoenicians; Minos himself was possibly a Phoenician who exercised an alien thalassocracy in Aegean waters.

But a reaction against 'Phoenicomania' has duly set in.[36] Some critics have objected that the arguments from place-names are hardly ever conclusive; others have pointed out that Homer nowhere suggests anything like a Phoenician trade-monopoly; others again argue that Thebes was a most unsuitable station for sailors from overseas,[37] and that the arrangement of the legend in which Cadmus is converted into a Phoenician was a mere afterthought.[38] But the chief weapon of the Phoenicophobes is derived from archaeology. The extreme rarity of material remains betokening Phoenician navigation before 1000 B.C., which stands in sharp contrast with the abundance of Minoan relics, can hardly be explained away as fortuitous. The scale on which excavation on the prehistoric sites of the Mediterranean has been carried out practically rules out such a play of chance. Hence it is now generally agreed that much of the credit which used to be given to the Phoenicians for opening up the Mediterranean must be transferred to the Minoans.

But with the decline of Minoan seamanship the Phoenicians found their opportunity. Though the Phoenician seaboard is not particularly well suited to navigation, being frequently buffeted by on-shore winds, it possesses two good harbours in Sidon and Tyre; and its hinterland makes up for scanty crops with abundant ship-timber. The first clear evidence of Phoenician sea-voyages is found on an Egyptian fresco of c. 1475 B.C., which depicts a vessel manned by sailors of an unmistakably Semitic type.[39] Further testimony from a papyrus of c. 1200 B.C. shows that by then the Phoenicians dominated the commerce of the Levant.[40] Their appearance in the Aegean may be assigned to a similar date. In the *Odyssey* the Phoenicians are represented as touching in at Greek ports soon after the fall of Troy (traditional date c. 1190 B.C.; in any case before 1100 B.C.); and there is no reason to suppose that Homer is here transferring the conditions of his own age (c. 800 B.C.) to the period of the Trojan War. On the

THE ANCIENT EXPLORERS

other hand it is not necessary to suggest that the Phoenicians supplanted the Minoans in the Aegean or the western Mediterranean as early as 1400 B.C.[41] The sack of Cnossus and the establishment of Achaean overlordship in Crete which took place about that time did not necessarily entail the cessation of Minoan seafaring. Indeed sporadic finds of imported pottery from the eastern Mediterranean of the period 1100–800 B.C. (mostly in eastern Sicily and Etruria) indicate rather some lingering traffic with the Aegean area, and especially with Crete and the Cyclades.[42]

In the western Mediterranean the Greeks credited the Phoenicians with having occupied the best parts of Libya and Iberia before the age of Homer (say, before 850 B.C.), and they dated their oldest colonies, Utica and Gades (Cadiz), at 1101 and 1110 respectively.[43] These dates may have been derived from Phoenician sources, and if, nevertheless, we decide to reject them as guesswork, we cannot similarly dismiss the references in the Old Testament to the Tarshish voyages of the Phoenicians, the oldest of which, undertaken by King Hiram of Tyre, must be placed not long after 1000 B.C.[44] On the other hand no Phoenician objects of any period have been found in eastern Sicily, and their occurrence in the western half of the island cannot be attested before c. 800 B.C.[45]

It is probable that the earliest Phoenician objective in the West was Spain rather than North Africa, for the Far West, with its great wealth of metals, no doubt exercised a greater attraction than the Middle West, which was merely agricultural land. Presumably the Phoenicians first followed an old Minoan sea-track from Sicily to Spain, and subsequently searched out the route along the African coast. From Utica, or from Carthage, a later settlement, which was founded c. 800 B.C.,[46] they eventually thrust across to western Sicily, and from this base they made their entry into the Tyrrhenian Sea, where they colonized Sardinia and visited the ports of Etruria.[47] But they left the head of the Gulf of Lions to the Greeks to explore and occupy, and their settlements on the eastern coast of Spain were probably sparse.[48]

The explorations of the Phoenicians, as described above, were on a more modest scale and of later date than was formerly believed. Even so, if we bear in mind their relatively small man-power, and their eccentric position at the extreme end of the Mediterranean,

THE MEDITERRANEAN AND THE BLACK SEA

which handicapped them as compared with the Minoans and the Greeks, their record of exploration is a notable one. And in the absence of any certain information as to the limits of Minoan sea-travel, we are entitled to assert that the Phoenicians were the first people, to our knowledge, to open up the Mediterranean from end to end.

6. THE ODYSSEY

At the beginning of the first millennium B.C. the inhabitants of Greece, out of whom the Greek nation was then being formed, had settled down to a stay-at-home existence. But the same causes that had lured the prehistoric Aegeans out upon the water also helped to cure their successors of their original sea-shyness.[49] Moreover the memory of past explorations probably never died out in the Aegean area. A reminiscence of these may be discerned in Homer's *Odyssey*, a poem whose kernel may be assigned to the ninth or eighth century. Not that Homer inherited much information about Mediterranean topography. In the Levant, Cyprus, Phoenicia, and Egypt are little more than names to him. In the western seas the scenes of Odysseus's adventures will not fit into any geographical framework. Despite the plausible and seductive identifications which have been proposed by Homerists ancient and modern,[50] the majority of them remain as it were in the air. Calypso's Isle, the Isle of the Sun, the lands of Circe and of the Cyclops can be located, if considered singly, on a whole variety of sites;[51] but to plot Odysseus's course from one to any other is little better than a game of chance.[52] Yet here and there Homer reveals a knowledge which is not born of mere fancy. His lotus plant is a real speciality of the coast of Tripoli;[53] his Aeolian Islands possess the distinctive features of the Lipari group;[54] his land of the Phaeacians is reminiscent in various details of Drepanum and the Aegates Islands off north-west Sicily.[55] But more important than these points of detail is Homer's general description of wind and weather conditions in Mediterranean waters. Unlike the story of the 'Egyptian Sindbad' or the Babylonian epic of Gilgamesh, in which the sea is little more than a stage property (p. 234), the *Odyssey* smells of brine and seaweed and ozone.

The truth to sea-life which we find in Homer suggests that he drew upon some tradition of actual seamanship. M. Bérard, who more than any other writer has taught us to appreciate the underlying realism of the *Odyssey*, contends that Homer's source was a corpus of Phoenician sailing directions. Given a Phoenician source, it seems in any case preferable to assume that it was a collection of sailors' yarns. M. Bérard himself admits that the *Odyssey* does not in the least resemble an official sailing-guide in point of form.[56] Besides, it is uncertain whether the Phoenicians had issued any 'Mediterranean Pilots' in Homer's day; and if they had, they would assuredly have kept them secret, as (on M. Bérard's own showing) the Portuguese and Dutch kept theirs in the sixteenth and seventeenth centuries. But is it necessary to postulate for Homer a Phoenician source of any kind? Is it not possible that he drew upon indigenous Aegean tales? The west coast of Asia Minor, which by common consent was the place in which the core of the *Odyssey* was composed, was also the chief repository of the lost Minoan culture.[57] The *Odyssey* therefore may be taken as evidence that the Minoan traditions of seamanship had not wholly died out in the Aegean area, but lingered on to be quickened into fresh life by the Greeks. This apparently haphazard way of filling the gaps in the settlement areas was not wholly due to the accidents of navigation. It is largely to be explained by the nature of the welcome which the explorers received from the populations in possession. Thus, although the coast of Etruria was occasionally visited by Greek traders of the ninth and eighth centuries, no attempt was made by them to find new homes here, or in the neighbouring Latium. The subsequent history of Central Italy sufficiently explains this aloofness.

7. THE GREEK COLONIAL MOVEMENT

Be it cause and effect, or mere coincidence, it was in the age of Homer (*c.* 800 B.C.) that the Greeks, as we may henceforth call the inhabitants of the Aegean lands, took up the heritage of the Minoans and engaged in their first naval ventures. By a fortunate chance the Phoenicians at this time became preoccupied with self-defence against the war-lords of the Asiatic hinterland, and could only offer a limited and belated

THE MEDITERRANEAN AND THE BLACK SEA

resistance to the intruders upon their thalassocracy. In the eighth, seventh, and sixth centuries the Greeks were able to carry out a movement of colonial expansion by which they became wedded for better for worse to the sea.[58] This movement, moreover, was closely bound up with voyages of discovery which preceded their earliest settlements and were in turn called forth by these.

According to a common tradition the Greeks owed much of their success in finding good sites for settlement overseas to the oracle of Apollo at Delphi. It is true that previous to the foundation of a colony Greek emigrants made a point of consulting an oracle, and there are actual instances in which the priests of Apollo had a voice in the selection of the site, as at Cyrene (pp. 33–4) and at Byzantium. But as a rule the oracle merely sanctioned a choice of site already accomplished.[59] The real work of exploration should rather be credited to individual merchants or fishermen or corsairs, who put to use some lingering memory of Minoan sea-travel, or spied out a Phoenician track, or, trusting to luck, were blown by a discerning wind, or drifted on a sagacious current, to places which repaid a visit. In view of the large number of Greek colonies which were primarily fishing settlements, it may be suggested that many eligible sites were discovered by trawlers venturing out beyond their usual sphere of operations. As might be expected, Greek adventurers succeeded no more than modern discoverers in always happening straightway upon the nearest or best spot for settlement; the process by which they filled the vacant spaces on the Mediterranean seaboard was anything but systematic.

Of Greek exploration in the Levant scarcely any information has reached us. In view of the decline of Egypt in the first millennium B.C., intercourse with this country was now of less importance than in the days of the Minoan thalassocracy. On the other hand Egypt was no longer difficult to find. Cyprus, which was still inhabited by descendants of the Minoan or Achaean colonists, formed a convenient half-way house from Greece; and it is not unlikely that piratical descents on the Nile delta, such as had been a feature of the period of Achaean ascendancy, went on being made by stray corsairs from the Aegean, so that the routes to Egypt were never wholly lost out of sight.[60] In any case, relations between Greece and Egypt had been

31

resumed by 650 B.C., for at this time we find Greek traders and soldiers of adventure established in a colony at Naucratis on the Delta.[61]

A much fuller record of the exploration and settlement of the region of Cyrene has been preserved for us by Herodotus. This district is encompassed on either side by sterile land and is flanked to westward by a gulf, known in antiquity as Syrtis Major, where a shallow coast, on-shore winds, and a comparatively strong tide created risks that were unfamiliar to Mediterranean seamen.[62] Hence it appears not to have been frequented by the Minoan and Phoenician seafarers, and its discovery and colonization by the Greeks was in the nature of a real adventure. The gist of Herodotus's story is as follows.[63] The king of the Aegean island of Thera, while consulting the oracle at Delphi on quite another matter, was bidden to found a city in Libya (Africa). The people of Thera took no notice of this advice, 'for they did not know Libya, where in the world it was, and durst not send colonists to an unknown goal'. For seven years after this no rain fell on Thera. The Theraeans again consulted the oracle and received the same reply. So they sent messengers to Crete to inquire whether any Cretan or visitor to Crete had been to Libya. In one of the Cretan towns these found one Corobius, a purple fisher, who said that he had been driven by winds to Libya and an island named Platea. The envoys took Corobius in their pay and brought him to Thera. Under the guidance of Corobius a small advance party presently left Thera for Platea, but merely stayed long enough to drop their pilot, who was to act as their caretaker while they returned to Thera for reinforcements. A further batch of emigrants was now enrolled by the community of Thera, which ordered a compulsory ballot among the eligible men. The new draft set out on board of two men-of-war, but chanced upon some inhospitable part of the Libyan mainland, from which they soon put back to Thera, 'for there was nothing else they could do'. On their return home, however, they were greeted with showers of stones and ordered to double back. On their second journey to Libya they found their course to Platea, where Corobius in the meantime would have starved, had not a vessel bound from Samos to Egypt touched in at Platea and revictualled him. For two years the colonists tarried at Platea. Thereupon they left *en masse* for Delphi, where they announced to Apollo their disappointment in Libya. But Apollo retorted that they had not been to Libya (i.e., the mainland). At last the emigrants recognized that Apollo 'would not let them off from the colony'. Returning to Africa, they shifted their quarters to a more fertile island named Aziris, where they stayed seven years. After this interval some unusually friendly natives invited them to a better and roomier site on the mainland. Here the wanderers ended their journey and founded the city of Cyrene (c. 650-630 B.C.).

The temporary settlements at Platea and Aziris illustrate the familiar rule that emigrants who are not sure of their welcome in a foreign land will alight, if possible, on an off-shore island or a peninsula. Thucydides noted that the Phoenicians had followed this rule in Sicily. He might have added that this was also a Greek practice, and have prophesied that modern colonists would equally adopt it.[64]

The general belief of the Greeks, that the era of colonial settlement in the West was opened with the foundations of Naxos (under Mt Etna), may be accepted in this sense, that Naxos was in all probability their first landfall in Sicily, for the snow-laden peak of Etna would offer them an even surer mark than Cape Cod did to the Pilgrim Fathers. But finds of Greek pottery in the Bay of Naples and on the Etruscan coast, which date back at least as far as 800 B.C., are plain evidence that Greek seamen had re-discovered the Straits of Messina long before 735 B.C., the traditional foundation-date of Naxos; and the archaeological record indicates that by 750 B.C. at the latest a permanent Greek settlement had been made at Cumae (just outside by Bay of Naples).[65] But in general the Greek foundation dates (e.g., 734 B.C. for Syracuse and c. 705 for Tarentum) accord well with modern research on the sites.[66]

From Sicily and South Italy the Greeks were slow to adventure themselves on the 'Spanish Main'. The discovery of the Straits of Gibraltar by a Greek merchant, which did not happen till c. 650-630 B.C., was the result of a chapter of accidents. The successful explorer was a Samian named Colaeus, the selfsame man who had rescued a marooned sailor on the island of Platea near Cyrene (p. 31). Proceeding from Platea to Egypt, Colaeus was turned back by a persistent easterly wind which drove him willy-nilly past the Straits and clear

THE ANCIENT EXPLORERS

of the Phoenician station at Gades, but allowed him to put in at Tartessus.[67] Returning home with a cargo of Spanish silver, Colaeus again eluded the Phoenicians and set up in Samos as a nabob.[68] His colossal good luck could not fail to put other Greek fortune-hunters on his track; indeed, the real credit for opening up the farther West belongs not to Samos but to the neighbouring city of Phocaea. This small but resolute community, the 'Amalfi of ancient Greece', was not content to rely on haphazard individual effort, but organized collective expeditions, and instead of using ordinary merchant vessels, equipped its explorers with warships. Thus armed, the Phocaean mariners could defy what Matthew Arnold has euphemistically called the 'indignation' of the grave Tyrian traders.[69] They established a regular sea-track from Sicily or the Bay of Naples past the Balearic Islands, and thence probably to the Puenta de Yfach, a bold promontory of south-eastern Spain which matches Gibraltar as a landmark.[70] This route was consolidated with a chain of settlements, most of which betray their Phocaean origin by the Ionic suffix -ussa to their names. The chain included Pithecussa (Ischia, near Naples), Ichnussa, i.e., Sardinia,[71] Pityussa (Iviza), Melussa and Crommyussa (Mallorca and Menorca?).[72] The presence of Phocaeans in Sardinia is also indicated by an archaic Ionic inscription at Olbia[73] and pottery of sixth-century style.[74] The last stage of the Phocaean track was marked by a colony at Maenace, a little to the east of the Phoenician settlement at Malaga.[75] This Phocaean outpost was the westernmost of Greek cities.

It was probably in the last years of the sixth century, and after the opening up of the route to Tartessus, that the Phocaeans proceeded to explore the Gulf of Lions. It is not certain whether they reached the Gulf by coasting along Italy, or, as seems more probable, by way of Sardinia and Corsica. With a good eye for position they made a settlement on the site of modern Marseille, the city of Massilia, which soon grew out of a hamlet of fishermen into the chief bulwark of hellenism in the West (*c.* 600 B.C.).[76] The story of the romantic circumstances in which Massilia was founded has fortunately been preserved. 'Phocaean sailors alighted in the Gallic Gulf near the mouths of the Rhône. Captivated by the natural beauties of the region, they returned home with a report of what they had seen and asked

THE MEDITERRANEAN AND THE BLACK SEA

for reinforcements. The leaders of the second expedition were Simus and Protis. These went to solicit the friendship of the native king Nanus, in whose territory they desired to found the city. As it happened, the king was engaged in arranging the wedding of his daughter Gyptis. After the fashion of his tribe, he proposed to have her choose her husband at a banquet and hand her over to him on the spot. All the suitors were invited to the feast, but the Greek strangers also were called in. When the maid was introduced and bidden to serve with water the guest of her choice, she passed over all the rest and handed the water to Protis.'[77] It need not be objected that this story is too pretty to be true. If Pocahontas was smitten with a paleface, why should not Gyptis fall in love with a Greek vagabond? Massilia in turn became the starting-point for further exploration in the Gulf of Lions and down the east coast of Spain to Denia or beyond. By 500 B.C. a belt of Greek colonies extended from Nice and Hyères to Denia, or even to Alicante.[78]

The Ionian Greeks completed the circle of discovery in the farther West by following out the coast of North Africa toward the Straits of Gibraltar, and occupying stations along it. The evidence for this is derived from surviving Ionic names of small towns and islands along the western portion of the North African coast.[79] The exact situation of these places and the date of their occupation can no longer be ascertained; but it is clear that they marked stages on an alternative Greek route from Sicily to Tartessus. Thus in the course of the sixth century the western Mediterranean bade fair to become a Greek lake.

But before 500 B.C. the Greeks had lost touch with the farther West. They did not widely follow the Phocaeans's good example of corporate action. On the other hand the Phoenician settlers at Carthage took a leaf out of the Phocaean book and organized collective resistance against foreign interlopers. They took over the guard of the Straits of Gades; they secured to themselves all the south of Spain, the Balearic Islands, and Sardinia; they foiled a last colonizing effort of the Phocaeans in Corsica by defeating them in a set naval battle (*c.* 535 B.C.).[80] In the area of Phocaean thalassocracy Massilia and the settlements along the Gulf of Lions alone maintained themselves. The shrinking of the Greek geographical horizon in the West may be measured by comparing the knowledge of the geographer Hecataeus (*c.* 500 B.C.) about Spain and north-west Africa, and the blank ignorance of the historian Herodotus (*c.* 450 B.C.) about these regions of the farther West.

We have already noticed that the Phoenicians did not frequent the Italian coasts. Until the fifth century B.C. these remained mostly under the control of the Etruscans. It is uncertain whether the Etruscans were natives of Italy or immigrants who came from Asia Minor early in the first millennium.[81] Of their immigration, if such it was, no details can be furnished. In any case they maintained a local thalassocracy and discouraged alien exploration in their waters. It is probable that they did not allow the Greeks to proceed beyond the Bay of Naples, and that the vast quantities of Greek pottery discovered in Etruscan tombs of the seventh and sixth centuries had been transhipped at Cumae on to Etruscan vessels.[82] The opening up of the west coast of Italy by the Greeks may be dated from the battle of Cumae (474 B.C.), when the Syracusans destroyed the Etruscan naval power. Henceforth Syracusan and Massilian traders could join hands in the ports of Rome, whose growing ascendancy in Italy was probably not regretted by the Greek mariners.

For reasons not yet well understood, the Adriatic Sea long remained a backwater.[83] The Phoenicians apparently by-passed it altogether, and it was left to the dauntless Phocaeans to open up its recesses (probably *c.* 600 B.C.).[84] By 500 the Greek colony of Spina (near the Po estuary) had achieved a temporary prosperity as a port of entry into Etruscan territory; and speculations as to the identity of the Po with the mysterious amber river Eridanus suggest that at this date the Greeks had an inkling of its importance as an avenue into the interior of Europe, and onward to the Atlantic and Nordic faces of Europe.[85] By 500 some unknown Greeks had also pried into the cul-de-sacs of the Dalmatian coast, as may be gathered from the knowledge which the geographer Hecataeus could show about it.[86] In the fourth century the Syracusan tyrants patrolled the Adriatic and established colonies at Ancona and Hadria (near the Po estuary).[87] But the principal stream of Greek traffic was directed henceforth up the Dalmatian coast, where the wine merchants found some of their most profitable markets (see also p. 142).

In the exploration of the Mediterranean the Romans played no direct part, but they rendered it possible for the full fruits of previous discovery to be reaped. They rounded up the Dalmatian pirates and destroyed the Carthaginian trade-monopoly, so that from 200 B.C. every part of the Mediterranean was accessible to all comers. During the crises that befell the Roman Republic in the first century B.C. the security of the seas was often jeopardized; but under the emperors it was established more fully than at any later period previous to the nineteenth century. Apart from the inevitable risks of wind and weather, Mediterranean navigation in the first two centuries A.D. was ceasing to be a venture and was being reduced to a routine.

8. THE BLACK SEA

In spite of its proximity to the Aegean Sea, the Black Sea does not appear to have been frequented by sailors from the Mediterranean until the end of the prehistoric period. This may have been due in part to its more forbidding climate, to its summer fogs and winter barriers of ice. But the chief reason for its seclusion undoubtedly lay in the double lock of the Dardanelles and Bosphorus. Navigation up these sea-rivers is impeded by strong currents (rising to 5-6 knots), which shoot like torpedoes off the headlands and form troublesome eddies in the backwaters, and by a north-east wind which reinforces the rush of water for nine months of the year.[88] In addition, as recent history has once more shown, these straits cannot be forced against the resistance of the powers that control them.

The first apparent evidence of exploration by Aegean adventurers in the Black Sea is to be found in the legend of the Argonauts. According to this tale the Thessalian prince Jason won his way into the Black Sea and followed its south coast as far as the Caucasus region at an interval of two generations before the Trojan War (say 1250 B.C.). But the Argonaut legend, as is now generally agreed, is a composite structure and reflects the experience of several centuries.[89] If we peel away the later accretions and lay bare the Homeric core, all that we find is a curt allusion in the Iliad to Jason's lordship of the Aegean island of Lemnos, and to an obscure passage in the *Odyssey* relating how the good ship Argo, 'that lay at all men's hearts', slipped through the Drifting Rocks by an uncertain course to a goal not specified.[90]

Further proof, if not of actual discovery in the Black Sea, yet of a desire to explore it, has been found in the story of the Trojan War. The Achaeans, so some scholars contend, sought not so much to enter Troy as to get past it, so as to open up trade with the Black Sea.[91] But the *Iliad* contains not an inkling of this purpose, indeed it nowhere mentions the Black Sea or its approaches. Moreover a mercantile theory of the Trojan War is not required. If we are not content to accept Homer's sex motive, we can adequately explain the war as the most memorable of various descents upon Asia by which Achaean captains of adventure acquired booty or territory for themselves.[92]

In the *Odyssey*, as we now possess it, there is no hint of a Black Sea. Nevertheless it is held by a good many scholars that the scene of Odysseus's adventure was originally laid in those waters, and that their transference to the West was an afterthought.[93] This theory is based on supposed resemblances between these localities and various places in or near the Black Sea, and on a reference by Homer to a 'Cimmerian' folk, whom it is tempting at first sight to equate with the Cimmerians of the Crimea. But the identifications of Calypso's Isle, the Land of Circe, etc., with Black Sea sites are not a whit more convincing than any others; and the Crimea was certainly not the home of the Homeric Cimmerians, who are described as living in fog and gloom, and hidden from the sun's rays. Besides, it is on the face of it most unlikely that any poet or editor should have shifted the scenes of Odysseus's travels *en bloc* from one end of the Greek horizon to the other.

Evidence of prehistoric intercourse between the Aegean and the Black Seas has also been found in the archaeological record. It has been suggested that the gold of Mycenae came from the Urals,[94] its silver from the Upper Euphrates, the 'miltos' or red paint with which the 'cheeks' of Homeric ships were rouged from the region of Sinope,[95] and the lapis lazuli pearl at Dimini (near Jason's Thessalian home) from some Black Sea entrepôt.[96] Again, the so-called 'Mycenaean' pottery from Akalan near Amisus has been accepted as proof of an Aegean export trade to the Black Sea before 1200 B.C.[97] But none of these arguments carries much weight. The Aegean peoples could procure gold from Egypt,[98] silver from Spain; their lapis lazuli might have travelled from its home in Turkestan by way of Babylon, where this jewel was in great demand;[99] the red paint of Homeric ships was probably not the cinnabar of Sinope (a criminally expensive object for such a purpose), but common red lead. Of the published sherds from Akalan, one is of the eighth-century 'orientalizing' type, the others are of the debased 'sub-Mycenaean' variety (*c.* 1200–800 B.C.): none seems of earlier date than the Trojan War.[100] It is also noteworthy that the prehistoric cultures of South Russia show little affinity with Minoan civilization, and that Troy has yielded more evidence of contacts with the adjacent continents than with the Aegean area.[101]

On the other hand we may accept a Greek tradition of early settlements in the Black Sea by Carians from south-western Asia Minor,[102] for along its coasts we find names with the distinctively Carian suffix -essus.[103] The Carians were a sea-faring people as early as the second millennium B.C.; but their best known exploits, their raids on Egypt[104] and the cruise of Darius's admiral Scylax in the 'Red Sea' (pp. 78–9), belong to the seventh and sixth centuries. Their voyages to the Black Sea are difficult to date, but may be tentatively assigned to the period between the fall of Troy and the Greek explorations (*c.* 1200–800 B.C.). It is not unlikely that they brought the Aegean pottery mentioned above to Akalan.

But whatever the date of the Carian settlements, the effective discovery of the Black Sea was the work of the Greeks. No records of their first voyages in these regions have survived, indeed our literary evidence is almost limited to the foundation-dates of their colonies. These suggest that intensive occupation of the coast began about 650 B.C., and this conclusion has been borne out by excavation on the north and west coasts.[105] But the records of individual towns suggest that tentative settlements were being made by 750 B.C., and the excellent situation of most of the Black Sea towns shows that their definitive foundation must have been preceded by a long and searching reconnaissance.[106] If any conclusion can be drawn from the mention of the Ister (Danube) in the Hesiodic poems,[107] we may infer that Greek exploration of the Black Sea had begun by 800 B.C. The primary object of Greek exploration was probably not the wheat of the Russian 'black earth' zone, important as this eventually became,

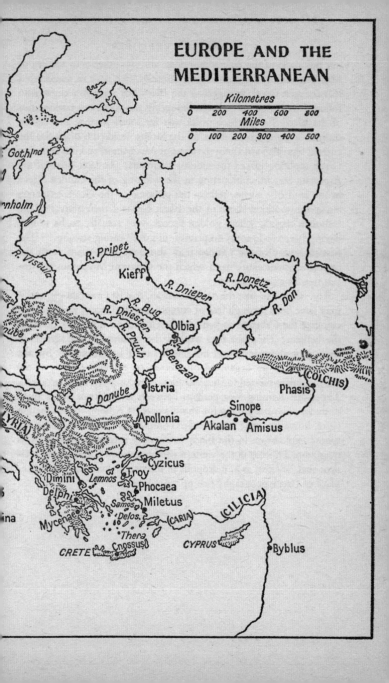

THE ANCIENT EXPLORERS

but the metal of the Asia Minor coast and the fisheries of the Sea of Marmora and of the Russian river estuaries.[108] It may be conjectured that the first Greek voyagers to the Black Sea were fishermen who found by repeated visits that in the spring the north-east trade-winds of the Black Sea approach make way for south-westers, and that in summer a night breeze blows up the Straits. Inside the Black Sea the current sets eastward along the coast of Asia Minor, and it is here that Greek tradition placed the oldest settlements. The later form of the Argonaut legend, culminating in the winning of the Golden Fleece in the land of Colchis, reflects the progress of the Greek explorers along the southern shore of the Black Sea and their arrival in the Caucasus region, where golden fleeces were actually to be found, having been made so by suspension in gold-bearing streams.[109] The first discovery of the Crimea may also have been made from the opposite Asiatic shore from which the Crimean Mountains are discernible on a clear day.[110]

The Black Sea, like the Mediterranean, had its pirate shores, and it may have been through fear of corsairs that the Greeks were slow to colonize the Caucasus region and the Crimea.[111] Similarly, dread of the bellicose Thracians may explain the tardiness of the Greeks in settling on the magnificent but unsafe site of Byzantium.[112] But the explorers had no organized opposition to encounter, such as the Carthaginians offered to later intruders in the western Mediterranean. The worst fighting was perhaps between rival Greek bands from different cities scrambling for eligible sites.[113] Among the competing cities, however, none had a record comparable with that of Miletus, whose exploration in the Black Sea is only surpassed by that of its neighbour Phocaea in the western Mediterranean. It is no mere accident that Miletus, as a metropolis of discoverers, was also the birthplace of Greek geography (see pp. 161-2, 226).

CHAPTER 3

THE ATLANTIC

1. PREHISTORIC VOYAGES

A SEAMAN passing from the Mediterranean into the Atlantic enters upon a different and less hospitable world. In place of the steady breezes of a Mediterranean summer, he may at any time encounter a heavy gale or lie becalmed in a fog. If he hugs the coast, he may be cast up by an on-shore swell or become the plaything of the tides. If he stands out to sea, he may lose sight of land for ever.[1]

In spite of these disadvantages and dangers, Atlantic navigation dates well back into the prehistoric era. The natives of western Europe probably learnt the rudiments of seamanship on navigable rivers and estuaries, of which the Atlantic seaboard has no lack, or in the backwater of the Baltic. As shipwrights they equalled the Mediterranean craftsmen. Their coracles, though designed for paddling on rivers, were strong enough for the open sea.[2] But in the Bronze Age (after 1800 B.C.) they had learnt to fashion sailing vessels of Mediterranean type.[3] The seaworthiness of the ships used by the Veneti of Brittany impressed Julius Caesar: with their high bows, designed to meet Atlantic rollers, they were forerunners of the shapely Viking craft.[4] The injunction of the Codex Theodosianus (A.D. 438), forbidding Roman subjects to teach shipbuilding to barbarians, came 2,000 years too late to have effect on the Atlantic peoples.

The chief stimulus to primitive travel on the Atlantic was the quest for metals and amber. In the Bronze Age Ireland exported gold to Britain and the Continent, and imported amber.[5] From Jutland the earliest consignments of amber for southern Europe were sent by sea.[6] Tin from Cornwall, perhaps also from Brittany, was traded to Scandinavia; and it is possible, though not certain, that the Mediterranean derived its supplies of this metal since 2000 B.C. from Atlantic sources.[7] The prehistoric carrying business between Atlantic and Mediterranean seems to have been concentrated in the hands of the people of Tartessus, which was the farthest goal of early Mediterranean traders (pp. 25, 34).

Homer's land of giant Laestrygones, where the paths of day and night lie close together, has often been located in the far north Atlantic; but no satisfying explanation for these words has yet been offered.[8] On the other hand a vague inkling of sub-Arctic Europe may perhaps be detected in his reference to the Cimmerians who lived at Ocean's marge in fog and gloom. This information might have reached the early Greek world by way of the Tartessian voyagers to the Atlantic tin-lands.[9] It is also not unlikely that the mythical river Eridanus, which the Greeks always associated with the production of amber and at first imagined as flowing northward through western Europe, derived its name from the Rhine and its association with an amber isle (Heligoland) at its estuary from the Elbe.[10] But the first definite information about Atlantic lands derived from a *Periplus* (or Sailing Direction) written by a Massilian captain *c.* 525 B.C., which partly survives in the doggerel poem of a late Roman author, Festus Rufus Avienus.[11] According to the *Periplus* the Tartessians plied regularly to 'Oestrymnis' (Brittany); but apparently they left the cross-Channel traffic to the Oestrymnians.[12] The British Isles were also introduced to the Mediterranean world by the Massilian seaman. Unfortunately his account, as reduced to poetry by Avienus, reads perilously like nonsense:

> But hence [from Brittany] in two suns' time a boat
> May travel to the Sacred Isle [i.e., Ireland]:[13]
> (Thus 'twas yclept by men of old.)
> It lieth in betwixt the waves,
> And occupieth many acres.
> And in it roomily resideth
> The folk of the Hibernians.
> And eke hard by there stretcheth out
> The isle of the Albiones.[14]

It will be noticed that Avienus disposes of Britain in a contemptuous aside. But it would be rash to infer from this that Ireland loomed larger than Britain in the imagination of the Massilian writer.

2. MIDACRITUS

About 600 B.C., when the Phocaeans forced their way to Tartessus (pp. 34-5), the king of that city received them with high favour, showering his Spanish silver upon them and inviting them to settle in his dominions.[15] Did he allow them any share in the Atlantic trade? No hint of such a privilege is contained in our Greek authors, but a faint memory thereof may be discerned in a stray remark of Pliny: 'Midacritus was the first to import "white lead" (i.e., tin) from the Tin Island.'[16] Midacritus, as his name declares, was plainly a Greek, and presumably a Phocaean.[17] But what precisely was his exploit? Pliny's words perhaps mean no more than this, that Midacritus went to Tartessus to meet a cargo of tin which the Tartessians had brought from farther afield. But this would not have been a deed to resound through the ages: compared with Colaeus's discovery of Tartessus and his huge haul of silver (pp. 33-4), it would have been quite a commonplace performance. It therefore seems justifiable to infer that Midacritus fetched the tin from Tin-land, by which may be meant either Brittany or Cornwall.[18]

Further evidence of Greek cruises in the Atlantic comes from Avienus. This author applies to C. Roca (near Lisbon) the name of Ophiussa, which is typically Ionic; and it is not unlikely that the whole of his description of the coast to C. Aryium (Ortegal) is drawn from the Massilian Periplus.[19] But these early Greek intrusions upon the Atlantic were but tentative and short-lived. By 500 B.C. the Ocean gates had been bolted against them by the Carthaginians,[20] and to these belongs the credit of first having explored the North Atlantic effectively.

3. HIMILCO

The Phoenicians probably reached the Straits of Gibraltar soon after 1200 B.C., and established the colony of Gades about 1100 B.C. (p. 28). There is no evidence of their having ventured out as yet into the Atlantic, so as to break in upon the commerce of the Tartessians. But the ostentatious friendliness of the king of Tartessus to the Greeks was tantamount to a breach with the Phoenicians.[21] It is remarkable that after 500 B.C. Tartessus passes out of notice. In all probability the Carthaginians, having driven the Greeks off Spain and isolated the Tartessians, proceeded to destroy their city.[22] Henceforth Gades was the key of the Atlantic, and its Carthaginians became the pioneers of Atlantic discovery.

The first recorded voyage of a Carthaginian captain in the northern Atlantic is the cruise of Himilco. This is known to us by nothing more than a passing allusion by Pliny[23] and a few perplexed verses of Avienus, which tell us that Himilco undertook a four months' cruise along the old Tartessian trade-route and on the way was beset by calms and shoals, by entangling weeds and inquisitive sea-monsters.[24] Avienus states that Himilco sighted Brittany, but leaves us to guess whether he passed on to the British Isles. All that we can say is that on a four months' journey there was ample time for him to visit Britain, and that he had an obvious interest in opening direct relations with the Cornish miners.[25] Nevertheless it is fairly certain that Himilco did not explore the continental coast beyond Brittany.[26] According to Avienus this coast lay deserted in Himilco's day, as a result of Celtic migrations, which implies that the Carthaginian seaman avoided it.

Himilco's adventures with shoals and weeds have been taken to prove that he visited the Sargasso Sea by the Azores, where marine plants luxuriate on a sunlit ocean sill.[27] Such an excursion could no doubt be fitted into Himilco's time-table. But it is most unlikely that he made search for the Azores, which in his day were almost certainly unsuspected: his discovery, if such there was, must have been accidental.[28] Besides, there is no need to go to the Sargasso Sea for the impediments to navigation which Himilco experienced. If allowance be made for a little seamanlike exaggeration on his part, all the phenomena described by him may be located near Gades. In summer the belt of calms often rises to the latitude of Gibraltar,[29] and between here and C. St Vincent shoals and banks of weed may still be found.[30] At four days' sail from Gades fishermen used to frequent a bank overgrown with marine plants and peopled with tunnies 'of incredible obesity'.[31] According to Pliny schools of whales used to calve in the recesses of the 'Ocean of Gades', and cachalots followed in to prey upon them.[32] Thus there is no adequate reason to suppose that Himilco was blown out to the Azores.

The date of Himilco's cruise may be fixed approximately. Pliny states that it occurred 'when the Carthaginians were at the height of their power',[33] which must mean before their disastrous defeat by the Syracusans at the battle of Himera (480 B.C.). But seeing that the Carthaginians would not want to lose time in entering upon the heritage of Tartessus, we may place Himilco's venture soon after the destruction of that town, i.e., about 500 B.C.

Himilco's discoveries gave the Carthaginians control of the tin trade at least as far as Brittany. But it is likely that the natives of Brittany retained a share of the cross-Channel traffic in their hands. At any rate the Carthaginians were not wholly satisfied with the results of Himilco's cruise, but at some later date set themselves to discover a direct route from Spain to the tin mines. This they eventually accomplished by standing out northward from Corunna.[34] The 'Cassiterides', or Tin Islands, which were the goal of their endeavour, may be identified with the Scillies, which produced very little tin, but might serve as a depot, just as St Michael's Mount did in the days of Pytheas.[35] But more probably the 'Islands' were the stannaries of the Cornish mainland. It would have been quite in keeping with the practice of ancient explorers, and of later ones to boot, if the Carthaginians on first impression had mistaken the Cornish peninsula for a cluster of islands, and had not troubled themselves to verify this assumption by following out the Cornish coast to right and left.[36]

4. PYTHEAS

For two centuries the gates of the Atlantic were kept closed by the Carthaginians against the Greeks. But about 300 B.C. they were thrust ajar for a while and gave passage to a Massilian captain named Pytheas, who has the best claim among ancient travellers to rank with the great discoverers of modern times.

For a man of his eminence, we know strangely little about Pytheas.[37] From the fact that he finds mention in the works of Dicaearchus, a pupil of Aristotle, but not in those of Aristotle himself, we may infer that his explorations fell between 322 and 285 B.C. His success in eluding the blockade at the Straits points to 310–306 B.C. as the date of his venture, for in those years the Carthaginians were engrossed in the defence of their own city against the Syracusans. But we can only guess at the scale of his expedition, at the number and the tonnage of his ships. The primary object of his cruise was no doubt to break the Carthaginian monopoly of Atlantic trade. Being a person of slender means, Pytheas could hardly afford to equip an expedition without some hope of a commercial return, and his visit to the Cornish mines betrays his mundane interests. But the really distinctive feature of his travels is that they also served a scientific purpose. Pytheas was a skilled astronomer. He had observed that the Pole Star was not situated at the Pole,[38] and had calculated the latitude of Massilia within a few minutes of the correct figure.[39] During his voyage he studied the tides, and he took further observations of latitude which gave the astronomer Hipparchus his reference points in plotting the map of northern and central Europe.[40] In scientific discovery his journey was more fruitful than any preceding the age of Henry of Portugal.

The route followed by Pytheas can only be fixed by a continual process of dead-reckoning from the shamefully few bearings which ancient writers have provided.[41] It is particularly regrettable that our chief source, Strabo, had succumbed to the higher criticism of Polybius, and thought himself entitled to dismiss Pytheas in a few contemptuous allusions. Our remaining information comes mostly from Diodorus and Pliny. These two compilers excerpted the historian Timaeus, who was a contemporary of Pytheas and a painstaking scholar; but they took their notes from him in a very slipshod manner.

From Gades to C. Ortegal Pytheas's track is unmistakable. At the latter point our uncertainties begin. He is next heard of on the Breton coast and at the isle of Ushant.[42] But did he arrive here by following the long in-shore route, or by cutting across the Bay of Biscay? Some modern scholars have used an ambiguous passage of Strabo in support of the latter view.[43] But it is most improbable that any ancient explorer, however versed in sun and stars, should have made an open-sea journey of 300 miles on a maiden trip. Besides, Strabo atones for his obscurity in one place by stating in another that Pytheas reported easier travelling along the north coast of Spain when proceeding towards 'Celtice' (France) than Ocean-ward.[44] This clearly shows that Pytheas kept close in. The same conclusion emerges from a speculation of Timaeus, that the Atlantic tides were due to the spates and ebbs on the rivers of western Europe.[45] Evidently Timaeus had heard something of the big rivers of the Atlantic seaboard, and for this information he was almost certainly beholden to Pytheas.

The next stage on Pytheas's journey admits of little doubt. In order to visit the Cornish tin-mines, his natural route would lie along the old-established track from Ushant to Land's End. Since Pytheas gave a measurement of the distance from Belerium (Land's End) to the French coast,[46] it may be assumed that he made a direct passage to Cornwall. His impressions of the Cornish miners were summed up by him as follows:

The natives of Britain by the headland of Belerium are unusually hospitable, and thanks to their intercourse with foreign traders have grown gentle in their manner. They extract the tin from its bed by a cunning process. The bed is of rock, but contains earthy interstices, along which they cut a gallery. Having smelted the tin and refined it, they hammer it into knuckle-bone shape and convey it to an adjacent island named Ictis. They wait till the ebb-tide has drained the intervening frith, and then transport whole loads of tin on wagons.[47]

From Land's End Pytheas followed the British coast right round to his starting-point. This achievement is not mentioned explicitly by any ancient writer, but is plainly implied by the fact that he described Britain as a triangle and furnished measurements for its three sides.[48] The order in which Diodorus, who is our informant on this point, mentions the faces of Britain creates a slight presumption that Pytheas sailed round against the clock.[49] No ancient author states whether Pytheas visited Ireland. But even if he never set foot on its shores he could not have failed to sight them from the Mull of Galloway. There is little doubt that Pytheas provided the information which enabled the geographer Eratosthenes (c. 235 B.C.) to locate Ireland in its proper position relative to Britain.[50]

From the British coasts Pytheas made occasional excursions up-country. So much may be gathered from Polybius's sarcastic remark, that Pytheas claimed 'to have walked all over Britain'.[51] It also follows from the detailed descriptions which he gave of the inhabitants, of which a few extracts are preserved in Diodorus. Pytheas noticed that the natives lived a primitive life, gave combat with chariots in their infrequent wars, and stacked their corn in underground silos. This information has all the appearance of being derived from flying visits into the interior of Britain.

Britain is triangular in shape like Sicily, with three unequal sides. It extends obliquely along Europe. The promontory nearest the mainland, called 'Cantium', is said to be distant about 100 stades [c. 11 miles], at which point the sea forms a current; the second cape, known as 'Belerium', is stated to lie four days' sail from the continent; the third is known to jut out into the open sea and to be named 'Orca'. The shortest of the sides is reckoned at 7,500 stades [c. 820 miles], the one that stretches along Europe; the second, from the Strait to the apex, at 15,000 stades [c. 1,650 miles]; the remaining one at 20,000 stades [c. 2,200 miles], so that the whole circumference of the island is 42,500 stades [c. 4,670 miles]. The inhabitants of Britain are said to be sprung from the soil and to preserve a primitive style of life. They make use of chariots in war, such as the ancient Greek heroes are reputed to have employed in the Trojan War; and their habitations are rough-and-ready, being for the most part constructed of wattles or logs. They harvest their grain crops by cutting off the ears without the haulms and stowing them in their covered granges; from these they pull out the oldest and prepare them for food. They are simple in their habits, and far removed from the cunning and knavishness of modern man. Their diet is inexpensive and quite different from the luxury that is born of wealth. The island is thickly populated, and has an extremely chilly climate, as one would expect in a sub-Arctic region. It has many kings and potentates, who live for the most part in a state of mutual peace.[52]

'Above Britain', i.e., at its northern extremity, Pytheas observed the sea running 'eighty cubits high'.[53] This statement has usually been referred to the tidal range of ebb and flow. But at the northern end of Scotland the tides only rise ten or twelve feet; even in the Bristol Channel they barely attain one-half of eighty cubits, and the Bristol Channel is not 'above Britain'. There can be hardly any doubt that Pytheas witnessed one of those gales in the Pentland Firth, in which the wind and an opposite tidal race upheave solid billows to a height of sixty feet, and columns of spray rise hundreds of feet above these.[54]

In the north of Scotland Pytheas gathered information about an

'Island of Thule', which created a greater stir among ancient geographers than his report on Britain. His description of Thule is as follows: It is the northernmost of the British Isles, at six days' sail to north of Britain proper, and at one day's distance from the frozen 'Cronian' sea. Round Thule itself 'there is neither sea nor air, but a mixture like sea-lung, in which earth and air are suspended; the sea-lung binds everything together'.[55] At midsummer the path of the sun as seen from Thule coincides with that of the Bear round the heavens ($66\frac{1}{2}°$); at night the sun retires to its 'resting-place' for a mere nap of two to three hours. For lack of the crops and cattle of more genial lands, its inhabitants subsist on wild berries and 'millet', which they thresh in covered barns because of the continual rains. From the plentiful honey of their bees they prepare mead for beverage.[56] Pytheas, it should be noted, expressly acknowledged that his information about Thule was at second-hand. All that he claimed to have seen for himself was the sea-lung: perhaps this was the cause of his not making a voyage to Thule in person.

Where are we to find Thule? Despite the authority of Ptolemy and some modern scholars,[57] we must look farther than the Shetlands. Perhaps we need not press the objection that in the Shetlands the sun's setting time never falls below five hours, for Pytheas's informants had no means of measuring time accurately, and their 'two or three hours' may have been no more than a conventional expression for 'a short while'. But by no standard of reckoning could the Shetlands be described as 'six days' sail' from the British continent.

Modern opinion in general is divided between Iceland[58] and Norway.[59] At first sight Thule cannot be Norway, which is not an island, British or otherwise, and does not lie to north of Britain. But on further inspection these difficulties fade away. If Pliny could write about the 'island' of Scandinavia,[60] we need not be surprised that Pytheas's informants might have described Thule in similar terms. Again, however much we may dissociate Norway from Britain, to a man in Pytheas's position it would appear as an extension of Britain, just as the Tin Islands appeared to the Carthaginians who reached them from Corunna as satellites of Spain.[61] Lastly, ancient writers commonly lacked precision in their indications of direction (pp. 16–17). If Strabo was frequently out of his bearings by 45°, *a fortiori* the rude folk who described Thule to Pytheas might have used the term 'north' to include north-east. On the other hand several features in Pytheas' account of Thule point to Norway. Under ancient sailing conditions six days would be a fair allowance for the journey from the north of Scotland. Again, Norway extends northward into the Arctic Circle and southward within the range of bees. Lastly, the 'sea-lung' in which modern scholars have recognized jelly-fish, phosphorescent medusae,[62] and ice-sludge,[63] and Polybius a *reductio ad absurdum* of Pytheas's entire story,[64] may be taken as an apt description of the sea-fogs off the upper Norwegian coasts, in which all elements are held in solution in a clammy and impalpable moisture.[65]

Similarly, in view of the ancients' lack of precision in measuring distance by sea, it is not a material point that Iceland would lie somewhat more than six days' sail from Britain, and still less weight attaches to the statement that the Irish monk Dicuil found Iceland uninhabited, for the island might well have become desolate during the thousand years that separated him from Pytheas. A seemingly more serious objection has been advanced by critics who maintain that Iceland lies beyond the range of bees. This contention, if correct, would indeed be fatal to Iceland's claims. But clover and heather now grow in Iceland, and travellers have observed the presence of wild bees.[66] In ancient times, therefore, the natives may have gathered sufficient honey for their mead. Furthermore, dense fogs, in which frozen air is drawn from the relatively warm sea, so as to form a soft and treacherous layer of drift ice, also envelops mariners in the latitude of Iceland.

On present evidence, therefore, we may hesitate to make a final decision between Iceland and Norway.

From Britain Pytheas re-crossed the Channel, presumably on the same track as on his outward journey, and skirted the continental coast 'beyond the Rhine to Scythia' and 'as far as the river Tanais'.[67] In the course of this voyage he passed a broad estuary of 6,000 stades (660 miles) along the territory of the 'Guiones?' or 'Gutones?'; and he saw, or heard of, an island variously called Abalus, Balcia, Balisia, Basilia, or Baunonia, on which amber was cast up by the spring tides and collected by the natives for fuel. According to one account this island lay in the aforesaid estuary, one day's sail from the mainland;

THE ATLANTIC

according to another it was situated 'in the Ocean, opposite Scythia above Gaul'.[68]

These discordant data raise one greater and one lesser problem, the name of the amber isle and the extent of Pytheas's cruise along the European coast. A comparison of the various ancient texts suggests that the island bore the native name of Abalus (Celtic for 'apple'),[69] that the Greeks variously transcribed this into Balcia, Balisia, or Basilia, and that Baunonia is a copyist's blunder. But the question that matters is not what the island was called or miscalled, but where it lay. The mention of Scythia and the Tanais (the Don) might suggest that Pytheas entered the Baltic Sea and followed it as far as the Vistula (which on this hypothesis was mistaken by him for the Don). Moreover, if we read 'Gutones' and identify these with the Goths, Pytheas's estuary must have lain in the eastern Baltic, which was the early habitat of that tribe,[70] and may be imagined as running from Rügen to Memel or Libau. The amber island in this case would be Bornholm. On the other hand we must beware of attaching too strict a sense to the term 'Scythia'. In Pytheas's age the Mediterranean peoples had not yet made the acquaintance of the Germans; hence they extended 'Scythia' to all that lay beyond 'Celtice'. Still less should we lay any stress on the allusion to the Tanais. This occurs in a jocular remark of Polybius, to the effect that 'Pytheas returned all along Europe from Gades to the Tanais'. This was plainly nothing more than a loose expression like our 'Dan to Beersheba' or 'China to Peru'. Again, the best manuscripts of Pliny read 'Guionibus', not 'Gutonibus', and this probably should be expanded into 'Inguaeonibus', a group of peoples who dwelt in the north-west of Germany.[71] Finally, the amber island is located 'above Gaul', and therefore must have lain in the Atlantic. In all probability, then, Pytheas's estuary was the Frisian bight from Texel to Jutland, and the amber island Heligoland, which no doubt was at all times the depot for Atlantic amber. The Frisian bight, it is true, does not extend over anything like 660 miles, but, as we shall see presently, Pytheas's measurements of distances were often much overstated. We may conclude, therefore, that Pytheas sailed to the Elbe estuary, and since the amber island was no doubt the object of his quest, we may assume that he put out from this point to Heligoland. But the manner in which the natives disposed of their amber suggests that in Pytheas's day it had ceased to be an article of commerce.

From this point Pytheas returned to the Mediterranean by a direct journey along the European coast: so much can be inferred from Polybius's remark, quoted above.[72]

We have left to the last a fundamental question. Did Pytheas really carry out the journey which we have detailed, or did he make up a sailor's yarn out of borrowed scraps and his own exuberant fancy? Was he a predecessor of Othere, Leif Ericsson, and Columbus, or of the Zenos and of Frederic Cook? In antiquity, as we have seen, his story had a mixed reception, and charges were made against his good faith which require to be examined.

The criticisms directed against Pytheas fall under three main heads, his description of the Ocean, his account of Thule, and his estimates of distance. Under the first head comes Pytheas's observation that it was faster sailing from C. Ortegal to the Gulf of Gascony than in the opposite direction. This, however puzzling, is true. The Gulf Stream sends an arm along the northern coast of Spain, and this current is reinforced at times by an on-shore swell.[73] Another point against Pytheas, had he really said so, would be his assertion that the ebb and flow of tides synchronized with the new and the full moon. But here we almost certainly have a misunderstanding on the part of our informant Plutarch.[74] Presumably what Pytheas observed was that the new-moon and mid-moon tides are the highest.

Of the objections to Thule, the most emphatic was that it usurped Ireland's rightful place in the cold as the northernmost of habitable lands.[75] This count in the indictment was pressed by Strabo as being quite conclusive. And so it was, but against Strabo. A more wary critic, Polybius, refused to swallow Pytheas's sea-lung, at which modern critics have also strained.[76] This was a reasonable ground for scepticism; but, as we saw on p. 52, it really enhances Pytheas's credit. It might further be urged against him that millet does not thrive in high latitudes. But in all probability Pytheas's error merely consisted in not recognizing oats as such – a natural mistake for a stranger from Mediterranean lands, where this grain was not much cultivated. Let us also bear in mind that Pytheas did not profess to have seen Thule, but merely passed on what others had told him.

THE ATLANTIC

The most serious charge against Pytheas is his habitual overstatement of distances. Apparently he measured four days' sail from Belerium (Land's End) to France, and 'several days' sail' from Cantium (the South Foreland) to the Continent. But it is not certain that in these cases he was measuring across to the nearest point in Europe.[77] Again, we cannot be sure of his terminal points in estimating the arc of the Frisian bight. But the real crux of this question lies in the dimensions which he gave to Britain. Reckoning its sides as 7,500, 15,000 and 20,000 stades (825, 1,650, and 2,200 miles) in length, he computed that Britain had a perimeter of 42,500 stades (4,675 miles), i.e., more than double its actual circumference.[78] It has been suggested in Pytheas's defence that his own estimates of distance were given in days' sailings, and that some later Greek writer applied a false scale in converting days into stades.[79] But what conceivable scale could have given double value to Pytheas's measurements? At best, Pytheas must have grossly overstated the size of Britain. But before we discredit Pytheas on this charge, let us remember that ancient seafarers had not even a tolerably accurate device for reckoning naval distances (p. 15). Herodotus exaggerated the length of the Black Sea, Pliny the Sea of Azov, and Nearchus the distance from the Indus to the Persian Gulf (p. 15), in much the same proportion as Pytheas swelled out Britain.[80] Yet nobody would maintain that the Greeks had not explored the Black Sea and Sea of Azov, or that Nearchus burnt his boats and sneaked home by land.

The indictment against Pytheas therefore does not carry much weight. On the other hand we have set forth in our narrative various data regarding seaways and currents, the natural features of countries and the habits of their populations, which were not only correct, but could only be discovered by personal investigation. The story of Pytheas needs correction on points of detail, but it is a record of actual travel.

On the assumption that Pytheas kept close in to the coast, except for a double open-sea passage between Ushant and Land's End, and similar trips between the Elbe and Heligoland, his entire cruise from Gades to Gades extended over 7,000 to 7,500 miles. If the achievement of Pytheas, who hardly ever lost sight of land, cannot be compared for boldness with that of Columbus, yet in point of distance

THE ANCIENT EXPLORERS

traversed it exceeded the memorable voyage of A.D. 1492. But the very magnitude of Pytheas's discoveries prevented their due appreciation among his countrymen. Moreover Pytheas, unlike Columbus, had no successors. No other Greek interloper followed in his wake, for after him the gates of the Atlantic were again held closed for some 150 years.[81]

5. OTHER VOYAGES IN THE NORTH ATLANTIC

The fall of Carthage in 146 B.C. opened a new era in Atlantic navigation. With the destruction of the Phoenician thalassocracy the freedom of the seas, both inner and outer, was definitely secured under Roman rule, and the surviving Phoenician merchants of Utica or Gades had to meet foreign competition on even terms. The manner in which they lost their monopoly of Atlantic trade has been described by Strabo:

> The Tin Islands are ten in number, and lie in a cluster to the north of the port of the Artabri (Corunna), in the open sea. One of them is deserted; the others are inhabited by men in black costume, with tunics reaching to their feet and girdles round their chests, who stride about with long sticks, like the Furies in tragedies. They subsist mostly on the cattle they graze. They also possess mines of tin and lead; in exchange for these and for hides they accept from the merchants pottery, salt, and bronzeware. Formerly the Phoenicians plied this trade alone, keeping their route from Gades hidden from all men. When the Romans shadowed one of their skippers, in order to discover their marts, the skipper deliberately cast up his vessel on a sandbank. Having thus lured his pursuers to the destruction which he had likewise courted for himself, he saved his own life, and received back on public account the value of the cargo which he had sacrificed. Nevertheless the Romans went on trying till they had spied out the route. Later on, when Publius Crassus made the crossing and ascertained that the mines lay near the surface and the natives were peaceable, he opened up this sea for all to work to their hearts' content, although the distance was greater than to Britain.[82]

Two points in this story call for comment. There is no need to assume that the interlopers were Romans from Rome. These only participated in overseas ventures to the extent of financing them

(p. 94). It is quite possible that the rivals of the Phoenicians were, as before, of Greek nationality. In the second and first centuries B.C. the Greeks of South Italy took the lead in the foreign trade of the peninsula, and being under Roman protection they were wont to style themselves 'Romans'.[83] It is among these that one would naturally look for the Paul Prys of the Atlantic traffic. Secondly, we must forego the temptation to recognize in Publius Crassus the right-hand man of Julius Caesar in the Gallic Wars. Caesar's own commentaries belie this identification. Not only do they fail to mention any cross-Channel exploration by a Publius Crassus, but they expressly state that when Caesar invaded Britain in 55-54 B.C. he had no first-hand information about that country except from Gallic traders and from another officer, one Gaius Volusenus.[84] Besides, Caesar asserts that the tin of Britain came from the inland, thus betraying a profound ignorance of the Cornish stannaries.[85] Such a misconception precludes the idea that any staff officer of his could have discovered Cornwall by an oversea route. Conversely, Strabo's description of the stanniferous area as a cluster of islands distinct from Britain shows that his informant's range of knowledge was limited to the old open-sea route of the Carthaginians from Spain, which dropped plumb on Land's End and stopped dead there. No officer of Caesar could have discovered Cornwall without also finding out that it was an integral part of Britain. The explorer Crassus can hardly be other than the grandfather of Caesar's lieutenant, who was governor of Farther Spain in 96-93 B.C. and fought out a war with rebellious tribes of the Atlantic seaboard.[86] But whatever the date at which the traffic from Corunna to Land's End attained its zenith, it did not outlast its historian Strabo by much. In the first century A.D. the opening up of new tin mines in Spain itself reduced the Cornish workings to inactivity and deprived Crassus's route of all commercial importance.[87]

The above extract from Strabo shows that the Carthaginian and 'Roman' traffickers in tin contributed but little to the general opening up of Britain. Thanks to this, and to the widespread discredit into which the story of Pytheas's travels had fallen, Britain still remained a mystery land in the days of Julius Caesar, and his military invasions in 55-54 B.C. were in the nature of an exploration.[88] Caesar received an early reminder of the risks which beset adventurers on unknown seas. Unaware of the daily variations in the Atlantic tides, he beached his vessels on the Walmer shingle, only to find that a flood-tide had carried them inland and knocked them about severely.[89] In 54 B.C. Caesar marched through Kent into Hertfordshire and Essex, and no doubt saw more of the interior than Pytheas. In one respect he was able to correct Pytheas, in that he ascertained the true measurements of the British coastline. But this was an isolated success which he obtained by careful questioning of the natives, and not by direct exploration. In general he added little to the existing stock of information on Britain, and he gave currency to an obstinate error in maintaining that one side of Britain faced north, while another looked 'westward towards Spain'.[90] Yet Caesar's conquests had at least the effect of throwing open the Channel to all comers. By the time of Augustus four lines of traffic between Britain and the Continent were in regular operation: from the Gironde, the Loire, the Seine, and the Rhine (by way of Boulogne). The fresh knowledge acquired by the traders of the Augustan age is reflected in Strabo's description of Britain:

> The greater part of the island is level and under timber, but it contains many hilly districts. It produces corn, cattle, gold, silver, and iron. All these are exported, and also hides, slaves, and powerfully built dogs for the chase. The men are of higher stature than the Gauls, and less fair-haired, but their flesh is more puffy. Here is proof of their size. We ourselves saw in Rome British striplings who overtopped the tallest men there by a clear half-foot; but they were bandy-legged and badly proportioned in all their limbs. Their customs resemble those of the Gauls, but are more simple and primitive. Some of them have plenty of milk, but cannot make it into cheese for lack of knowledge; they are likewise ignorant of garden cultivation and other husbandry. They are governed by hereditary kings: in war they mostly rely on chariots, like some of the Gauls. Their strongholds are forest enclosures: they fence off a wide ring with hewn timber and build huts for themselves, and stalls for their cattle, for a short season. Their climate is rainy rather than snowy. In the open air the mist holds on for long, so that in the course of a whole day the sun will only be visible for a few noontide hours.[91]

Occasional merchants also found their way from Gaul to Ireland; but they evidently did not follow the Irish coastline for any distance,

THE ATLANTIC

for they reported that it extended twenty days' sail from east to west;[92] nor did they acquire more than a vague knowledge of the interior, as will appear from the following extract from Strabo:

> Ireland is a large island, of greater length than breadth, extending along the north side of Britain. We have nothing definite to say about it, save that its inhabitants are wilder than the Britons, being cannibals and coarse feeders, and think it decent to eat up their dead parents and to have open intercourse with women, even with mothers and sisters. But for these statements we have no trustworthy authority.[93]

The following passage from Pomponius Mela, who wrote some fifty years after Strabo, shows but a slight improvement upon him:

> Above Britain lies Iuverna, of nearly equal extent, but of oblong shape, with two sides of same length. Its climate is unfavourable for the maturing of crops, but there is such a profuse growth of grass, and this is as sweet as it is rich, that the cattle can sate themselves in a short part of the day, and unless they are kept from continual browsing on the pasture they will burst asunder. The cultivators of the land are uncouth and ignorant of accomplishments beyond all other races, and are utterly lacking in any sense of duty.[94]

The exploration of the British inland was achieved by the Roman generals who were engaged on its conquest from A.D. 43 to 84. Most of these merely picked their way from point to point according to the exigencies of each campaign. But the last and greatest of them, Gaius Julius Agricola, opened up the whole stretch of country from York to Perth. Agricola's fleet at the same time followed out the east coast as far as the Orkneys.[95] From two obscurely worded passages in Tacitus it has even been inferred that the fleet circumnavigated Britain. But all that can safely be inferred from Tacitus's actual words is that the ships cruised along the entire eastern coast of Britain and rounded its northern end.[96] We do not know where Trutulum lay, and cannot be sure whether the 'nearest' coast of Britain should be taken for the south or the east coast: perhaps Tacitus merely means that the fleet made a double journey along the east coast. In any case, its cruises do not appear to have added much to the existing stock of geographic knowledge. From the works of Pomponius Mela and Pliny, who wrote during the early stages of the Roman conquest, we know that the Orkneys, the Shetlands, and the Hebrides had been placed on the Roman maps before the period of Agricola's governorship.[97] On the other hand Agricola's campaigns did not prevent the geographer Ptolemy from miscalculating by several degrees the latitude of northern Scotland and thus being reduced to imagining it as extending in an eastward direction, as if it were bent back on a hinge inserted at the Forth estuary.[98]

It was the intention of Agricola to conquer and open up Ireland as well as Britain, and it is possible that he went so far as to send a detachment to reconnoitre.[99] But before he could attack in force he was recalled from his command, and his successors left Ireland severely alone. Nevertheless the conquest of Britain facilitated mercantile intercourse with Ireland. Writing about A.D. 150, the geographer Ptolemy showed a fair knowledge of the entire Irish coast.[100] He not only enumerates several of the rivers on the coasts facing Gaul and Britain – the Barrow (Birgus), Moore's Avoca (Oboca), the Liffey (Lupias), the Boyne (Buvinda), and the Lagan (Logia), but also has the more remote Shannon (Senus) on his list. His names of Irish tribes are for the most part not difficult to identify. Most significant perhaps is his mention of Dublin (Eblana), which suggests an early connexion between this town and Roman Carnarvon (Segontium). The interior remained almost wholly unknown; but one late Roman author had found out that there are no snakes in Ireland.[101]

The continental coast of the North Sea was explored by the fleets of Augustus, in connexion with that emperor's attempt to extend the Roman frontier from Rhine to Elbe. In 12 B.C. his stepson Drusus sailed out of the Zuyder Lake (as it then was) by a specially constructed canal, captured most of the Frisian islands, and established a base at Borkum for operations along the Ems.[102] In A.D. 5 a Roman flotilla similarly supported the army of Tiberius on the lower Elbe and made a reconnaissance off Jutland as far as C. Skagen, thus considerably outsailing Pytheas in this direction.[103] But on several occasions the Roman squadrons suffered from gales on the Frisian coast, whose ill-defined shores and half-concealed sand-banks constituted a novel kind of danger to Mediterranean seamen.[104] It is probable that after the recall of the army to the Rhine in A.D. 16 the Roman fleets made no further excursions beyond that river. There is no evidence of Mediterranean sailors pushing on from Jutland to Norway, or turning C. Skagen and entering the Baltic. Had such explorations taken place, it would hardly have been possible for Pliny to describe Scandinavia as an island of unknown size in an archipelago of an Ocean bay, and Mt Sevo (in South Sweden?) as 'not less high than the Rhipaean range';[105] or for Ptolemy to transfer Pytheas's Thule to Shetland;[106] or for the writers of *mirabilia* to ignore such a natural wonder as the Maelstrom.

6. THE WEST AFRICAN COAST

Atlantic exploration in antiquity was mainly directed to the European seaboard. But the record of travel along the African coast, if briefer, is hardly less remarkable. In the central Atlantic the conditions of navigation differ considerably from those of European latitudes. Once a ship approaches the Canaries, it leaves the zone of westerly gales and is carried along by a north-easterly trade wind which persists through the summer. Beyond C. Verde it may encounter tornadoes, but it will have no lack of sheltered stations. On the other hand the Moroccan shore, with its cliffs and breakers, has a dangerous approach;[107] to the south of Mt Atlas the Sahara extends to the sea over ten degrees of latitude; and beyond the Senegal the sturdy negro tribes would be a fair match for white men without firearms. In the Middle Ages explorers were hindered by some strange fatality from doubling C. Bojador,[108] and it required all the crusading (and slave-trading) zeal of Henry of Portugal to break this spell. Moreover West Africa could offer to the ancients no such unique merchandise as the tin of the northern Atlantic.

In antiquity the West African coast appears to have remained totally unexplored until the arrival of seamen from the Mediterranean; and among the Mediterranean peoples the Phoenicians alone became familiar with any part of it. Nevertheless it is not certain whether these were the first to adventure themselves on the central Atlantic. As we saw on pp. 44–5, it is possible that a Greek pioneer named Midacritus preceded the Phoenicians in the northern Atlantic. With Midacritus we ought perhaps to couple a Massilian seafarer, Euthymenes. The one sure fact about Euthymenes is that he claimed to have seen a river on the west African coast whose waters were being driven back by an on-shore wind, and that in this river he espied crocodiles, which suggested to him that his estuary formed the upper reach of the Nile. We need not hesitate to accept Euthymenes's claim and to identify his river with the Senegal, where crocodiles still lie in wait for the unwary. But into what period does Euthymenes's journey fall? It has been thought that he was a contemporary of Pytheas and eluded the Carthaginian blockade of the Straits simultaneously with his Massilian compatriot. But it is equally possible that he escaped the blockade by living before its establishment. Euthymenes's theory about the source of the Nile suggests that he belonged to the sixth century, when speculations on this interminable river were rife among Greek geographers. An early date for Euthymenes is also indicated by the fact that he, unlike Pytheas, attracted no attention among the later Greek writers.[109] We may tentatively assign his voyage to the middle of the sixth century, i.e., soon after the cruise of Midacritus. But Euthymenes, like Midacritus, is little more than a name, and his voyage, whatever its date, had no practical consequences. The effective opening up of the West African coast was undoubtedly the work of the Phoenicians.

The earliest Phoenician cruises in the central Atlantic have been dated back by M. Bérard to the second millennium.[110] On this assumption he provides an ingenious explanation of the *locus desperatus* in Homer about the Ethiopians 'who are sundered in two parts, some near the setting sun, some near its rise'.[111] He locates the eastern Ethiopians on the upper Nile, the western ones in Senegambia, where presumably Phoenician mariners had discovered them in pre-Homeric days. But the Phoenicians probably did not commence the Atlantic explorations until about 500 B.C. (pp. 45–6). And in any case there is another explanation for Homer's bisected Ethiopians, namely that the eastern section were the Negroes of the Somali coast, whom the Egyptians had reached by sea long before Homer's day (pp. 74–5), and that the western moiety were the Sudanese, whose land stretched westward *ad infinitum* from the Nile valley.

THE ATLANTIC

7. HANNO'S CRUISE

If the first Phoenician voyages to West Africa did not precede the destruction of Tartessus, they probably followed it by no long interval. Their principal explorer in these regions, Hanno, was known as a contemporary of Himilco,[112] whose cruise cannot be dated much later than 500 B.C. (pp. 46–7). But since the primary object of Hanno's expedition was to establish colonies along the Moroccan coast, we may assume that this part of West Africa at least had been previously visited by Carthaginians.[113] The earliest of these visits may be placed close to 500 B.C., Hanno's voyage at any rate before 480 B.C.

The expedition of Hanno was probably the greatest achievement of Carthaginian seamanship, and is certainly the best known of all ancient voyages of discovery by sea. On his return to Carthage Hanno posted up a brief account of his adventures in the temple of Moloch, and this record by the explorer's own hand has survived in a Greek translation,[114] which was probably made in the first instance at the instigation of another African explorer, the historian Polybius.[115] Hanno's account, it is true, is not free from obscurities, and modern scholars are still at variance in estimating the range of his cruise and in fixing the intermediate stages: some contend that he did not get past C. Nun in Morocco,[116] others that he made the Camerun district, or even the Gabun estuary.[117] Hanno's story may best be told in his own words:[118]

The Carthaginians commissioned Hanno to sail past the Pillars of Heracles [the Straits of Gibraltar], and to found cities of the Libyphoenicians [Phoenicians residing in Africa]. He set sail with sixty vessels of fifty oars and a multitude of men and women to the number of 30,000, and provisions and other equipment.

After putting out to sea and passing the Pillars we sailed beyond them for two days. Here we founded our first city, which we named Thymiaterium [mod. Mehedia]. Under it a wide plain opened to view.

Thence we stood out westward and made C. Soloeis, a densely wooded Libyan promontory [C. Cantin].

Having founded a temple of Poseidon at this point we sailed on for half a day to east, until we arrived at a lagoon full of high and thick-grown cane. This was haunted by elephants and multitudes of other grazing beasts. [The marshes of the river Tensift.]

We skirted the lagoon for about one day's journey. Then we founded sea-side towns which we named Carian Fort [Mogador], Gutta, Acra [Agadir], Melitta, and Arambys.

Putting out from that point we reached a big river flowing from Libya, the Lixus. [The Draa, which still carries an imposing volume of water when in spate.] Lixite nomads pastured their flocks on its banks. We made friends with them and stayed with them for a time.

Beyond these dwelt inhospitable Ethiopians. Their land is infested with wild beasts and is broken up with high mountain-chains [the anti-Atlas], from which the Lixus is said to flow. These highlands are inhabited by a freakish race of men, the Troglodytes [Pygmies?], who are said by the Lixites to run faster than horses.

From them [the Lixites] we took interpreters and coasted along the desert [the Sahara] southward for two days. Thence we turned back east for one day. There we found at the top of a gulf a small island, with a circuit of five stades [*c*. half a mile]. Here we founded a colony named Cerne. We estimated from the distance traversed that it lay in a line with Carthage[119]; for it was the same distance from Carthage to the Pillars and thence to Cerne.

From that point we sailed through the delta of a big river named the Chretes, and came to a lake containing three islands larger than Cerne[120]. From there we accomplished one day's sail and arrived at the head of the lake. Beyond this a very high range of mountains rose like a tower. This was peopled by swarms of wild men in beasts' skins, who drove us off with stones and would not let us land [the Guanches].

Sailing on from that point we came to another deep and wide river, which was infested with crocodiles and hippopotami. Thence we turned back to Cerne.

From Cerne we sailed south for twelve days, skirting the land. This was peopled all the way with Ethiopians, who ran away from us and did not stay. Their tongue was unintelligible to us and to the Lixites in our company.

On the last day of the twelve we made fast under a high wooded range, with varied and fragrant trees [C. Verde].

It is now our turn to put back to Cerne in order to locate it. This is the most difficult problem in Hanno's story. Cerne has been variously placed near C. Juby, at Herne (near the Rio de Ouro), at Arguin

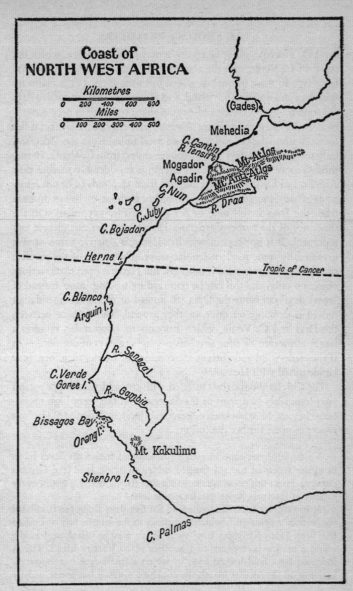

(near C. Blanco), and at Goree. We may rule out Goree, which lies south of C. Verde, and the region of C. Juby, which is to be sure from two to three normal days' sail from the Draa, but contains no wide bay and no off-shore island, and is much more than twelve days sail from C. Verde. Herne and Arguin alike are islands in deep easterly recesses. But Arguin measures about $2\frac{1}{2}$ miles in circumference; it lies fourteen normal days' sail (c. 800 nautical miles) from the Draa, and only seven to eight days' sail (c. 400 sea-miles) from C. Verde. Herne, though more than five stades in circuit, is considerably smaller than Arguin, and lies the required distance from C. Verde (c. 675 miles = 12 days' sail); moreover it has an excellent harbour.[121] But its distance from the Draa is full nine or ten days' sail (c. 525 miles). On the similarity of the names Cerne and Herne it is safer not to base any argument. It is not certain whether Herne is a native name or was invented by some modern scholar-sailor. Thus none of the possible sites fulfils all the required conditions. But Herne is open to one serious objection only, and this can be removed by reading 'nine' instead of 'two' days' sail from the Draa (θ' instead of β'). This emendation indeed is desirable on quite another ground. The distance between the Draa and C. Verde, which amounts to 1,200 miles, requires a larger allowance of time than the 2 + 1 + 12 days of our text, even if the trade-wind and current are considered. Cerne, therefore, may be identified with Herne.[122]

The Chretes and the river infested with crocodiles and hippopotami are presumably two arms of the Senegal. But the 'very high mountains' beyond have not been satisfactorily located. This part of Hanno's report requires further discussion.

We doubled the cape in two days' sail and found ourselves in an immense recess of the sea fringed with low-lying land [the Gambia estuary]. From this point we could see fires flaring up by night in every quarter at intervals, some greater, some less.

Here we drew water and sailed on for five days along the land until we reached a great gulf, which according to the interpreters was called the West Horn [Bissagos Bay].[123] In it lay a large island, and in the island a marine lake containing another island [Orang Island. This is indented by a land-locked bay, in which a small eyot is ensconced]. Landing on the smaller island, we could see nothing but forest, and by night many fires being kindled, and we heard the noise of pipes and cymbals and a din of tom-toms and the shouts of a multitude. We were seized with fear, and our interpreters told us to leave the island.

We left in a hurry and coasted along a country with a fragrant smoke of blazing timber, from which streams of fire plunged into the sea. The land was unapproachable for heat. [This phenomenon, which has been mistaken for a phosphorescent sea and for the play of lightning in a tornado, plainly was an extensive grass fire, which would appear to move like a stream as the wind drove it along. The streams, it is true, could not actually have reached the sea because of the mangrove swamps which fringe the coast,[124] but on a low-lying shore they might appear to do so, as seen from shipboard. The same awe-inspiring sight was reported on this coast by later discoverers, who clearly recognized that it was caused by blazing grass.[125]]

So we sailed away in fear, and in four days' journey saw the land ablaze by night. In the centre a leaping flame towered above the others and appeared to reach the stars. This was the highest mountain which we saw: it was called the Chariot of the Gods. [The mention of a very high mountain and of a shooting flame have been taken as evidence that Hanno saw Camerun Peak in eruption. This peak, with its 13,370 feet, is the only very high mountain in West Africa, and it has been active in recent years. But it is hardly credible that Hanno should have reached Camerun in the time indicated, for after passing C. Verde he would be moving into the belt of tropical calms, and on rounding C. Palmas he would be retarded by the strong Guinea current. Hanno's time-table cannot be adapted to such a long voyage without such drastic alterations of the text as most editors would shrink from. The actual data of time indicate that Hanno's mountain was Kakulima in Sierra Leone, which has only 2,910 feet, but shows up high in a flat country. A grass fire on its flanks could not be accurately described as leaping to the stars. On the other hand the fire of a volcano would certainly not set the land ablaze for a distance of four days' sail.]

Following the rivers of fire for three further days, we reached a gulf named the Southern Horn [Sherbro Sound]. In the gulf lay an island like the previous one, with a lake, and in it another island. [Macauley Island, which enfolds a smaller island in a tongue of sea, as at Orang.] The second island was full of wild people. By far the greater number were women with hairy bodies. Our interpreters called them Gorillas. We gave chase to the men, but could not catch any, for they all scampered up steep rocks and pelted us with stones. We secured three women, who bit and scratched and resisted their captors. But we killed and flayed them, and brought the hides to Carthage. [It is disputable whether these were a primitive race of human dwarfs, or, as seems more likely, chimpanzees, a tailless tribe, and eminently anthropoid, whose males are better runners than fighters. They were certainly not the powerful brutes whom we call gorillas.[126] In that case all the scampering would have been done by Hanno's party.]

This was the end of our journey, owing to lack of provisions.[127]

From the above account it appears that Hanno sailed as far as Sierra Leone, within eight degrees of the Equator. His outward journey extended over some 3,000 miles. If we assume that his predecessors sailed as far as the fringe of the Sahara, but no farther, Hanno discovered in one summer more of the West African coast than the sailors of the later Middle Ages in 150.[128] His cruise, therefore, is quite comparable with that of Pytheas. But, like Pytheas, Hanno did not have any successors. It is true that Carthaginian traders followed on his track and established a traffic in gold,[129] for which they would probably have to go as far as Cerne. From a remark of the Roman historian Caelius Antipater, that 'he had seen a merchant who went to Ethiopia for the sake of trade',[130] we may infer that this traffic persisted to the fall of Carthage in 146 B.C. Shortly after this event the Greek historian Polybius borrowed a fleet from Scipio Aemilianus and examined the West African coast as far as a river Bambotus, which was full of crocodiles – plainly the Senegal.[131] But if Polybius hoped thus to secure the gold trade for the Greeks and Romans, he was disappointed.[132] Whereas the fall of Carthage led to increased traffic in the northern Atlantic, it seems to have made a sudden end to West African commerce. The Greek traders, so far as is known, wholly ignored Polybius's researches; and the Roman soldiers in conquering Morocco (c. A.D. 42) made no use of a fleet in the Atlantic to support their operations by land. To judge by the sad confusion in Pliny's and Ptolemy's descriptions of Western Africa, these writers had little to go upon except a dim memory of Hanno's cruise.[133] Great as this explorer's achievements were, they did not make West Africa safe for geography.

8. THE CANARIES AND MADEIRA

In threading the West African coast the Carthaginian seamen could not fail to notice Fuerteventura, the easternmost of the Canary group, which is visible from C. Juby. We may therefore take for granted that they discovered the Canary archipelago. But there is no record of their visits, or of Punic settlements on the islands, which had no special attractions for mere merchants; and if any knowledge of the Canaries leaked through to the Greeks, it only did so in the form of vague rumours about certain 'Fortunate Islands' in the far West (pp. 235, 242). Neither Greeks nor Romans had sufficient faith in this land of bliss to make search for it after the fall of Carthage; and thus the effective discovery of the Canaries was left over to a king of Morocco named Juba (*c.* 25 B.C.–A.D. 25), who was the author of a book on African geography and had probably read the surviving Punic records of travel. Juba fitted out an expedition which brought back detailed and accurate information about the Canaries.[134] It enumerated six out of the seven habitable islands and characterized them correctly, duly mentioning the nuts and palms of 'Canaria' (Grand Canary), and the streak of cloud which the trade-winds condense on the Peak of Teneriffe ('Ninguaria', or 'Snow Island'). Its sailing directions, as reproduced by Pliny, were unhelpful: 'from the Purple Islands (presumably situated near the Straits) proceed 250 miles to W., then 375 miles to E.' But this is probably one of Pliny's blunders of transcription. Juba's party found traces of past habitation, but no living creatures except dogs, goats, and lizards. These remained the sole tenants, until the Guanches came in from the African mainland,[135] for Juba's expedition did not lead up to a settlement by Mediterranean peoples.

While no doubt exists as to the discovery of the Canaries in antiquity, the exploration of the Madeiras is open to dispute, yet may be regarded as probable. It is difficult not to recognize the main island of the Madeira group in a story of Diodorus, which relates how a Carthaginian vessel sailing along the African coast was blown out to sea, and at several days' sail from the mainland happened upon a large-sized island with fertile plains, well-wooded mountains,

THE ANCIENT EXPLORERS

navigable rivers, and a perfectly tempered climate. The knowledge of this discovery was about to be put to use by the Etruscans, who were powerful on sea at that time. But the Carthaginian government, which was generally friendly to the Etruscans, yet not in such measure as to allow them a share in Atlantic traffic, headed off the projected settlement and did its best to suppress information about the new health resort.[136] But that Madeira has no navigable streams, the above description fits in singularly well. Its size c. 40 × 15 miles, would entitle it to be called big on the Greek or Carthaginian standard; and the false detail about rivers, which the explorers (or the arm-chair travellers who passed on their tale) might have inferred from the regular summer rains of Madeira, is hardly sufficient to discredit Diodorus's narrative. The discovery is dated by the reference to Etruscan sea-power, which falls within the narrow period 535 B.C. (battle of Alalia) and 474 B.C. (battle of Cumae).

There is no further evidence of visits to Madeira until 80 B.C. At this time, according to Plutarch, some sailors from Gades found their way to certain 'Happy Isles', and on their return to Spain reported their adventure to the Roman exile Sertorius. The isles were two in number, at a distance of 10,000 stades (1,100 miles) from Africa; they enjoyed shelter from north and east winds, but stood open to balmy breezes from west and south which brought a gentle and fertilizing rain.[137] The distance here given suits the Azores better than Madeira, but it does not exceed the correct measurement (700 miles from Cadiz) so far as to damage Plutarch's story fatally. On the other hand the Madeira group does in fact contain two inhabited islands, and the climate of Funchal could not be better described than in Plutarch's words. We need not doubt that Sertorius's informants re-discovered Madeira; but their venture led to nothing. Sertorius, who had thought of retiring to Madeira, found other occupation, and his project of a settlement died with him.

While we admit the discovery of Madeira, we must probably reject a like claim for the Azores. We have already dealt with the assumption that they were visited by Himilco (p. 46). We may confidently dismiss the theory that they were the Cassiterides.[138] The reputed number (ten) of these coincides with that of the Azores. But the Cassiterides lay to north of Corunna, the starting-point of the tin merchants, not to south-west.[139] It is alleged that Phoenician gold coins with a horse or horse's head engraved upon them were unearthed in 1749 on Cuervo Island. Gold pieces of this type were indeed struck by Carthage.[140] But it is a sound rule to discredit insufficiently authenticated coin-finds, and adequate corroborative evidence for the Cuervo haul is lacking. The known facts indicate that the Azores were not discovered till the Middle Ages.[141]

9. ALLEGED VOYAGES TO AMERICA

Did any ancient sailor cross the Atlantic from end to end?[142] Contact between Europe and America has been inferred from similarities between American and Mediterranean place-names, from resemblances between ancient Mediterranean culture and the civilization of Mexico, and from reputed finds of Greek or Phoenician coins on American soil. But the homonyms are conspicuously few, and were they more numerous they could still be explained as mere coincidences. The similarities of culture can safely be attributed to parallel development. The supposed ancient coins mostly fail to materialize when search is made for them; the few which have been traced are certified forgeries.[143]

The only other evidence of transatlantic voyages is derived from some casual anecdotes in ancient writers. According to Pliny and Mela, Metellus Celer, who was proconsul of Gaul in 60–59 B.C., received from the king of the Celtic tribe of Boii a present of some 'Indians', who were said to have been cast up on the German coast.[144] Could these unfortunates have been Indians of the red variety? It is conceivable that a kayak should have been blown across the Ocean without being swamped,[145] and that Metellus should have mistaken Redskins for Hindus. But the whole story can be better explained on the supposition that the shipwrecked men were real Hindus (p. 198). A yet stranger tale was recounted to the antiquarian Pausanias (c. A.D. 170), à propos of satyrs in art and real life.[146] A ship belonging to one Euphemus of Caria was once upon a time blown through the Straits of Gibraltar and across the Ocean to an island inhabited by Redskins with horses' tails, who frightened the new-comers away by their menacing attitude, and behaved with abominable lewdness towards one unlucky member of the ship's company that had been unchivalrously abandoned to the satyrs. Could these beasts have been natives of the Antilles? According to the Spanish descubridores the Antillians were of red hue, wore (detachable) horse-tails, and had very disgusting habits. It is just possible that the 'levanter' of the western Mediterranean should have carried off Euphemus as it undoubtedly bore away Colaeus (pp. 33-4), and that the same persistent breeze which wafted Columbus to America should have driven a previous traveller on the same latitude. But the whole of Pausanias's story excites suspicion by its vagueness. When did Euphemus live? How many days did he take to sail across? How did he contrive to return to Europe? Such detail as is supplied is of just that sensational sort with which sailors of all countries and ages impart a caerulean colour to their yarns. In all probability Euphemus the Carian was nothing more than a nasty-minded relative of Plato's figment, Er the son of Armenius. There is no need to transfer to him the credit which is usually reserved for Columbus.

CHAPTER 4

INDIAN WATERS

I. GEOGRAPHICAL

THAT part of the Indian Ocean which is called by this name in modern atlases and maps hardly comes within our story; the two parts of it which concern us are generally represented under different names. Thus the irregularly shaped expanse of waters bounded by India on the east, Arabia on the west, and southern Persia and Baluchistan on the north is called the 'Arabian Sea', while the two sides of a triangle of which the eastern coast of India is one and the coastline of Burma and Malay is the other form the 'Bay of Bengal'. To the south of both seas lies the 'Indian Ocean' proper, being roughly south of a line drawn from Cape Guardafui, the easternmost point of Africa, through Cape Comorin, the southernmost point of India, to the northern end of Sumatra. The Arabian Sea is brought near to the west by its two long and narrow inlets, the Red Sea and the Persian Gulf, and is marked by the regularity with which moderate winds called monsoons (Arabic *mausim* – season) blow between east Africa and north-west India. From November to March the wind blows steadily from the north-east, being steadiest and strongest in January, while from the end of April to the beginning of October the wind blows from the south-west. In the Bay of Bengal the winds are irregular. The currents of the ocean are controlled by these monsoons, while the tides about Arabia and north-east Africa vary generally between rises of 5½ and 8 feet.

Of these eastern waters one small part, the Red Sea, has taken its name from the usage of the Greeks. The waters of this gulf move with much irregularity, and make navigation hard and perilous; storms are frequent and the summer climate is most unpleasant. The Gulfs of Suez and Akaba at the northern end, where only are tides really seen, are difficult to navigate because of awkward winds. Still, from June to August a useful north-west wind blows down the sea throughout its length.[1] The Greeks at an early date applied their word *erythra*, meaning red, to this narrow sea between Egypt and Arabia. But in course of time they stretched this name so as to cover not only the Red Sea but also what is now the Arabian Sea and also the Persian Gulf, and in this chapter, when the word Erythraean occurs, it is used in this extended Greek sense. Some Greeks said that since the sea was not red, the name perpetuated that of king *Erythras*, a Persian who held sway along the Persian Gulf.[2] In modern times it has been plausibly suggested that this story hides a truth which may be that Phoenicians ('Red', that is 'Sunburnt' men), who were settled in the Bahrein Islands of the Persian Gulf, sailed, and even migrated, round the coasts of Arabia, leaving their name in several places.

2. THE EARLIEST EXPLORERS

Naturally, the earliest explorers of the 'Erythraean' Sea were the more advanced of the peoples who dwelt on, or could easily reach, its coasts: the Egyptians, the East Africans, the Arabians, the civilized races of Mesopotamia, the Phoenicians, the peoples of the Iranian plateau, and the Indians all knew its waters, and of their explorations we shall have something to say, though we are in effect talking about their own sea. Until the time of Alexander the Great, Greek expansion took place along uninterrupted sea-lines in the Greeks' own area of the Mediterranean Sea, and did not extend to waters or lands on the farther side of intervening peoples. Only after Egypt became a Greek country did eastern waters begin to attract the Greeks.

Were the Phoenicians the first to make long sailings in eastern waters? The latest discoveries of archaeology lead us to believe not. Thus the predynastic Egyptians used boats like those used in the Persian Gulf, and the first dynasty used seals like the early Sumerian. Again, a curious carved handle of an Egyptian stone knife, dating from predynastic times, that is, before 3500, shows us early Egyptians who sailed in high-prowed, high-sterned ships, and fought with men who worshipped Babylonian gods of Sumerian date.[3] A little later we find, round about 3000 B.C., the metal-work of Egypt (especially of the sixth dynasty) and that of Ur distinctly like each other.[4] This looks like communication between two cultures by sea, as well as by land, before any race had grown to power in any part of coastal

Arabia. But this power soon came and expressed itself in men of 'Himyarite' stock, who, after any Phoenician movement in eastern waters had been completed, settled round the coasts of Arabia and set up astonishing barriers of commercial secrecy against all traders of other races. Some scholars indeed think that the origin of the name 'Red Sea' as applied by the Greeks to the whole bulk of oriental waters is really 'Himyar' or 'Red', the eponymous hero of these Arabs.[5] At any rate, the voyages of Egyptians and Mesopotamians, the former using Sauu (Kosseir), the latter using Eridu (Abu Shahrein), even in their greatest periods, became restricted in their length, with certain exceptions of rare occurrence. Thus the land of Punt or Puoni (Somaliland west of C. Guardafui and perhaps the coast of Yemen also), and 'God's Land' (south-west Arabia), first visited under Sahure of the fifth dynasty (*c.* 2965–2825), were for many ages the general limits of Egyptian exploration by sea, which seems, however, to have reached Socotra, for a tale of the thirteenth dynasty speaks of Socotra as 'island of the Genius', Pa-anch, ruled by the King of Punt (see pp. 233–4). But after the fall of the twelfth dynasty (*c.* 2212–2000) Egypt had much trouble till the glorious eighteenth dynasty (*c.* 1580–1322) reached a culmination in the great expedition beyond Bab el-Mandeb sent by queen Hatshepsut to Somali and Socotra, included in the land of 'incense-terraces', and along the coast of Arabia to Hadramut, and even beyond to the neighbourhood of the Kuria Muria Islands (Gnbti) and Dhofar. The reliefs of the temple at Deir el-Bahri, recording Hatshepsut's great penetration, show us that hitherto the lands outside the Red Sea were known chiefly by hearsay and the precious things which men desired were paid for only by much money. For the god Amon Re says: 'No one trod these incense-terraces, which the people knew not; they were heard of from mouth to mouth by hearsay of the ancestors. The marvels brought thence under thy fathers, the kings of Lower Egypt, were brought from one to another, and since the time of the ancestors of the kings of Upper Egypt who were of old, as a return for many payments: none reaching them except through carriers.' But Amon Re led the expedition and brought back many wares of worth. The cattle on the reliefs are not the humped of Somali, but those of south Arabia, while the incense trees shown are like the leafy ones of Dhofar rather than the barer ones of Somaliland. Under this dynasty the voyages to Punt at least (taken as being the coasts of the Gulf of Aden) were made every year, and often under the nineteenth and twentieth dynasties (1346 onwards).[6] By this period the limits of exploration were reached.

Again, for many ages, especially in the first half of the third millennium, the voyages of the people of Sumer, Akkad, and Lagash, from their Mesopotamian port Eridu, extended to Magan (El Hasr), Melukhkha (Oman), and other regions on the southern shore of the Persian Gulf, but no farther, though the monarchs of Ur seem to have been forward in trying to push right round Arabia so that the name Magan came to be applied to a coast of the Red Sea, while to the Assyrians Melukhkha had come to mean Ethiopia. Again, in the Babylonian story of Gilgamesh (of Elam?), the farthest places reached, 'Isle of the Blest', and 'Waters of Death', would seem to be Socrota ('Dvipa Sukhadara' – island abode of bliss), and the Straits of Bab el-Mandeb; and there are traces of an Elamite migration round Arabian coasts.[7] Still, all through the Arabians had maintained a stiff barrier along their outer coast. Did they also, and the Phoenicians, as has been maintained, use, for coasting at any rate, the monsoons, so as to reach Sofala and Rhodesia in South Africa, and the marts of India to the east? In the days of King Solomon (tenth century) the Phoenicians took part in voyages to Ophir between Sheba and Havilah. Ophir has been identified with regions of East Africa, India, Ceylon, Australia, and such remote and impossible countries as New Zealand and South America, but it should be sought somewhere along the south coast of Arabia, whither the Arabians brought gold from more distant Havilah, which we may place in north-east Arabia,[8] rather than down the east coast of Africa, though the renewal of the voyage every three years suggests a trading expedition like those which Arabians even now take from the Red Sea to Muscat and thence to Malabar and back to Africa and as far south as Madagascar or even Sofala,[9] and none can dispute the influence, and even more, of Arab races along the east coast of Africa. In due course the ancient Arabians left traces of their tradings in Baluchistan, Makran, and along the north-western coast of India.[10] Can we go farther and trace Phoenicians to Sumatra and even claim for them a colony at Shantung about 680 B.C.?[11] It is safer not to go so far as this, but to admit Arabian coasting voyages as far as north-west India and beyond Guardafui, and complete control of the Persian Gulf trade by Phoenicians before 650 B.C., and to conclude that in general the peoples round the Erythraean Sea for centuries met each other on more or less short voyages and did not as a rule wander far and wide on longer travels. In this way, indeed, the Arabians and Phoenicians may really have done something to spread, as Professor Elliott Smith propounds, on its long slow journey eastwards to America, the inspiration rising from the great stone monuments of Egypt.[12] Many smaller half-way voyages were taken along the coast between the Indus and the mouth of the rivers Tigris and Euphrates, in connexion with the commerce of the Hittites and of the Sumerians, Assyrians, and Babylonians, through the Chaldaeans, a trade existing about 3000 B.C., but much increased under Sennacherib, and revealed in discoveries made at Ur, for example, and in the palace of Nebuchadrezzar, and also in the Indus valley,[13] a trade which must have been like that shown by Ezekiel, by the Greek writer Theophrastus, and by an anonymous Greek writer of the Roman Empire. The voyages made by Indians such as we hear of in the *Rig Veda*, in *Jataka* stories, and in the *Mahavamsa* of Ceylon, were probably coasting voyages of this kind, and as is shown by the *Puranas* (which we cannot date) there was an age-old coasting commerce between north-west India and eastern coasts of Africa, indicating a use of the monsoons in the simplest way, that is with the wind straight behind the ship.[14]

3. THE EARLY GREEKS AND PERSIANS. SCYLAX

But it was undoubtedly the Arabians, chiefly in coasting voyages, who were the most persistent in Indian waters, visiting on secret carryings East African, South Asian, and Indian coasts with fair frequency, at the time when the Greeks were developing their culture in the West. No Greek appeared in the 'Erythraean Sea' before the time of Darius. The Greeks were a Mediterranean people with a Mediterranean mind; their world was Greece and coasts of the Mediterranean Sea. When a foreigner barred the way, there the Greeks stopped. They might desire to increase their knowledge, but there was little economic motive to push them on, for travelling and the purchase and carriage of wares were costly; nor was there a demand in the West for eastern luxuries until the rise of Rome to imperial power. So knowledge and wares of the East came to them by way of carrier-peoples. Whereas in the tenth century Hebrews and Phoenicians were sailing to Ophir, the Greeks of Homer knew nothing beyond Thebes of Egypt and Joppa. In the East the Arabians, stretching from Suez to the Euphrates behind the localized Hebrews, blocked their way. Above all, Egypt, as it were the gate to the Red Sea, was until the middle of the seventh century a closed land, and even the rise of Greek Naucratis on the western channel of the Nile, and the opening up of Egypt[15] (though not the Red Sea) to Westerners by Psammetichus I, Apries, Necho, and Amasis II, and the report of the circumnavigation of Africa by Phoenicians in the time of Necho (see Chap. 5) did not bring the Greeks into eastern seas, though that monarch made good use of the Red Sea.[16] The Greeks might desire to find out what lay in the distance towards the east, but until tales of the wonders, real and unreal, of India came to their ears, they had no encouragement to push through the barriers of intermediary peoples.

The first nation to draw the attention of the Greeks more actively towards the East was the Persians, though for the most part, being a land-power, they naturally taught the Greeks more of eastern lands within their empire (Chap. 7) than of eastern waters. Yet it was at Persian instigation that the Erythraean Sea was first navigated by a man whom we can name. Curiously enough, he explored not from west to east, but from east to west. This man, a Carian of Caryanda named Scylax, was sent with other trusted persons, probably Ionian Greeks, about the year 510 by Darius Hystaspis (who wished to know where the Indus flowed into the sea), from a town near the modern Attock, to explore the river Indus to its mouth and then to coast along the shores westwards, and, avoiding the Persian Gulf, which the Persians already knew about well enough, to sail round unknown Arabia into the Red Sea where they held Egypt. He was ordered to study the coast-line and to report upon it. Entering the Indus from the Kabul river, Scylax carried out what he had undertaken and after two and a half years' adventurous voyaging, and doubtless trading, came to Arsinoe, near the modern Suez. But the Greeks did not follow his example, though Darius, perhaps as a result, added Sind to his

INDIAN WATERS

Punjab conquests, and certainly made use of Indian waters, reopening the old canal between the Nile and the Red Sea. It is possible, says a hieroglyphic inscription of Persian date found along this canal route, 'for ships to sail direct from the Nile to Persia over Saba'. Darius claims that ships passed from Egypt to Persia round the Arabian coasts. Owing chiefly to lack of details in Herodotus's account of Scylax, who came from a place so close to that historian's own birthplace, doubts have been thrown upon the reality of Scylax's voyage. But the evidence of those who doubt is entirely negative. Since Nearchus (about whom we shall tell shortly) wrote his account of exploration from the Indus to the Euphrates in Ionic Greek, it is likely that the Ionians really possessed a report of Scylax's voyage. The length of time taken by Scylax might easily have been due to his being delayed by an adverse monsoon. His name was remembered so well that it was tacked on to a much later coasting record which had nothing to do with the Erythraean Sea. We must remember too that in those days there were no newspaper-presses and book-presses which would have circulated thousands of copies of Scylax's work in a short time.[17]

How much of Hecataeus's writings about the Indus districts, and how much of the travellers' tales which Herodotus gives about South Arabian trade, comes from Scylax, we cannot tell.[18] But of exploration by Greeks along Arabian and African coasts there is little to record, probably because little was done. In this respect there is almost a blank until we come to the explorations, conquests, and never completed commercial schemes of Alexander. The Persians, still holding Egypt, traded round south Arabia from Egypt's ports at Kosseir and Suez; but how much they sailed themselves on eastern waters, and allowed Greeks to do so, we cannot tell. After 448, when peace was made between Persia and the Athenian Empire, a few Greeks seem to have reached the chief port of Yemen outside the Red Sea. Thus, while Aeschylus (before 448) gives us merely the red-bottomed stream of the Red Sea, Herodotus (soon after 448, when he travelled in Egypt) learnt of the shape of that sea, of various tales which the Arabians told to all western visitors to prevent them from finding the real eastern sources of precious gums and aromatics, and of long-living Ethiopians, some of whom were Somali who dwelt outside the Arabian Gulf at the ends of the earth on the borders of the 'Southern Sea'. Aristophanes in 414 knew of a 'happy city by the Red Sea', which Euripides before 406 knew as 'Arabia the happy' (eudaemon), not far from sea-washed Asia. Later on this town (today Aden), with its district, was named definitely 'Arabia Eudaemon'. Euripides also echoes hearsay reports of the Somali coasts. Neither he nor Aristophanes had entered Asia or Egypt and yet they could write about these things in plays.[19] But of East Africa, and of the Persian Gulf, no Greek yet knew, and many believed that the Erythraean Sea was a lake which joined Africa to India, the Nile to the Indus.[20]

4. NEARCHUS; FROM INDUS TO EUPHRATES

We can trace no further Greek exploration along the coasts of Arabia and Africa for two generations. Nearly all that Greeks learnt about Arabia, Iran, and India came from the Persians, and the success of Ctesias's book on India (see Chap. 10) shows that they could not tell true stories of distant lands from false ones. But the conquests of Alexander opened up entirely new prospects. As we read in Chapter 7 the outward campaigns of the great Macedonian opened up Asia by land. But when he had explored the Indus, he decided to explore and open up a route by sea from the Indus to the Euphrates. For this enterprise he chose Nearchus, and we are fortunate in having a reproduction of Nearchus's own report embodied by Arrian who wrote centuries later. Again it is exploration from east to west.

With 150 ships and crews, 5,000 all told, and, besides these, armed units for protection, put under his command, Nearchus was to expect provisions from Alexander's army as it marched parallel with the exploring fleet through Baluchistan, but it turned out that after the first part of their course the voyagers had to land and search constantly for food and drink along the coast, fish-meal and wild dates being almost their only fare. Nearchus was to wait until the north-west monsoon, now known for the first time by the Greeks, began to blow, but was forced to leave a month early because of hostility on the part of the Indians. After a start late in September 325, the gusts of the changing monsoon caused twenty-four days' delay at Crocola (Karachi), whence the voyagers reached the river Arabis (Purali, if not the Habb) in five days. As they sailed along the coast of the Oritae, halting twice, they lost three ships in a great gale, but the crews were saved by swimming. On reaching a part of the coast (Ras Kuchri or Kachari) near Cocala (not far from Gourund), Nearchus anchored, landed, fortified a camp, and obtained touch with Alexander's officer Leonnatus, and so was able to replenish his supplies. With his ships again in good repair, he sailed with a fair wind to the torrent Tomerus, that is the Muklow or Hingol. Here the Greeks landed in spite of opposition on the part of about 600 real savages; for near the mouth 'dwelt men in stuffy huts; who when they saw the fleet sailing towards them were astonished, and extending themselves along the shore made a battle-line with an eye to fighting against any who disembarked. They carried short spears about 6 cubits long; the points were not made of iron, yet the sharpened ends, heat-hardened with fire, were just as useful'. On defeating them, Nearchus was able to take a closer look: 'The captives were hairy over their heads as well as the rest of their persons, and had nails like wild beasts; at least they were said to use their nails like iron tools, and to kill fish by tearing with them, and to cut up woods of the softer sorts. The other kinds they cut with sharp stones. For they had no iron. For clothing they wore the skins of wild beasts, some indeed wearing the thick skins of the bigger fish' (porpoises or whales, not fish). Onesicritus's report of the voyage shows that the fleet noticed that some of the primitive men along the coasts of Gedrosia (the modern Makran) and also Carmania were 'turtle-eaters', who used the flesh of turtles for food, and their shells for hut-roofs.

After more repairs, Nearchus sailed on the sixth day and soon reached Malana (Ras Malan), and from here as far as Cape Jask it took him twenty days to coast along the lonely but not waterless coast of the savage Fish-eaters of Gedrosia. On the first day's sail Bagisara (C. Arabah), on the third Calama (by the small river Kalami) was reached, where they received food from the natives. About 100 stadia out lay the island Carbine, reported to be uninhabited:

The natives said that this island was sacred to the Sun and was called Nosala, and that no man was willing to put to shore on it; whosoever came to land there in ignorance, vanished. Nearchus indeed says that one boat with a crew of Egyptians vanished not far from this island,

and the chief officers of that ship strongly urged about this, that they vanished having put to shore on the island through folly. Nearchus sent a thirty-oared ship to sail in a circle round the island, with orders not to bring it to land, but to shout to the men as they sailed as close by as possible, and to call out the name of the helmsman and any other well-known name. But as none heard, he says he sailed to the island himself, and forced the unwilling sailors to land; and that he stepped on shore himself and proved the story about the island to be a fable.

He heard also that it was once inhabited by one of the Nereids who made love to all who landed, turned them into fish, and threw them into the sea. But the Sun in pity and anger made them men once more.

This enchanted Mermaiden's Island, found by Nearchus, is Astola, about twelve miles out from the Kalami, and it is still a centre of sun-worship. In 1744 Captain Blair was warned by the natives of Pasni that the island was enchanted, and that a ship had been turned into a rock there. Nothing daunted, he landed, found many fine turtles (which were probably the reason why the people of Pasni did not want him to land) and saw a rock which from afar looked like a sailing ship. Until 1820 a terribly savage race of piratical inhabitants used to destroy, with all on board, ships which came near.

Before reaching port Mosarna (Pasni town and Ras Jaddi) the grain-supplies of the fleet were running short, and wild goats were seized at Cysa. A native guide Hydraces was obtained at Mosarna, and the voyage was for a time less trying after that. Fertile land and not altogether savage folk were found near the inland village Barna (Ras Shamal Bunder). On putting in at port Cophas (near Ras Koppah) for water, the fleet were scornfully surprised to see there, just as one can today, fishermen paddling in canoes instead of rowing in boats with oars like respectable Greeks, and off Cyiza a curious thing befell them. 'Nearchus says they saw in the east water blown upwards from the sea, as though it were carried along by a blast of a whirlwind. The astonished seamen asked the guides of the fleet what was and whence came this happening, and they answered that it was whales which, moving through the sea, blow the water aloft; whereupon the oars fell out of the astonished seamen's hands.' Nearchus encouraged them to attack the animals head-on, as it were in a battle; so they bent to their oars, and 'by the time they were coming near to the beasts, they shouted aloud to the full capacity of their heads' (we should say 'lungs'); 'the trumpets blared; and the rattling of the oars sounded afar. And so the whales, now seen off the very prows of the ships, plunged astonished into the depths, and not long afterwards rose up again off the sterns, and blew up great masses of sea again'. They learnt that at some places whales were sometimes cast up on shore and left their bones to become materials for natives' houses. Soon the problem of supplies grew so urgent that it was found necessary to land in the bay of Gwattar, and attack a small city which looked promising. As Nearchus approached, the inhabitants in friendly manner gave them toasted tunnies, cakes, and dates, but he started to deal deceitfully with them, and held part of the wall. Then 'an interpreter with Nearchus's fleet proclaimed that they were to give corn to the army, if they wanted to keep their city safe; but they denied having any, and attacked the wall', without success. Nearchus was allowed to search the place, and his scouts 'showed him much meal pounded out of roast fish, but of wheat and barley very little. For it happened that their grain-course was made from fish, while they used bread-cakes as it were for dessert'. Nearchus made the best of what was available, and set sail again. He put in again at several places, but as the want of good food grew worse, he became afraid lest landing might be followed by desertion. The sailors ate fish-meal, heads of wild dates, a little corn and date-fruit and seven camels found in miserable villages, until they reached Cape Jask and the much more fertile land of Carmania (Kerman) where fruit and corn were obtained. Little indeed had the Fish-eaters provided, who for the most part did not even build boats, but made only huge nets out of date-tree bark, in which they caught fish. The smaller ones they ate raw as soon as taken from the water, the larger and tougher they dried in the sun and made into a kind of bread. Prickly-crabs, oysters, and cockles were also taken, and salt sauce. Only a few ate grain-bread as a savoury with their fish; only the 'richest' could use the bones of whales to build their houses with; most of them had to build theirs with fish bones.

The voyagers soon saw a great headland which they learnt to be Cape Maceta, that is Ras Mussendam of Arabia. Nearchus rejected Onesicritus's suggestion that the fleet should cross and sail along the

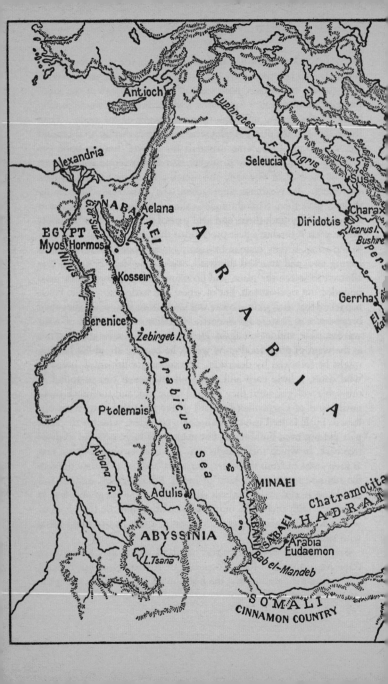

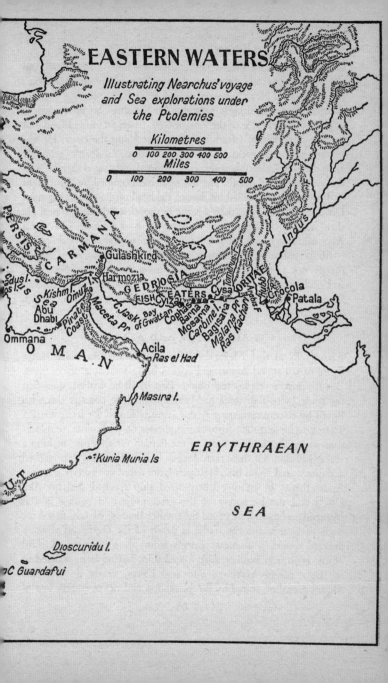

Arabian shore of the Persian Gulf, which was now to be entered, and the fleet came to rest by the river Anamis (Minab) in fertile Harmozia (near Ormuz). Nearchus went inland from here, meeting joyfully Alexander at Gulashkird; returning to the ships he proceeded to explore the north side of the Persian Gulf, passing Ormuz Island, but putting in at Kishm Island, and elsewhere along the coast. He observed the shoals, rocks, and reefs, and on one sandbank (probably Bassadore on Kishm) their ships ran aground and caused three weeks' delay. Noting full details, like the headland occupied by Abu Shehr or Bushire, Nearchus led his fleet to the head of the gulf at the mart Diridotis at what was then a separate mouth of the Euphrates. He then went back and up the Pasitigris to Alexander at Susa.[21] An Alexandria was founded at Charax to open up the new trade-route.

About half a year had been spent upon this exploration, which though not a great one, reflects credit upon the ability of Nearchus, and made Greeks realize the Persian Gulf as an inlet of the Erythraean Sea. As we shall see, it was not destined to have very brilliant results.

5. ALEXANDER AND ARABIA

Arrian tells us that Alexander, after his return from India, had a desire to sail round Arabia and Africa, being insatiable of conquest. But commerce was his real object. First of all he wished, now that the route Indus-Babylonia had been explored, to connect this with Egypt by circumnavigating Arabia, already famed for its aromatics. Therefore he sent men from the Euphrates on preliminary explorations along the Arabian coast of the Persian Gulf, before making a real expedition. Of these men Archias in a thirty-oared ship reached and discovered Tylos Island (Bahrein), but dared not go farther; then Androsthenes in another thirty-oared ship reached perhaps Abu Dhabi and the beginning of the great projection ending in Ras Mussendam beyond, surveyed the smaller Bahrein islands, noted the trade of Gerrha and the traffic in pearls of the Gulf, and wrote a detailed report with measurements from place to place. A third officer Hieron in another thirty-oared ship, having been ordered to sail right round Arabia to Suez and Egypt, nearly doubled Ras Mussendam, but, scared by the barrenness and never-ending line of the coast, went back and reported that Arabia was nearly as big as India. Besides these, ships were sent out from Suez down the Red Sea, probably to meet Hieron. They reached Yemen outside the Straits, but went no farther through lack of water (so their leaders said); nevertheless they learnt the names of the prosperous Arabian tribes of Yemen and Hadramut, and their trade in aromatics, and raided some unguarded incense-trees.[22] Alexander was preparing for his main effort when he died in June 323. Had he lived longer the full use of the monsoons might have been discovered now instead of more than three centuries later. From now onwards the motive of gain encroaches upon those of conquest and knowledge, and our story resolves itself next into the efforts of the Ptolemies to pierce the Arabian barrier in the Indian Ocean, chiefly for commercial reasons.

6. THE PTOLEMAIC GREEKS IN THE 'ERYTHRAEAN' SEA

(a) The Exploration of the Red Sea and the Gulf of Aden

The Greek dynasty of the Ptolemies in Egypt was soon firmly established by Ptolemy I Soter (son of Lagus), 323–285, with Alexandria as a centre of government, culture and commerce. He had been with Alexander, and had seen the Indus mouth, and Egyptians had sailed with Nearchus; men might hope therefore to be able to trade from Alexandria to Patala on the Indus. But they soon found that the barrier of the Arabians was impassable, and there was not yet a Roman demand for eastern luxuries and Roman money to back enterprises. Ptolemy I built large ships and sent an admiral Philo down the Red Sea to explore. This man seems to have gone some way down the African side of the sea, and brought back chrysolites from Isle Zebirget which he discovered.[23] He probably was the first to inspire the Egyptian government with a desire to obtain elephants (for use in war) and ivory from the districts south of Egypt by voyages down the Red Sea – a desire which was fulfilled by Soter's successors. The second Ptolemy (Philadelphus), 285–246, took up the matter of Egypt's southern relations seriously. He opened up trade between Egypt and the Sabaean Arabians of Yemen and with those who held the Somali coasts, and began the penetration of Africa south of Egypt inwards from the Red Sea. To help these objects he founded new trading ports along the Red Sea from Suez to Ras Benas, establishing well-provided routes to them from the Nile, and sent out Satyrus to explore the shores of the 'Trogodytes' on the African coast with the object of fixing on suitable stations from which to conduct elephant-hunts. As a result, between Ras Benas and Bab el-Mandeb a number of stations were set the first being Ptolemais, where Eumedes netted some elephants. Philadelphus had great trouble in inducing African elephant-eaters to refrain from joyously chopping buttock-steaks off living elephants for food, in order that he might capture the beasts alive and uncurtailed. Besides this, it was apparently he who sent out from the Nile canal and Suez Aristo to explore the Arabian coast to Bab el-Mandeb before sending fleets to the Hedjaz coast and Al-'Ula (278–277). Pythagoras too was sent on a similar exploration and examined some islands off the Arabian coast.[24] A few men sent out by Philadelphus (for state enterprises were a feature of the early Ptolemies) began to explore the 'Cinnamon country' – that is, the Somali coast west of Guardafui,[25] but this region was really discovered through the increased efforts of Ptolemy III Euergetes I (246–221), who made the elephant-hunts a military organization. He put down piracy in the Red Sea, and sent out Simmias to re-explore the western coast inside the sea and to examine the Somali coast outside with a view to elephant-hunts, and possibly the Arabian coast with a view to trade.[26] Another Philo was sent to 'Ethiopia', and after him, as Strabo tells us, Peitholaus, Lichas, Pythangelus, Leon, and Charimortus, who all shared in setting up hunting-stations and giving their names to geographical features both inside and outside the Red Sea, and all set up pillars and altars on the Somali coast showing the farthest points reached by them. This idea of giving one's name to places caught on, and soon a number of features in the Red Sea received names of their explorers or first surveyors. From an inscription we know that Lichas was sent out twice, the second time at any rate by Ptolemy IV, on elephant-hunts. More interesting still is the inscription which is now in the British Museum. It tells how, after 209, under the fourth Ptolemy, Alexander, second in command of an elephant-hunting expedition conducted by Charimortus (mentioned above), and a subordinate named Apoasis and his soldiers, gave thanks to Ares, Victory-Bringer and Good-Hunting-Giver. Loans were

INDIAN WATERS

sometimes raised for visits to Somaliland.[27] At one time Euergetes was master of Babylonia, and sent some ships exploring from the Euphrates through the Persian Gulf on the way towards India in trade rivalry with the Seleucids, yet they do not seem to have gone outside the Persian Gulf, but simply to have reached El Katr, and to have sailed across to the pirate coast leading out to Cape Mussendam.[28]

The death of Euergetes marked the end of the best period of Ptolemaic history. The Arabian coast had been explored from Akaba to Aden, but not beyond, though a good deal was known about Hadramut, and the African coast was known from Suez perhaps as far as Guardafui.[29] But this temporary breaking into the Sabaean barrier was soon at an end, for the Sabaeans did not want the cinnamon-trade of the Somali to slip away from them. No Egyptian Greek had yet pushed as far as India since Alexander's campaigns.

(b) Penetration to India and East Africa

The degeneration of the Ptolemaic house, and the internal and external troubles of Egypt, did not prevent further progress during the second century, owing to the fact that the growing luxury of the West impelled Greeks in Egypt more and more towards commercial explorations, elephant-hunts being gradually neglected. Under the kings of the second century merchants frequently visited for trade all the regions of the west coast of the Red Sea and took voyages outside it; many interesting details were thus learnt about the inland and other tribes – the Fish-eaters, the Root-eaters, the Seed-eaters, the Elephant-eaters, the Ostrich-eaters, the Locust-eaters, and others. Even Socotra was heard of, but was not yet explored or named,[30] nor had distances been measured between Bab el-Mandeb and Guardafui. It was on these voyages to the Somali coasts and on the huntings after elephants that men now confirmed what Philadelphus's explorers had already vaguely reported, namely the flowing of the Atbara from Lake Psebo (Tsana) and the flooding of the Nile through rains in the Abyssinian hills, visible from the Red Sea. But it was along the Arabian coasts that really new progress was made. Here the Egyptian Greeks were now not only trading with the ports of the wealthy Sabaeans of Yemen, and meeting there Indians of North India, and Parthians of Persia, but also sailing beyond along the coast of Hadramut (the

THE ANCIENT EXPLORERS

famous Frankincense country), and beyond to the Kuria Muria Islands and Masira, which had been discovered just before Agatharchides wrote (about 120).[31]

But had anyone reached India yet from Egypt? So far as we can tell, one man only – namely Eudoxus of Cyzicus, one of the most enterprising explorers of antiquity. He came to Egypt after 146, when Euergetes II was reigning, and being a man who liked the idea of exploring unknown lands, was entrusted by the king with an important voyage, namely to India, the route to which, as we are expressly informed, Euergetes did not know.[32] We are told by Poseidonius, as reproduced by Strabo, that 'an Indian happened to be brought to the king by the guards of the Arabian recess' (Gulf of Suez?), 'who said that they found him cast ashore alone and half dead, but who he was and whence he came they did not know because they did not understand his language. The king handed him over to persons who would teach him to speak Greek. And when he learnt how to do this he related that while sailing from the Indies he missed his way and, having lost his fellow-voyagers through starvation, was safely carried to Egypt. He promised to be a guide on a voyage to the Indians for men chosen beforehand by the king. Eudoxus became one of these. He sailed with gifts and came back with a return cargo of aromatics and precious stones . . . but he was deceived in his hopes; for Euergetes robbed him of all his cargo. But when that king had departed this life, his queen Cleopatra succeeded to the monarchy; and by her Eudoxus was sent out a second time with greater provision. And when he was coming back he was carried astray beyond Ethiopia by winds', which would be the north-east monsoon.[33] Strabo then proceeds to relate how Eudoxus landed at several places, returned to Egypt, was again 'robbed', and decided to find a route to India from Spain round south Africa so as to avoid the exactions (which we know to have been legal) of the Ptolemies. This part of Eudoxus's adventures finds a place in another chapter (pp. 124 ff.).

It is not definitely stated that Eudoxus ever reached India at all, but in the face of no contrary evidence we may conclude he did. At any rate the scorn which Strabo pours upon the finding of the Indian and upon the voyages is unmerited. There is nothing to show that, by 'Arabian recess', the Red Sea as a whole is not meant, and a small

INDIAN WATERS

Indian ship might easily have come to grief, especially since the Arabians were excluding Indians from the Red Sea, and thus once inside any Indians would have to be independent of help. As we point out elsewhere the death of Eudoxus on his last voyage caused all information about his career to be oral only, not written. We may take it then that he reached not only Indians along the Arabian coast, but by two coasting voyages, one about 120 and the other about 115, India itself, probably the Indus region only; that on the second return he was blown away by the north-east monsoon in winter some way south of Guardafui, but managed to come back safely to Egypt. It is possible that he sighted Socotra: at any rate it must be due to Eudoxus that Artemidorus, writing about 100 B.C., could know that the African coast turned round at the 'Southern Horn', that is Guardafui, though he stated[34] that nothing was really known of the coast beyond that headland. That the Arabian barrier was not quite so strong as it was is shown by the fact that about 115 the chief power passed to the Himyarites, and about the same time Hadramut became a weaker kingdom.

(c) *Final Efforts of the Ptolemies. Socotra, North-West India*

In the last century B.C. the degenerate house of the Ptolemies allowed Egypt to slip into the hands of Rome, yet did not cause Egyptian trade to suffer much. But in oriental waters, though the Greeks were allowed to coast along Arabia, the rise of Arab-African communities (partly piratical) on the Somali coasts prevented them from exploring these in greater detail, so that with isolated exceptions they did not reach as far as Guardafui. On the other hand, and coasting along Arabia all the way from Bab el-Mandeb to Mussendam (but not on into the Persian Gulf) is reflected in the medley of fragments of περίπλοι (see p. 224) presented to us by Pliny through the medium of Juba and other compilers who drew their material from sources of the last century B.C.[35] The Himyarites were doubtless powerful enough by about 50 B.C., and against them, and the piratical Nabataeans in the north of the Red Sea, the Ptolemaic government seems to have acted by means of military officers whose sphere of action went well outside Bab el-Mandeb.[36] Well might Apollonius and others, of whom inscriptions tell us, give thanks to the gods that they had been saved from the Red Sea and Arabia and the Trogodytes,[37] and sometimes not even twenty Egyptian ships a year dared to sail outside Bab el-Mandeb at all.

The main discoveries made were in two directions. A little before the reign of Ptolemy XI Auletes (80–51, in exile 58–55), a few Greeks reached Socotra and explored it, naming it Dioscuridu Nesos from Dvipa Sukhadara, and it is possible that the Greeks who, Cosmas says, colonized that island were sent out by Auletes. The colonists remained part of the mixed population, but fell under the sway of Hadramut.[38] About the same time the Greeks, in their sailings along Arabia, discovered that Acila (beyond Ras el-Had, and probably to be identified with Kalhat near Sur) was a port for sailing to India, and to that land they began to sail in small coasting vessels, but very rarely, as is shown by the fact that even at the beginning of the Roman Empire no further details were known about the coast of Iran and about India beyond the information bequeathed by Alexander, Nearchus, Onesicritus, and Megasthenes, three centuries before, with the exception of Barygaza or Broach, which must be regarded as the farthest limit reached by Greeks in the Hellenistic era.[39] Inscriptions of the reign of Auletes show that before about 65 the protection of Egyptian commerce by military force was being extended into the 'Indian Sea',[40] which reflects the new goal. The exclusion of Egyptian Greeks from the Persian Gulf all this time was due to the policy of the Seleucids, and then of Parthia, as we shall now show.

7. THE SELEUCIDS AND THE ERYTHRAEAN SEA

After the Seleucid monarchy had solidified out of the chaos following Alexander's death, the Greeks of Asia, having only Arabian barriers in the Persian Gulf to deal with, might have established a brisk trade in Greek hands between the Euphrates and the Indus, as Alexander had wished. But in fact this did not come about, due partly to Gerrhaean and then Parthian influence in the Gulf, though Greek exploration and commerce were not wholly wanting. Seleucus Nicator made an honest attempt to carry out the schemes of Alexander, and his subjects seem to have made one or two voyages from the Euphrates to the Indus in connexion with spice-trading,[41] probably

INDIAN WATERS

after his peace with Chandragupta in 302, but there was very little Greek traffic along Nearchus's route. When Antiochus III restored Seleucid power he advanced in 205–204 along the Persian Gulf to attack the powerful Gerrhaeans in their salt-built towers and houses, but did not quell them. On the other hand Antiochus IV Epiphanes (176–164) about 165, irritated by the annoyances of intermediaries both Persian and Arabian, examined in detail the southern shore of the Gulf, and it was he probably who sent Numenius with a fleet to Cape Mussendam to deal with the troublesome Persians. Pliny shows that the survey, starting at Charax, extended along the southern shore and included islands near it, past Gerrha, until rocky shallows stayed the advance somewhere west of El Katr.[42] All this time the Seleucids had not desired to sail round Arabia and thus open a route which would at once let in their rival Ptolemies. What they wanted was to find the trade-centre of the north-western part of the Arabian Sea – which, as we saw, was Acila.

After the death of Epiphanes the Seleucids began to lose their empire; the Parthians became powerful and gradually won control of the Persian Gulf through their vassals of Persis and of the districts where the Euphrates and the Tigris flowed into it, so that probably only the Greeks of Seleucia were able to do much trading in it. But these seem to have explored beyond El Katr along the district of Oman, learning that Omana was not on the Carmanian side, as had been thought, and correcting ideas about other places.[43] They may have pushed round the Arabian peninsula to Acila.

That Seleucid Greeks ever reached India it would be unwise to deny. But they did not go often, nor did they learn any more than the Ptolemies. The opening up of Indian coasts to western trade was still to come. It is possible that the ever-rising power of Rome, now paramount even in the eastern Mediterranean, and all that this meant, tended to turn the attention of men away from oriental waters. Be this as it may, the establishment of the Principate by Augustus in 27 B.C. turned the minds of Egyptian Greeks towards the East again in a manner quite unknown before.

8. THE ROMAN EMPIRE. HIPPALUS AND THE MONSOONS

The whole western world was at peace at last under one man; Rome's power reached to the Euphrates, and to the borders of Arabia, and she ruled (after 30 B.C.) Egypt. She did not, like the earlier Ptolemies, encourage or direct commercial expeditions. But the new conditions of peace and security which rose under the monarchy of Augustus fostered a fresh spirit of enterprise, having an effect rather like that of the Peace of A.D. 750. The compelling motive of exploration in the Roman era was an unmixed love of gain. Under the Ptolemies scientific curiosity had fought hard with mere acquisitiveness, but it was now definitely superseded. The opportunities for gain arose from the rapid growth of wealth in Rome and the western provinces, which led to a demand for oriental luxuries on a scale unknown before. The Romans did not participate directly in the eastern trade, which they left in the hands of Greeks and others, but they backed the Greek merchants with their prestige, and eventually also with their capital. Greek seafarers therefore showed a new readiness to make long journeys into the unknown East.

Clear evidence of this renewed enterprise is furnished by Strabo, who informs us that in the reign of Augustus as many as 120 ships would put out in a single year from Myos Hormos and Berenice for north-east Africa and India, and that some even sailed as far as the Ganges.[44] In the reign of the same emperor we find the first embassies passing between India and Rome. One mission from Broach brought snakes and a boy born without arms; another from south India offered pearls, precious stones, and so on, and gave Roman subjects permission to visit the courts of Indian kings.[45] These embassies, followed or preceded, we may be sure, by others on the part of the Romans, greatly increased trade between the West and the East, for the evidence of literature and archaeology alike records a large inflow into the Mediterranean of Arabian, African, and Indian products – spices, perfumes, gums, pearls, ivory, woods, precious stones, and so on. The second emperor Tiberius was worried over the fact that Roman money was beginning to leave the empire in quantities in order to pay for the gems and silks beloved by girls and women, and

INDIAN WATERS

even by men.[46] He tried in vain to stop this outflow, for nothing on this earth can prevent a woman from wearing what she pleases of any material at any price, or regulate the number of pearls which may hang round her neck.

The increased trade with the East under the early Roman emperors was essentially an overseas traffic. This fact was due, not merely to the natural advantages of the sea route over the land journey across Asia, but to its greater freedom from political disturbances, particularly the continuous rivalry and occasional open hostility of the Parthians. But even the sea-trade was subject to great and constant checks. No Roman subject at any rate, and perhaps no Arabian or African, knew how to use the monsoons properly in the Indian waters, so that the voyages to India were made by lengthy coastings all the way. This meant that a Roman subject going to India had to run the gauntlet of the impositions of a pair of independent Arab sheikhs, first at Arabia Felix or Eudaemon (Aden), and next at marts of the Hadramut coast in South Arabia.[47] Beyond that the officials of Parthian and Persian potentates annoyed him all the way from Khor Reiri to Sonmiani Bay. It was natural that under these conditions Arabian and Parthian subjects should remain middlemen for a large part of the Indian sea-trade. Augustus made at least one effort against the Arabians by force of arms. In 25 B.C. an expedition under Aelius Gallus, which advanced towards Mar'ib, as we tell elsewhere (pp. 192-3), possibly startled the Sabaean folk of South Arabia.[48] About the year 1 B.C. a scheme for the circumnavigation of Arabia caused a survey of both sides of the Persian Gulf for Augustus by Isidore of Charax, doubtless by special treaty with Parthia, and seems to have resulted in a raid upon Aden by sea and the dismantling of that mart.[49] But the vital blow was delivered at some unknown date in the early years of Tiberius's reign by the peaceful action of a Greek merchant named Hippalus. This remarkable man became aware that India formed a peninsula jutting southwards into the vast waters of the Erythraean Sea; in the course of trading along its coasts he had observed correctly the shape of what we now call the 'Arabian Sea'; and he had formed an idea of the position of Indian ports.[50] At the same time he knew that from May to October a wind blew steadily across the sea from the south-west, while from November to March it blew with equal steadiness from the north-east. With these facts in his mind Hippalus sailed boldly one summer along the Arabian coast with the south-west wind behind his ship; he touched at no Arabian port except the now insignificant Aden, and as the coast receded beyond Ras Fartak he pushed straight ahead across the open sea. He found his theories correct, for his ship reached India near the mouth of the Indus. Further voyages on this system by Hippalus and imitators of him resulted in their reaching direct the Indian coast anywhere between the Indus and the Gulf of Cambay, and merchants returned in winter time by using the north-east counter-wind. Hippalus had achieved fame: his name was given to the south-west monsoon, to a cape on the African coast, and to part of the Arabian sea near the Persian Gulf.

Yet his work was but one step. The parts of India which produced most of the precious stones and spices now in demand were in the south. The voyages made with the monsoon behind the ships brought these direct to only two great marts and both of them were in the north – Barbaricon (Bahardipur?) on the Indus, and Barygaza (Broach) on the Nerbudda in the Gulf of Cambay. Not only was the Nerbudda difficult to find and to enter, but the Rann and Gulf of Cutch were very dangerous points to approach from the open sea. The alterations in the coast caused by time have made the dangers less, but the tides of the Nerbudda are still strong, and pilots are often asked for at the mouth. Both parts of the Rann, says the nameless Greek author of the *Periplus*, are 'marshy seas, with shallowing and continuous sandbanks reaching far from the land, so that often, when the mainland is not yet in sight, ships run aground in shallows and if they are dragged farther in they perish'. The Gulf also was dangerous: 'Those who happen upon the beginning of it, if they double backwards a little to the open sea, escape unhurt. But those who are shut up in the very depths of it perish; for the waves are big there and very heavy, and the sea rough and troubled, with whirlpools and violent eddies.' Unexpected rocks too made anchoring difficult. The traders from Egypt therefore sought a safer line of crossing; and they found it early in the reign of the emperor Claudius, by coasting to Ras Fartak as before and then cutting across to the Indian coast near the modern Jaigarh, or perhaps Rajapur, whence they could proceed north or south as they pleased, returning in winter from the same place. But

INDIAN WATERS

even this achievement was not enough. The Tamil land of south India, which was the chief source of the coveted luxuries, still lay far off, and in between lay coasts infested by pirates, as they ever remained until the British took strong action. Moreover, ships even of today are warned to avoid the coast of India during the summer monsoon lest they be blown on to it when coasting. It was thus desirable that a new route should be found. At some time in Claudius's reign[51] a freedman of one Annius Plocamus, who had farmed from the Roman state the right to collect the customs dues of the Red Sea, was sailing round Arabia (on what business we can only guess), when he was caught by a monsoon of which he knew nothing, and was blown to Taprobane (Ceylon), one of whose kings sent in return an embassy to Claudius under a Rachias, that is, a Raja. This accident may have given the Indiamen a very strong hint. About A.D. 50 a nameless merchant, bolder than the rest, by ordering his helmsman to pull constantly on his rudder, and his sailors to make a shift of the yard,[52] found an open-sea route from the Gulf of Aden in an arc of a circle (bent northwards) to the south Indian coast, which he touched near the greatest of all Indian marts, the town of Muziris (modern Cranganore). He must have returned in winter by a corresponding route, on an arc bent to the south.

The track discovered by this nameless mariner was henceforth adopted by most of the Roman subjects trading in the Indian seas. The starting-place varied: some chose Hisn Ghorab, some the ruins of Aden, some C. Guardafui, some the mouth of the Red Sea; but in all cases the start from Egypt was made, whenever possible, in July, and the return trip began about the end of December. The Malabar coast was reached from the Gulf of Aden in forty days, the length of time taken being due to the size and heavy cargoes of the ships, which carried armed guards for protection. A voyage from Italy to India (with a land- or river-journey through Egypt) took about sixteen weeks. Lucian (*c.* A.D. 150) states that within two Olympiads, i.e., not more than eight years, anyone might make the trip from the Pillars of Heracles to India and back three times over, spending sixteen months on each double voyage, and taking time to explore intermediate lands as he went. The same author expresses the hope that men might fly from Greece to India in a day.[53] In actual fact it was now usual on one double voyage in one ship to sail more than 5,500 miles. The distance from Aden to the Indus is about 1,470 miles; between Aden and Broach, 1,700; from Myos Hormos to Broach, 2,820; between Bab el-Mandeb and Cranganore 2,000; the journey from Suez to India, which may sometimes have been made without a stop,[54] would amount to 3,000. In any case, there can be no doubt that the three ways of crossing the Erythraean Sea all became well established. This triple choice of route is reflected in the variant forms of the story of St Thomas, one of which takes him to king Gondophares (in the Punjab), another to Jaigarh or thereabouts, a third to the vicinity of Cranganore.[55]

9. EXPLORATION TO THE GANGES; AND TO ZANZIBAR

The growth of trade which followed upon these discoveries is revealed in the increased size of the ships that were built for the Malabar voyage, and by the reports of merchants that larger ones still were needed for the cinnamon trade of the Somali coast.[56] It is also attested by the greater frequency of visits by Indians to Alexandria: one of these, named Sophon, has left a record of a trip in a temple at Redesiya.[57] But the chief proof is to be found in the great hoards of gold and silver coins of the Roman emperors in southern India, which were imported there in order to serve for the most part, not as currency, but as bullion to be weighed out in exchange for goods.[58] Already in the reign of Nero the so-called 'drain' of specie to the East, which has been a problem for nearly 2,000 years, was sufficiently large to cause misgivings.

We are fortunate in having a merchants' guide-book written during the latter part of the first century[59] by an unknown Greek trader, who reveals to us the extent of Greek travels in eastern waters during the years immediately following the discovery of the cross-passage to the Malabar coast. This nameless author describes each of the coast-regions of India to which the traders from Egypt plied. First comes the district of the Indus. Here there was the important mart called Barbaricon (Bahardipur) on the middle stream of its estuary, which in those days had seven mouths. At this place, under the control of a Saka king who ruled at Minnagara (Bahmanabad?), the Greeks ex-

INDIAN WATERS

changed products of the Roman empire for Indian, Parthian, Tibetan, and Chinese wares.[60] Avoiding the dangerous Rann and the equally formidable Gulf of Cutch, many traders visited another productive district in the Gulf of Cambay, where lay Barygaza (Broach) on the river Nammados (Nerbudda). Its Saka king Nambanos (Nahapana) provided a pilot-service for strangers wishing to have their ships towed up the river to the city. Says the anonymous guide-book:

The putting in and the putting out of ships are dangerous for the inexperienced and for those who are putting in at the mart for the first time. For when there is a pull of water at the time of flood-tide, there is no withstanding it, nor do the anchors hold against it; for which reason even big ships, caught by the force of it and turned broadside through the swiftness of the flow, run aground on the shoals and break up, while the smaller barks are even overturned. And some which have been turned on their sides in the channels through the ebb-tide, if you do not prop them up, when the flood-tide comes back suddenly, are filled with water by the first head of the flow. For so great are the forces of the uprush of the sea at the time of the new moon, especially during the flood-tide at night, that as soon as a putting in begins when the open water is calm, at once there is carried to the ears of those at the river mouth a noise heard far off like the shouting of an armed camp, and after a little the sea itself rushes on over the shoals with a roar.

Barygaza may have been the port through which a charming ivory statuette of Lakshmi, the Indian goddess of good fortune and prosperity, recently found at Puteoli in Italy, and of date before A.D. 79, was transmitted from Inner India.[61]

Nambanos also controlled the trade of other marts on the west coast by means of a sub-king, Sandanes (Chandaka).[62] The long stretch of coast farther south was infested by pirates (whom the Andhra kings were not strong enough to subdue) and was little visited,[63] but the Tamil coasts in the extreme south were quite secure. The chief marts were the Chera town Muziris (Cranganore), where resident Roman subjects had built a temple of Augustus, and the Pandya town Nelcynda (Kottayam) in Cochin backwaters; but most merchants preferred to unload at Bacare (Porakad).[64] Our author does not mention the Laccadives: yet these could not fail to be seen and noted before long. Beyond Bacare few traders went, but on the east side of

C. Comorin the Pandyan town of Colchoi (Kolkai) was regularly visited. Ceylon for the time being remained unexplored, because of the Tamil monopoly of trade with that island.[65] Farther along the east coast a fair number of Greeks visited marts of the powerful Chola kingdom, notably Camara (Kaviripaddinam) at the mouth of a branch of the Cauvery, Poduce (Pondicherry – see below), and Sopatma (Madras?). On this stretch navigation was aided by lighthouses. Still farther north the trade of Masalia (in the Masulipatam district), of Dosarene (Orissa), and of the Ganges mouths had been examined and to some extent shared in, but little exploration had been done beyond the Chola kingdom. The interior of India remained unexplored by the Greeks; but it is possible that the Saka king Nahapana may have allowed them to visit his capital, which was another Minnagara (Chitor?), and his great inland centre Ozene (Ujjain), where earthen imitations of Roman coins have been found, while Andhra kings seem to have let them come to Paethana (Paithan) and Tagara (Ter). The Chola capital Argaru (Uraiyur, now part of Trichinopoly)[66] was known if not visited. The southern Chinese too had been heard of by name (Sinae, Thinae, as the Chinese were known when approached by sea, the name coming from the Chin dynasty), but very few met the Greeks, usually at a mouth of the Ganges, probably at Tamluk. The great bend of the Bay of Bengal, and the marked though irregular monsoons that blow across it, had not yet been realized by the Greeks, and Malay was still thought by most of them to be an island. But this peninsula and Burma were beginning to take shape in their minds. Whereas Poseidonius in the Ptolemaic era had thought that Gaul lay opposite to India in the east, some men of Nero's age knew that beyond the Ganges existed a great headland which they called Chryse, that is, Malay, with part of Burma.[67]

On the African coast our guide-book makes it clear that Somaliland and its rising marts as far as the 'Cape of Spices' (C. Guardafui) were well explored, and that the Greeks had penetrated to the court of the Abyssinian king Zoscales (Za Hakale) at Auxume (Axum).[68] They had also discovered that beyond C. Guardafui the coast ran continually southward, and did not turn off to the west, as had been previously imagined; some of them had even sailed as far as the Zanzibar channel, noting Menuthias Island, apparently Zanzibar rather than Pemba.[69]

The coast of Arabia was also well explored, and Socotra received regular visits, especially on the return voyage from Malabar. Inland Arabia of course remained unpenetrated, as much of it still is at the present day.[70] The Gulf of Oman and the Persian Gulf were somewhat neglected, the reason being that the Parthians, who controlled both the coasts, were sometimes enemies and always rivals of Rome; yet Syrians, Jews, and Palmyrenes traded at the ports of these gulfs in some numbers.[71]

10. EXPLORATION TO CAPE DELGADO. GREEKS IN INDIA. PENETRATION TO CHINA

Under the emperors of the Flavian dynasty (A.D. 70–96) and in the first half of the second century A.D., the farthest limits of exploration in the Erythraean Sea were attained. At this period the general level of prosperity in Mediterranean lands stood higher than ever before, and consequently the demand for eastern luxuries was well sustained. The actual pioneers of eastern navigation continued as before to be drawn almost entirely from the Greeks, though their ventures at this stage may have been supported by Roman capital. On the African coast, these explorers now reached their farthest south. We have noticed cases of Greek voyagers, Eudoxus and the freedman of Annius Plocamus, making discoveries in the Indian Ocean in spite of themselves. A later instance of this kind is that of a skipper named Diogenes, who imitated Eudoxus by missing his course near C. Guardafui on the return from India and was blown down the east coast to the Zanzibar channel. Here he gathered and apparently followed up information about the true sources of the Nile and the great Alpine mountains near the Equator (see pp. 214–15). The same southern limit was touched by another merchant named Theophilus, but he was in turn outsailed by one Dioscorus, who went as far as C. Delgado.[72]

But east Africa attracted far less attention in the Greek and Roman world than India. The possibility of conquering this country was much talked of,[73] and the emperor Trajan, who penetrated overland as far as the Persian Gulf and sailed on its waters, remarked that were

THE ANCIENT EXPLORERS

he a young man he would continue his course to India.[74] The same emperor and also Hadrian and Antoninus Pius received embassies from Indian kings.[75] Under these conditions the Greeks, and also Syrian traders in growing numbers, came to know the western coast of India better than ever.[76] Not only did they reside at the ports, but they penetrated inland and visited most of the native rulers at their courts. Among their hosts in the second century were the Saka king Tiastanes (Chastana) at Ujjain; and Andhra king Baleocuros (Vilivayacura) at Nasik, where the Greeks made dedications in Hindu temples; his successor Siriptolemaeos (Sri Pulumayi, A.D. 138–70) at Paithan; Nedunj-Cheliyan at Modura, and other Tamil kings. At Nasik has been found red-glazed pottery, which may have come from the Mediterranean; and at Nevasa not only similar pottery, but also four Graeco-Roman jugs. Pieces of rouletted dishes with a Mediterranean look have been found in the Andhra towns at Brahmagiri, Maski, Kondapur, Amaravati, and Chandravalli, where silver denarii also of Augustus and Tiberius have turned up. More has been discovered at Sisupalgurh. Again, at a number of places, mostly in the Andhra empire, and especially at Kondapur, have been found earthen imitations of Roman gold and silver coins. The Saka and Andhra monarchs probably took adequate steps to deal with the pirates off the Konkan coast, and on one extant papyrus a scene between Greeks and Indians is apparently laid on the coast between Karwar Point and Mangalore. Use of harbours on the Konkan coast may well be the cause of two interesting modern discoveries at Kolhapur – of a good bronze statuette of the sea-god Poseidoro (Neptune), and of a Roman jug, made about A.D. 50. Not far north also, at Akota on the old site of Anbottaka, has been found a bronze jar-handle of a similar jug, adorned with a rough figure of Cupid.[77] The Greeks also explored extensively the districts of the Indus with its tributaries and divarications, and inland regions of northern India, whose capital towns were known by name if not visited. It is probably as the result of these maritime voyages that western influences affected Indian art at Gandhara, Aurungabad, Bagh, and Ajanta.[78]

We have some interesting details about the visits of Greek merchants to south India in Tamil poems of the earlier centuries after Christ. We are told that to Muchiri (Muziris) the Yavana, that is, the

Greeks, brought gold in large ships which beat the water into foam, and exchanged it for pepper which was carried away in large sacks, and for other rare things of the mountains and of the sea, given by the Chera king. A Pandya prince is asked by a poet to drink in peace the fragrant cool wine brought by the Greeks. Yavana of stern looks served as guards in the king's tent on the field of battle and at the gates of the Pandya capital Modura, in the reign of the second Nedunj-Cheliyan. Again, Greeks had their own quarter in Kaviripaddinam of the Chola kingdom, and Greek carpenters helped to build a Chola king's palace.[79] Kaviripaddinam (now Tranquebar) is the official mart called Khabaris by Ptolemy the geographer; he shows that nearly forty inland towns of the Tamils were known to Greek traders, who thus learnt, among other things, of the important beryl mines of India. Large quantities of Roman coins have been found near the mines of the Coimbatore district and in the marts from Cannanore to Pudukottai: hence 'dinar' (denarius) occurs in ancient Indian records. The region of Coimbatore inland is reached up two river-valleys, of the Ponnani from its mouth on the western coast, and the Cauvery from its mouth on the eastern. No one need doubt that this overland route was used by Greeks, not only to enter the area of the mines of Coimbatore, but also to reach Poduce (see below) and Khabari on the eastern coast, without sailing round Cape Comorin (the southern point of India). Other examples of vast peninsular corners cut off by travellers of ancient times are those of Jutland and of Cornwall.

The island of Ceylon, which had hitherto remained forbidden land, was now circumnavigated, and explorations were made inland, notably to two capital towns, Anurogrammon (Anuradhapura), and Maagrammon (Bintenna?).[80] The commerce which then sprang up was revived in the later Byzantine age: the copper coins which were then imported in large quantities are still being dug up and passed again into circulation.[81] Visits were also made to the Maldives; their number was estimated at 1,378, and nineteen were known by name.

The eastern coast of India, though never so much frequented as the western shores, was widely explored as far as the Ganges, and visits were paid to the courts of Chola and Andhra kings beyond Cape Comorin. Some relics of the bold Greeks who ventured to go as far as distant east India have recently been brought to light through discoveries made under the direction of Sir Mortimer Wheeler at the site called Arikamedu, two miles south of Pondicherry (called Poduce in the days of the Greek explorers). Here a busy port began to rise about the end of the last century before Christ, reaching its greatest importance from about A.D. 50 onwards. At an early stage in its history were brought the Graeco-Roman wine jars, of which at least one hundred and fifty bits were dug out; soon after came the Italian, usually red-glazed, 'Arretine' ware of the kind made at Arretium (Arezzo) between about 30 B.C. and A.D. 45. Of these have been found more than fifty fragments, including one on which is stamped C. VIBI OF (*Workshop of Gaius Vibius*), and three others bearing the potter's stamp. Mediterranean in aspect are the extensive remains of rouletted dishes; certainly Roman are some pieces of glass vessels and a couple of lamps; and there is at least one Roman gem-stone.[82]

During the reign of Hadrian, we may surmise, the great curve of the Bay of Bengal was realized by Greeks who were visiting the east coast of India. Some of these, desiring to know where the precious rubies came from, where lay the reputed lands of gold and silver beyond India, and where the Chinese lived who sent silk to the West, reflected that what Hippalus had done in the Arabian Sea might be repeated in the newly discovered bay. They found that monsoon winds did in fact blow across it, though not in such a regular fashion as on the western side of India.[83] To one Alexander is due the credit of having first used them, thus superseding the laborious coastal route from the Ganges. The earliest crossings were made as it were in a series of leaps: the first method was to sail from Curula near modern Karikal in a north-easterly direction to Palura (Ganjam?); thence straight across the north end of the bay to Sada (Thade near Sando way?) in Lower Burma; thence southward to Temala at C. Negrais, and so to Malay. Before long a somewhat bolder course was laid by skippers who turned due east from the neighbourhood of Masulipatam and made direct for Negrais. Later still a few adventurers, navigating their vessels in the same way as on the direct transit from Bab el-Mandeb to Malabar, sailed eastwards from the north end of Ceylon, between the Andamans and the Nicobars, and hit upon Malay, Sumatra, or the Straits of Malacca. Meanwhile Alexander had led the way into still remoter regions. Beyond Malay he had sailed across the

INDIAN WATERS

Gulf of Siam to Zabae (Kampot?) in Cambodia, and thence to Cattigara, where he found that he had at last reached the Chinese. The location of Cattigara is in dispute, but the evidence points to Hanoi in the Gulf of Tongking, Cochin-China, where a Roman coin has been found, and from Chinese records we learn that merchants from the Roman Empire were to be seen in Siam, Annam, and Tongking, but not normally east of the last-named region.[84]

Voyages so far afield as Cattigara were rare, as is most clearly proved by the fact that Ptolemy does not mention the Straits of Malacca or the Malay archipelago, and that while he names Java he leaves out of account Sumatra, the nearer and larger island. The marts of Burma and Malay which Ptolemy names, places corresponding with modern Ramu, Sandoway, Bassein, Rangoon, and Thatung, were the general limits of Greek navigation. It is very probable that the merchants avoided the Straits of Malacca by travelling overland across the neck between the Bay of Bengal and the Gulf of Siam. Yet other Greeks followed in the tracks of Alexander, probably carrying supplies for three years on board, which according to Chinese records was the rule with those who came to China by sea. Occasional traders visited southern Siam and Cambodia, coasted along Sumatra and Java, and charted the Nias, Sibiru, and Nassau Islands, besides collecting details about the interiors.[85] A few pioneers even sailed some distance up the coast of China and brought back the earliest first-hand information about that country.

The great Mekong river of Indo-China has revealed some traces of western intercourse. At Oc-ceo by the river's delta, diggings begun by L. Malleret have unearthed a gold coin of Antoninus Pius and one of his successor Marcus Aurelius, and some gems which have a Roman look; and near P'ong Tuk, forty miles inland up the river, has been found a good Roman lamp of bronze.[86]

Before these efforts intercourse between China and the Mediterranean lands had all been indirect, and the peoples of either region were only known to the other by hearsay, partly over the land-routes, partly through the Indians, and partly through the Sabaeans (who seem to have reached China early in the first century A.D.), but also through rare visits of Chinamen to north-east Africa and Axum. In A.D. 97 the Chinaman Kan Ying was persuaded not to sail from the

Persian Gulf to Egypt because the sea made a man long for home.[87] The Greeks and the Romans imagined the righteous Sinae,[88] the purveyors of the much prized silk, as living in a capital fortified with walls of brass, while the Chinese knew but dimly of the honest men of Ta-ts'in, and Chi'ih-San, names which they gave the Roman empire (particularly Syria) and to Alexandria respectively. But in the second century A.D. Greek merchants for a time established direct contact, and in October A.D. 166 the supreme achievement of western travel in the Far East befell. In that year, according to Chinese records, an 'embassy' (that is, a commercial mission) came from Antun (Marcus Aurelius) to the emperor Huan-ti, and from that time, so the annals say, dated direct intercourse between the Roman empire and China – a thing long desired, but hitherto prevented by Parthia. With pardonable pride the Chinese magnified the gifts brought by the Romans into 'tribute'. The record of this mission says that the Romans make coins of gold and silver, get a tenfold profit from sea-trade with Parthia and India, are honest and have no double prices, and have a full treasury. On this occasion the visitors offered ivory, rhinoceros-horns, and tortoise-shell; but, it was specially noted, no jewels or glass or other curiosities. We have other Chinese records of definite visits by Roman subjects, one for the year 226, another for A.D. 284. The former tells of Lun, we may suppose a Greek called Leon, who reached Cochin-China by sea and was given an audience with the Chinese Emperor; the latter notices a visit of Roman subjects bringing aghal-wood to the Emperor. These explorers soon discovered that the walls of the capital city of Thinae (in Theinni, probably in the Salween valley in Burma) were not made of brass; but they did not learn much about the interior of real China.[89] Neither did they open up the seas to the east of China, but went on believing that the country was bounded on that side by unknown land. Articles of Roman origin have been found in China; they are mostly made of glass. But some of the evidence for a western origin fails to satisfy.

11. CONCLUSION

After the second century A.D. the record of eastern exploration, as of most other forms of enterprise in the Roman empire, comes to an end. In addition to the other causes which made for the decline of expansive energy, the economic decay and loss of purchasing power among the Mediterranean peoples weakened the mainspring of eastern trade and travel. Henceforth Roman subjects ceased to ply to India and the Far East; the Arabians and still more the Axumites of Abyssinia, and then also in part the Persians, seized all the commerce and became the leading navigators of the Erythraean, and with the retirement of the Romans the Chinese came sailing even into the Persian Gulf.[90] It is hard to judge the significance of embassies received by Elagabalus, Aurelian, and Constantine the Great.[91] Under the late Roman emperors of Constantinople there was a partial revival of trade, but not of exploration. Individual Christian missionaries, with perhaps the example of St Thomas of an earlier age to spur them on, may have used the ocean route to eastern lands, but they did not explore new waterways. Christian settlements were made in south India and Ceylon, but their authors were in most cases Persian, not Roman rulers. Chinese coins drifted westward as far as the Persian Gulf, but not in the hands of Greek traders. Thus India once more became a land of fable, and the geographical lore gathered together by Ptolemy began to be dissipated.[92] Finally, the Arabian conquests of the seventh century interposed an impassable block between the peoples of Europe and the eastern seas, and not only India but Arabia became utterly remote. The re-discovery of the Indian Ocean began in 1487, when Pedro de Covilhão sailed from Aden to Sofala. But the really epoch-making exploration was that of Vasco da Gama in 1497–9, and as his voyage was round south Africa, it can hardly be regarded as a resumption of the ancient Greek ventures.

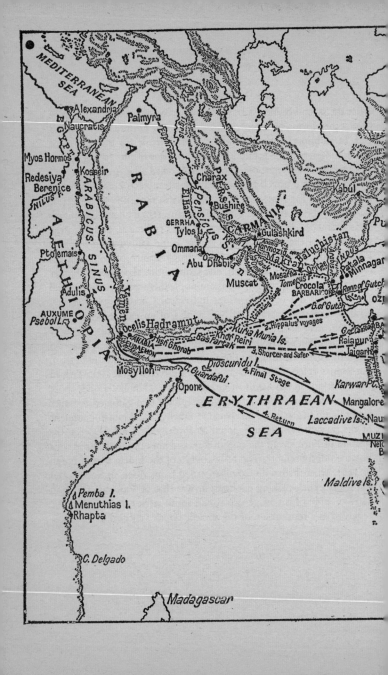

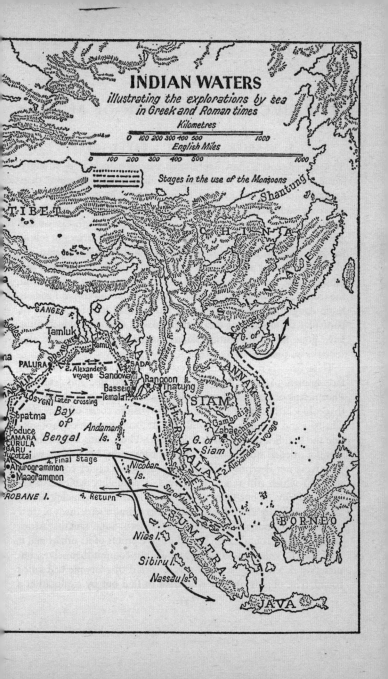

CHAPTER 5

THE CIRCUMNAVIGATION OF AFRICA

1. BARRIERS; OBJECTS; EARLY IDEAS

AFRICA presents against exploration barriers which were strongly felt in ancient times: (i) the never-ending coast-line, more than 15,000 miles long, which was found, by all who tried to trace it out completely in sea voyages, to stretch and stretch away ever south and south-west on the western side, ever south and south-east on the eastern side; (ii) the Sahara with its desert wastes, a barrier not felt in full by voyagers, but having an effect upon their minds against the idea of an easy or safe pushing southwards; (iii) heat, a barrier felt most on land, but also on the sea, for Africa lies almost wholly within the tropics; (iv) unhealthy climate, which within this tropical zone is more dangerous for the people of Europe than it is for others. In particular the malaria of the western coast is a scourge which must have gone far in preventing sailings towards the south. (See also remarks on pp. 216–17.)

These barriers stood against all races. In addition the Greeks had to face human barriers; thus the Tartessians perhaps, and later the Phoenicians of Gades and other places, and before 500 those of Carthage, blocked the way on all the north-western coasts of Africa (see Chaps. 2 and 3, pp. 35, 62), while the Arabians were in control outside the Red Sea on the eastern side. Both Phoenicians and Arabians were of oriental stock, both carriers of commerce, both secret when they were not deceptive. The result is that this chapter consists of a problem which still remains unsolved, a possibility which still remains unproved; but there is no lack of interest in glancing at the attempts made by Phoenicians and Greeks to find a settlement of their doubts. People came to take one of three views. Some said that Africa could be sailed round, and claimed to give proofs of it: others said it could not be sailed round; others would not commit themselves, but merely pointed out that there was no proof that anyone had sailed round. The objects of men who tried to find out by exploration a

THE CIRCUMNAVIGATION OF AFRICA

voyage from west to east were satisfaction of a desire for knowledge and adventure, and avoidance of the restrictions and exactions of whoever was holding Egypt against men who wished to trade with 'Ethiopia' (in the sense of north-east Africa) and Arabia; one man made India his goal, as we shall see (pp. 124 ff.). We may compare the attempts to work round Africa to India made in the fourteenth and fifteenth centuries so as to avoid Moslem hostility in Egypt and Arabia. As for the voyage from the east round to the west, there was no commercial goal except for the Phoenicians of Levant to find a new route to old haunts in Spain and north-western Africa.

In the days (c. 900–800) when the Homeric poems were written down men pictured the inhabited world as a round plane with an Ocean river flowing endlessly round it;[1] this idea was not at all connected with an 'Atlantic' or 'Indian' Ocean, but it influenced later Greeks in the belief that Africa had *sea* all round it. In Hesiod Libya is a mere isthmus over which a boat can be hauled.[2] For centuries it rested not with the Greeks, but with the Phoenicians, the Arabians, or the Egyptians to try, if they would, to settle the matter. It so happened that of these people one in the service of another made the first attempt.

2. THE PHOENICIAN CIRCUMNAVIGATION

(a) *The Evidence*

Egypt had been opened up[3] to Mediterranean peoples by kings Psammetichus (664–610) and Necho (610–594), and both Greeks and Phoenicians were admitted up the Nile valley, though apparently only the latter were allowed upon the Red Sea. If we are to believe Herodotus, some of these Phoenicians, about the years 600–595 B.C. when Necho was ruling Egypt, actually achieved for the first time[4] the circumnavigation of Africa, from east to west. This is what Herodotus says:

> Libya [that is Africa] shows that it has sea all round except the part that borders on Asia, Necho a king of Egypt being the first within our knowledge to show this fact; for when he stopped digging the canal[5] which stretches from the Nile to the Arabian Gulf he sent forth

Phoenician men in ships, ordering them to sail back between the Pillars of Heracles until they came to the Northern [Mediterranean] Sea and thus to Egypt. The Phoenicians therefore setting forth from the Red Sea sailed in the Southern Sea [Arabian Sea and Indian Ocean], and whenever autumn came, they each time put ashore and sowed the land wherever they might be in Libya as they voyaged, and awaited the reaping-time; having then reaped the corn they set sail, so that after the passing of two years they doubled the Pillars of Heracles in the third year and came to Egypt. And they told things believable perhaps for others but unbelievable for me, namely that in sailing round Libya they had the sun on the right hand.[6] Thus was Libya known for the first time.

That is the plain bald statement, attractive and disappointing, of Herodotus, which has caused every sort of written comment from annotation to hundred-page books. Who were Herodotus's informants? Phoenicians of Tyre of Egypt, or possibly Egyptian priests in the time of Persian domination in Egypt, talking with Herodotus some time after 448, about which year peace between Athens and Persia let members of the Athenian empire into Egypt, about 150 years after the alleged voyage. If the priests were the tellers, they would hardly be lying, for the story was one which gave the greater glory to Phoenician foreigners, but ultimately the whole thing rests on the word of Phoenicians from whom came some of the greatest explorers and the greatest liars in history. Herodotus, whose faith no one can doubt, believed the reputed circumnavigation himself and used it as evidence that the Indian Ocean and the Atlantic were one water, a belief held by most later Greek thinkers without necessarily believing Herodotus's account. Very few people have denied that such a circumnavigation was humanly possible; but none can rightly say that it has been proved on the above sole evidence. Men like Polybius (historian and African explorer) and Poseidonius (physical philosopher) doubted the achievement, and naturally many people today also doubt it. On the side of the doubters the strongest has been E. J. Webb, on the side of the believers W. Müller, who has dealt fully with the probabilities of the case, but still does not prove his point.[7]

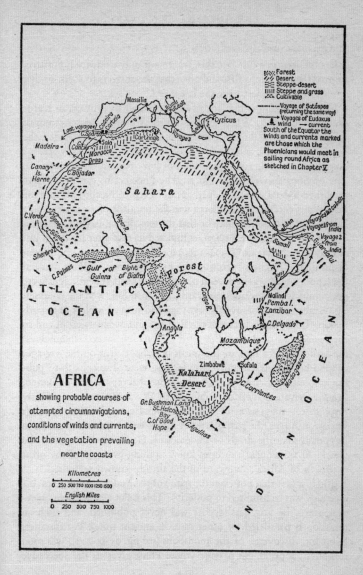

(b) Objections; and Answers thereto

It seems fairest in a case like this, where our only evidence is positive, to give the objections first and after that the answers to them so that in effect the defenders have the last word.

(i) Most striking is the lack of details[8] about the voyage in Herodotus's own report. Where are the travellers' tales, the descriptions of wild tribes, and of the great extension of the east coast of Africa, the winds and currents, and so on?[9] Was Herodotus unable to give them? Apparently so, otherwise he could not have resisted the temptation. The obvious answer to all this is: The Phoenicians can never have published a detailed report, perhaps not even to Necho; all that the informants could tell Herodotus was the remnant of a century and a half of oral tradition. It may be that the expedition, having been sent out by Necho late in his reign, completed its task after his death just as Henry the Navigator (1394–1460) died before he saw the fulfilment of his desire. Necho's successor, the busy Psammis, did not demand the report which was in the end never made;[10] subsequent reigns were not conducive to voyages eastwards, and then followed the conquest by Cambyses.[11] Moreover the voyage had proved useless as regards making traffic easier between the Red Sea and Mediterranean and so the details were neglected. But other explanations suggest themselves: the leader or secretary of the outfit may have died on the voyage; or the Phoenicians, commercially jealous, were always loath to publish accounts of their voyages, our one detailed report, that of Hanno, being a private dedication published to Western peoples apparently through the suggestion of Polybius several centuries after Hanno's voyage.[12] Again, the past glories of Egypt were of greater interest to Egyptians than the deeds of Phoenicians. Hence Herodotus and the Greeks of subsequent centuries knew nothing even about the great extension of Africa southward of the already known regions.

(ii) How is it that not even the name of the leader of so remarkable a voyage was transmitted to posterity? This is not a conclusive objection, since the voyagers' report was never published, and such an omission is paralleled by other cases in ancient times. For instance, Hippalus, discoverer of the monsoons (see pp. 95–6, 227), was soon forgotten; look at the passage where Pliny gives us the successive

THE CIRCUMNAVIGATION OF AFRICA

stages by which the Greeks discovered how to use these winds – no name is given. It might be said that were Herodotus's story an invention of the priests, as Gosselin suggested,[13] or of the Phoenicians, a leader would certainly have been named.

(iii) The Phoenicians must surely, if they really made the voyage, have noticed and reported even verbally the disappearance of the Great Bear and of the Pole Star by which they normally steered.[14] They must indeed have done both, but to Egyptians, if they were Herodotus's informants, the sun was a much more important heavenly body than the stars were, and so they preserved only the report about the sun. It has been argued that the priests, granted that they were Herodotus's informants, could deduce the point about the sun. We are not convinced that they would know all about the sun appearing in the north. They lived in Lower Egypt, outside the tropics, and what guarantee have we that they could imagine the sun to appear in the north on a voyage round Africa, when Herodotus, who had been to Assuan, could not swallow such a story?

(iv) Admitting that the Phoenicians were sent out by Necho, may we not claim that they came back without sailing round and lied as they so often did, inventing a plausible story about sowings and reapings, and talking about what they and learned men in Egypt perhaps knew happened to the sun under given conditions when seen by observants who had passed Bab el-Mandeb?[15] Hardly; how could they return to the Nile delta from the Red Sea unobserved? And if they lied, how is it that Herodotus's notice contains nothing that could possibly be called an attractive lie, but on the other hand one detail not only possible but probable, and another manifestly true? A fable-telling Phoenician would have decorated his tale with all manner of alluring falsehoods, or perhaps rather falsehoods calculated to deter, like those which the Carthaginians spread abroad about Himilco's voyage (pp. 46–7). The point about the sun will be explained later.

(v) Could the Phoenicians really have made this enormously long voyage of so many thousand miles in any sort of ship built in those days, and without the aid of the mariners' compass?[16] The voyage was certainly of great length, but instances taken from the Middle Ages, and even from Roman times, show that such a thing

THE ANCIENT EXPLORERS

was not impossible in small or smallish ships. In 1539 the Portuguese Diego Botelho sailed from Goa to Lisbon safely in a boat about five metres long and three broad![17] As for the matter of the compass, just as in the Indian Ocean the Greeks of Roman times found the monsoon winds an excellent 'compass' for crossing the sea direct even to south India, as we relate elsewhere (pp. 96 ff.), so in the Indian Ocean these Phoenicians of Necho's time found the actual coast of Africa (which obviously they were to follow) an excellent 'compass' for going round Africa – a thing which is in reality self-evident. The ancient sailors were well skilled in all the dangers of breakers and rocks in coasting. Again, at this period, it is the Phoenicians, if anyone, that we should expect to find making or trying to make this voyage. They were fearless and long-suffering seamen, and we know that they were bold enough to go into regions where hardly anyone else dared go. The Egyptians themselves were not seamen enough to try such a voyage, but Necho was just the sort of man to make others try it, and the Phoenicians were the people to choose for this, just as the Persians used them as a fighting sea-force.[18]

(vi) The time given for the voyage is too short.[19] Long as the voyage was, being in all more than 13,425 and less than 13,500 miles, if taken it could certainly be done within three years even at a slow rate of sailing (say between 20 and 25 miles a day, with two rests of six months each) provided that the voyagers started at the right time, when they would have winds and currents on the whole in their favour, as we shall describe below (pp. 120–1). We may compare the long-distance voyages from Egypt to Tongking taken by one Alexander under the Roman Empire as we have related.[20] That Scylax (pp. 78–9) took a shorter voyage in about the same time is accounted for by the fact that he must have traded with Arabian ports, if not for his own sake, at least for the sake of the Phoenicians whom Darius probably sent with him, and, besides, the Arabians and adverse monsoons may well have caused him great delays. Moreover we may add that Hanno, with not more favourable winds and currents, averaged 50 miles a day, and that Drake and Magellan, with hardly better ships, went round the world in three years, and that Vasco went from Malindi to Lisbon in eight months.

(vii) It is unlikely that men going round Africa, hitherto unknown, should be able to mark in time the changes of seasons in the tropics and in the southern hemisphere so as to be able to raise a crop, or having done that, to settle and grow corn in the face of savage peoples; since these Phoenicians (granted that they set out) sowed each 'autumn', they cannot have left the northern hemisphere.[21] Such argument is weak. The voyagers could observe gradually the climatic changes and even foresee them, and being in close touch with the coast all the time, could note the look of the land and perhaps the occupations of the people. At this time too there were no fierce Bantus pouring out of the heart of Africa, but only timid tribes whom they could easily impress. And again, 'autumn' was not confined to the months September, October, and November; when the Phoenicians decided it was convenient to land and sow, they called the season 'autumn'. As we shall see, the voyagers could certainly grow their corn as Herodotus states.[22]

(viii) Why was such a voyage as is alleged by Herodotus not repeated at a later date by Arabians, Greeks, or Romans? It must be remembered that no report had been officially published (though it may have remained a 'state secret'), and the bald statement about the sun, repeating what many knew to be a fact without the necessity of sailing far from Egyptian ports, tended to lessen the importance of the route, so that the Greeks soon came to believe that, in spite of the great distance, there was no great commercial goal along the route to make it worth while. Any Greek who tried to pierce the Arabian barrier into the Indian Ocean proper would find another one if he sailed round south Africa – the Carthaginian colonies on the western coast. As for the voyage from west to east, Phoenician and Carthaginian falsehoods and barriers stopped that also at the outset. This counter-argument we shall give in fuller form at the end of this chapter (see pp. 130–1). The great voyage of Pytheas in ancient times, and the discovery of America by the Northmen in the eleventh century, were likewise not repeated for ages; the latter being even forgotten until the nineteenth century.

(c) *A possible Reconstruction of the Voyage*

It will be seen that none of the objections is without an answer. Let us assume that the voyage actually took place as Herodotus said it did:

THE ANCIENT EXPLORERS

we can now sketch in some details about the way in which the Phoenicians performed the task, as has been done, for instance, by Rennell, Wheeler, and by Müller, the attempt of the last named being the most searching and satisfactory. Necho's object was to find out if, for the purpose of opening up a new sea-route instead of a canal-route to the Mediterranean, Africa could be sailed round; the voyagers were to look out for trade possibilities, but not to trade; the Phoenicians on their part desired to complete their knowledge of Africa and, thinking probably like Necho that Africa soon turned west after Bab el-Mandeb, hoped to reach their marts of the western coasts by a route which Greeks would always find it difficult to control unless they held Egypt. They may have been influenced by Arabian reports of Rhodesian gold.[23] Choosing therefore Phoenicians settled in the Nile delta or hiring them from Syria, Necho sent them from Suez or perhaps Kosseir on the Red Sea, probably in well-equipped fifty-oared ships or galleys loaded only with wheat of quick-growing kinds. Said Müller (more than a generation ago): 'The Phoenicians strove to draw the greatest possible advantage for their voyage from the atmospheric conditions of the Indian Ocean; but besides this they – the Englishmen of antiquity – devoted themselves to the principle that Time is Money as much as the practical sons of Albion do nowadays, and so, we may be sure, first entered on their journey when shipping voyages in the Mediterranean stopped.'[24] Knowing something of conditions as far as Guardafui they left, we may suppose, late in November, and rowed against a then unfavourable current of the Red Sea and the north-easterly monsoon as far as that headland. Turning south here with the coast they had the north-easterly monsoon and drift-current behind to the Equator, after which, in early northern spring, the south-easterly trade-wind blew at the side. Soon after crossing the Equator they noticed that what had been known as occasional in the southern parts of the Red Sea was now permanent – the sun was on their right hand, or rather, to the north of their ships, instead of being in the south, at midday. They had not expected this, and yet the farther south they went the stranger was the sight. Soon the swift Mozambique current swept them on, the Agulhas current taking them round the Cape, on the western side of which (perhaps in Great Bushman Land at St Helena Bay, south of the Kalahari desert) they landed during May of the first year in a subtropical climate where wheat could grow, sowed the grain in June, reaped in November, and set sail in December, being now in the second year of travel. When passing round the Cape they had seen the sun at midday about 50° from the zenith, very near the northern horizon. But now, as they were blown up the west coast by a favourable south wind and were helped along by the Benguela current, the wonder of the sun lessened. The Bight of Biafra was reached late in March, and when they reached the mouth of the Niger the sun was almost overhead. After this the displacement was much less and within the limits which they knew. A south-westerly wind and calms, in spite of the Guinea current, caused hard rowing along the coast to and round Cape Palmas (late in June), whereupon the north-easterly trade-wind and the ever-lessening Canaries current fought unsuccessfully against them all the way towards the Straits of Gibraltar; passing the tropic at the end of September, they landed again somewhere in Morocco (suitable for wheat) in November, sowed in December, reaped in June, and went on, quickening up in home waters. Once inside the Straits, the joy of home-rushing coupled with neutral winds and a favourable current made the last lap to the Nile easy.[25] Such is, with slight alterations, Müller's scheme, which fits into the probabilities better than the attempts of Wheeler, and even Rennell.[26] With some exceptions (for instance, the Mozambique channel) it has always been easier to sail from east to west than from west to east. Some ships were doubtless lost, but most of them came back safely.

Such may have been the voyage taken by these Phoenicians. But having heard both sides we must give the verdict 'Not proven', though in the face of no definite denial or impossibility the precise statement of Herodotus is in favour of the truth of the voyage. At any rate we can state with a feeling of certainty that Phoenicians were sent out by Necho with the object of sailing round Africa from the Red Sea to the Mediterranean, and at least sailed into the southern hemisphere.

Between his notice about the Phoenicians and his story of Sataspes, which we deal with next, Herodotus reports that the Carthaginians also said that Africa could be sailed round. He adds no details; his informants probably referred to the fact that since Hanno (not

THE ANCIENT EXPLORERS

mentioned by Herodotus) had reached Sierra Leone and Sherbro Sound, where the coast takes a decidedly eastern turn, the Carthaginians had on their own account become confident of the possibility of sailing right round.

3. SATASPES

But the voyage of Hanno[27] (between 500 and 480) was taken with the one object of colonizing the west coast, and had no connexion with sailing round Africa, so that it finds its place in Chapter 3 on Atlantic exploration. Curiously enough, the Persians seem to have entertained the idea under Darius, as is reflected in an obviously fictitious Magus (wise-man; or is it magus – swindler?) whom Heracleides[28] brought into his dialogue as having made the circumnavigation. At any rate we find a definite attempt made apparently on the mere whim of the foolish Xerxes (485-465). For quite worthy of belief is the curious account which Herodotus gives immediately after his report of the circumnavigation under Necho. It is a strange piece of Persian history heard by Herodotus while he was at Samos. An Achaemenian (that is, a Persian of the royal house) called Sataspes, son of Teaspes, was to have sailed round Libya from the Pillars of Heracles to the Arabian Gulf, but 'did not voyage round Libya when sent out to do this very thing, but, fearing the length of the voyage and the loneliness of it, came back again, and did not accomplish the task which his mother had laid upon him' with the consent of King Xerxes in expiation of a shameful act. 'Xerxes having agreed to these terms, Sataspes came to Egypt and having taken a ship and sailors from the people there sailed for the Pillars of Heracles. Having voyaged out through them and having doubled the headland of Libya called Soloeis' (Cape Spartel probably,[29] though the Soloeis of Hanno and the later so-called 'Scylax' is, as we have seen, Cape Cantin) 'he sailed towards the south, but having crossed over much sea in many months, because there was ever need of more and more voyaging he turned and sailed back to Egypt. And coming thence to king Xerxes' court he told him that at the farthest point of his voyage he sailed by dwarfed[30] men, wearing clothes made from palms, who, whenever the voyagers put the ship to land, fled towards the mountains, leaving their towns; the voyagers did them no harm as they went into the towns, taking only

THE CIRCUMNAVIGATION OF AFRICA

cattle away from them. The reason why he did not voyage right round was, he said, that the ship was not able to go on farther, but stopped.' Xerxes did not allow that he told the truth and caused Sataspes to be impaled as he had threatened. Herodotus then shows that his source of information was a good one – namely one of Sataspes's eunuchs who ran away with most of Sataspes's wealth, but was robbed by a Samian 'whose name I know but forget on purpose'.[31]

This is the first recorded attempt to sail round Africa from west to east, and there was no idea yet of reaching India in that way. It is possible that Sataspes sailed along the coasts of Morocco, Sahara, Senegal, and perhaps Guinea.[32] It is clear that he had gone southwards far enough to meet with negro tribes beyond the Sahara and civilized enough to dwell in 'towns'. They may have been Bushmen or Bosjesmans more widely spread than they are now. Xerxes may have heard of Necho's attempt;[33] we do not know. But it seems that Sataspes took only one ship, manned perhaps with Greeks and Phoenicians from the Nile delta, and sailed right beyond the limit of any Phoenician or Carthaginian settlements, which tended to exclude strangers, before landing for fresh food and water. He had much to oppose him, especially if he reached the Gulf of Guinea in which the comparative absence of winds and current of the Camerun coast, or the strong northerly current and south-easterly wind which prevail after the Gulf, may have caused him to say with truth that his ship would not go on.[34] Had his story been written down in full, he would now be a more famous man than Hanno. But of course he may have been caught anywhere south of Cape Verde in the doldrums. In the fourteenth century sailors often told how they just could not go farther than Cape Bojador.

4. THE GREEK OUTLOOK FROM THE FIFTH CENTURY TO THE SECOND. THE MYSTERIOUS SHIPS OF GADES

The Greeks now began to believe that it was impossible to sail round Africa, the nearness of Egypt to their own seas and the Persian domination in that land turning their minds to the Erythraean Sea more than to the Atlantic, and by Aeschylus's time lack of exploration under Persian rule caused the astounding belief of some that south of

Egypt near the unknown Nile sources east Africa could not be sailed round, but was joined to north India so as to make the modern Arabian Sea a lake.[35] The constant arrival of Indian wares on African coasts helped this view. Some seem to have believed that the west coast of Africa also could not be sailed round, but was blocked up.[36] Others, however, continued to believe that not far out of the Red Sea the coast of Africa turned away south-west and west to the Straits of Gibraltar, so that in their view all Africa was north of the Equator and formed a right-angled triangle, with a hypotenuse largely unknown. Hecataeus, Herodotus, and others were of this opinion,[37] maintained during several centuries during which very little exploration was made along either side of Africa (see pp. 68, 79–80). During the fourth century, at a period when the well-known West African mart Cerne (Herne – see p. 66) was a place to which the citizens of Athens or of western Greek cities were sending Attic pottery for ivory and gold, but only through the Carthaginians, some Athenians and others had wondered whether one could sail to Cerne from the Erythraean Sea round south Africa, but the historian Ephorus announced that any who tried could not go beyond certain 'columns' or small islands, because the heat was too great.[38] Who first caused this report we cannot tell; but it is fairly certain that the Carthaginians were deterring the Greeks on the western side by telling them, not without truth, that the Atlantic was muddy, shallow, full of seaweed, and very windless, as the so-called 'Scylax' and Aristotle of the same period show, especially south of Cerne, where no one could sail, though rumour had it that the sea beyond went all round to Egypt and that Africa was a great peninsula.[39] Aristotle further deduced a temperate zone in the southern hemisphere, but could not decide whether it was inhabited or not.[40] Alexander the Great by his Indian expedition disproved the supposed connexion between India and east Africa, and this seems (we are not certain) to have led him to a natural desire to sail round south Africa from the Red Sea to Carthage, which he thought would be a short voyage, but he did not live long enough even to begin preparations.[41] Gradually the curiosity of the Greeks was roused still further, and in 241 Rome laid Carthage low and in due course won Spain. Yet no one made the voyage, and one school of thought by the end of the century had come to regard the inhabited part of the earth as wholly within the northern hemisphere, cut off from the south by great heat; to say that men lived beyond the torrid zone was useless talk;[42] but wise Eratosthenes deduced the possibility of a circumnavigation of Africa from the likeness to each other of the tides of the Indian and the Atlantic Oceans. During the second century men speculated about unexplored Africa; men like Crates (died 145) imagined a separate world south of Africa; Hipparchus the astronomer and Polybius the African explorer show that many believed that Africa was not surrounded by sea but continued infinitely as land.[43]

The Greeks were not reaching beyond Guardafui on the one side, 'Ethiopia' and the 'Cinnamon Country' (Somali coast) being looked on as the southernmost limits of the inhabited world, nor were the Carthaginians on the Atlantic exploring beyond their existing trading-stations, as we show elsewhere. Now a very active part in this Atlantic trade was taken by the Phoenician city Gadeira or Gades (Cadiz) in Spain, and from the second century B.C. to the reign of Augustus (27 B.C.–A.D. 14) there were frequent rumours and even (men thought) tangible signs that ships of Gades were sailing round south Africa from west to east so as to trade with Somali and Arabia without submitting to the exactions of the Ptolemies in Egypt. Thus Caelius Antipater (c. 120) said he had met with a man who had sailed from Spain to Ethiopia (probably the man meant a region of west Africa) for trade.[44] Suddenly came a bold attempt to solve the problem, a little before the end of the century.

5. EUDOXUS

(a) The Evidence

Poseidonius (born c. 135), who disbelieved all the alleged circumnavigations of Africa (except by ships of Gades), made much of the attempt of Eudoxus, and Strabo later saw fit to give the substance of Poseidonius's account.[45] Eudoxus was a well-to-do native of Cyzicus who came as a sacred ambassador of his city to a festival of Persephone in Egypt during the reign of Ptolemy Euergetes II Physcon between 146 and 117. As we describe elsewhere, he made two successful voyages to, or towards, India; on returning from the second (taken between 117 and 108) he was caught by the north-easterly monsoon and blown a fair distance down the east coast of Africa south of

THE ANCIENT EXPLORERS

Guardafui. Wherever he was forced to land he made friends with the natives by sharing with them his bread, wine, and dried figs, which they did not have, and in one case taught them secrets of the use of fire.[46] He received water and guides in exchange and made a list of some of their words – a very modern idea which astonished Strabo. And he found an end of a wooden prow, with a horse carved upon it from a wrecked ship, and heard that it belonged to some voyagers from the West. Eudoxus took it back to Egypt, found Ptolemy Soter II Lathyrus reigning, and for the second time had to hand over his cargo to the government by law. Shipmasters of the market-place of Alexandria told him that the figurehead came from Gades. Smaller ships of the poorer sort were called 'horses' from the device put on them, and were used for fishing voyages along the west coast as far as the Lixus river (Wadi Draa). Strabo continues: 'Some of the shipmasters indeed recognized the figurehead as having belonged to one of the ships which had sailed too far beyond the Lixus river and had not come back safely.'

Eudoxus, convinced that Africa could be sailed round, and clearly chafing at the annoying exactions of the Ptolemies, and wishing to act as a free business man and seeker after knowledge, determined upon what no man had thought of before (so far as we know) – to sail round Africa in order to reach India and avoid the Ptolemies and the Arabians in doing it. Like nearly all Greeks who had ever thought about the matter, he did not realize how far south Africa really stretched. He went home to Cyzicus, 'put all his property on board ship, and set sail, He put in first at Dicaearchia' (Puteoli in Italy) 'and Massilia' (in Gaul) 'and at places all along the coast round Spain to Gades in order chiefly to get together a good cargo. And everywhere ringing out the details of his doings and making money in trade, he fitted out a big ship with two tow-boats like those which pirates use. He put music-boys and girls on board', suitable we may guess as gifts for the harems of Indian kings besides providing musical and other entertainment at sea, 'and also physicians', for malaria and other diseases, and perhaps by request of India, 'and carpenters besides', for repairs and perhaps at the request of India, 'and then at last set sail well in the open sea for India with constant west breezes. But when those who were with him were tiring of the voyage' (taken apparently

THE CIRCUMNAVIGATION OF AFRICA

well away from but parallel with the coast, to avoid any interference) 'he went with a fair wind towards the land, unwillingly and in fear of the ebbs and flows of the tide. And in fact what he feared did happen; for the ship ran aground, yet gently so that it was not broken up altogether, but the cargo was brought safely to land in time, and so were most of the timbers, out of which he put together a third boat about as large as a fifty-oared ship and sailed until he found himself among people speaking the same words as those of which he had made a list before,[47] and at once learnt at least that the people of that district were of the same race as the other Ethiopians and were neighbours of the kingdom of Bogus' (Bocchus). This means that Eudoxus had done nothing remarkable – he had gone a little south of Morocco, but perhaps no farther. It is clear that he was sent aground by the north-easterly trade-wind which blows from the north near the coast, and by a strong tide.[48]

Strabo goes on: 'So he gave up the voyage to India and turned back', apparently because he thought he had reached the neighbourhood of the place where he had found the figurehead and could return, make a fresh start with larger ships, and quickly repeat the now known voyage so far, before going on to India; 'and on the voyage along he saw and noted down a well-watered and well-wooded but unpeopled island,' perhaps the already discovered Madeira[49] (see Chap. 3, pp. 69–70) rather than any one of the Canaries, 'and having safely reached Maurusia' (roughly Morocco) 'sold his boats, betook himself on foot to the court of Bogus', failed to persuade him to take up this expedition on his own account, and was almost tricked into being marooned on a desert island. What island this was we cannot tell: it may have been Lanzarote or Fuerteventura, the nearest of the Canaries which Juba's explorers later found to have no trace of man except signs of buildings (see p. 69). It may have been the very island which Eudoxus himself noticed. He fled to Roman territory, which would be the allied territory of Numidia, and from this country he crossed, probably from its port Rusicada (Philippeville) rather than Hippo Regius (near Bona), to Spain, presumably landing at Carteia and going on to Gades once more. He and his men had tales enough to tell, and we seem to have them (much distorted in transition) in the remarks of Pomponius Mela[50] of the first century after Christ. What they told was the glimpses they had of tribes during the accidental sail down the east coast of Africa and also on the first voyage down the west coast. They spoke about mouthless folk and others dumb and others with lips stuck together, who would be timid peoples of the west coast and Hottentots or others who joined their lips together with some ornament; about noseless folk who might be south African peoples whose thick lips almost hide the flat nose; and about other strange tribes including dwarfed men. Still, nothing is proved, and we cannot very well take Eudoxus farther south than Poseidonius did. Strabo proceeds: 'Having built a round ship and a long ship of fifty oars so as to sail on the open sea in the former, and make the acquaintance of the land with the latter, and having put on board agricultural tools and seeds, and also builders' (because we may suppose he realized something of the great lengthiness of his project and planned, like Necho's Phoenicians, to raise crops, as well as to build huts and provide for wreckings) 'he set out again with an eye to the same circumnavigation as before, intending in case the voyage were delayed to winter in the island he had already examined, to sow and reap a crop, and then finish the voyage which he had at the beginning determined on.' Poseidonius added that probably the people of Gades and Spain knew what happened afterwards. 'How was it that' Eudoxus 'did not fear . . . to sail again along Libya, with provision enough to colonize an island?' exclaims Strabo. If he had a fear he mastered it, but this enterprising forerunner of Vasco da Gama and those with him were never heard of again, and their fate remained unknown. They may have been involved in the destruction of old Carthaginian settlements on the west coast by natives; or in the great spreading movement of the big black Bantus which seems to have taken place about this time; or, caught in one of the cyclones which sometimes rage near Africa,[51] the ships were wrecked with loss of all hands, or they may have met their fate in another way, for something had made Eudoxus sail well out to sea on the first outward and return journey – and this may have been hostility on the part not so much of possible last remnants of Carthaginian colonies or their destroyers, as of Bocchus of Mauretania, whose jealousy was perhaps still more roused when Eudoxus was bold enough to appeal to him at his court. Or Eudoxus may have been

THE CIRCUMNAVIGATION OF AFRICA

marooned in a mutiny, like Hudson in 1610, and, very nearly, Columbus. At any rate Eudoxus and all his crew perished to a man. They did not succeed in finding the passage round.

(b) Some Objections against the Account, and Answers to them

Strabo scoffs at Poseidonius for doubting the older stories of circumnavigation and yet (if we may use a modern phrase) accepting this 'Andersen's fairy tale' (that is, believing that ships of Gades had sailed round), and suggests that Poseidonius or others invented it. Strabo's doubts about the first voyages (those to India) have little weight, but about the African part he makes some points: 'The man who recognized that figurehead – wasn't he a wonderful fellow, and the man who believed him a more wonderful fellow still – one who with a hope such as this went back to his own city and then emigrated from it to the regions outside the Pillars?' There are indeed objections to the story: (i) The alleged identification of the ship's prow; but firstly this was a mistake of Eudoxus, and not confined to him alone among Greeks; the reigning Ptolemy did not know of any Spanish ships having sailed through the Nile canal, and so Eudoxus thought (perhaps misinterpreting a native word) it must have come the supposedly short way round Africa. It has been suggested that the fragment, like the others we mentioned above, was indeed Spanish and had been washed all round Africa; which is impossible, especially from west to east, because of the Agulhas and Mozambique currents.[52] Secondly, when the skippers in Alexandria heard Eudoxus's belief they were so tickled that they went one better and almost named the ship. (ii) The alleged discovery of men on the north-western coast of Africa speaking a language like one spoken on the upper eastern coast. But it is possible that there were, just as there are now, African races on both these coasts speaking languages the same as or similar to each other. Nor need the languages have been the same; even today there is a tendency among sailors to class all foreign tongues together as 'French', if not 'double-Dutch'. (iii) The absence of any real detail about geography, African races, and commerce of the north-west coast, Carthage not having been long destroyed. But we may answer that on the earlier part of his first African voyage Eudoxus, like Sataspes, did not land except when forced; an experienced man of his type might well be bold enough to sail out to sea to avoid being pushed ashore or being harassed by unfriendly folk; that with his companions he met death on a last voyage before he could publish any account of his explorations; and that the details which we have look much like what he and those with him told by word of mouth in between the voyages. (iv) No real reason is given for his turning back on the first voyage from Spain. This omission seems to be due to the oral nature of the whole tradition, and we have tried to give his reason above.

Moreover, strongly in favour of the story are the following points: (i) Poseidonius was reporting what had happened recently in his younger days and did not make Eudoxus succeed; (ii) the alleged robberies of Eudoxus by the kings of Egypt are a true touch; they show an innocent ignorance on Poseidonius's part of the fact that in state enterprises of Egypt the king claimed a right to the cargo. We can picture Eudoxus in anger saying: 'It is sheer robbery. ... The Pot-bellied (Physcon) and the Chick-Pea (Lathyrus) robbed me. ... I will go to India round Africa!' (iii) The story makes him call at just the likely places for a Westerner collecting a good export-cargo for eastern regions – the best ports in Italy, Gaul, and Spain.[53] (iv) On the whole the tradition is free from exaggeration and falsehood. We can believe it. A similar failure befell the next explorers known to have tried to find the route to India round Africa. For in A.D. 1291 Ugolino Vivaldo and Guido Vivaldo, having prepared two galleys for this purpose, commerce and to a less extent proselytism being the objects, sailed from Genoa in May and reached Cape Nun, but were heard of no more.[54]

6. SUBSEQUENT OUTLOOK OF THE GREEKS AND ROMANS

The story nears its end, for after Eudoxus's failure men began to despair of sailing round Africa, and with the coming of the Roman Empire attention was drawn in more obviously profitable directions. However, the phantom ships of Gades (a town now becoming Romanized) seem to have still stirred the minds of men as conducting heroes of navigation round south Africa. But, with Carthage gone,

THE CIRCUMNAVIGATION OF AFRICA

went also the prosperity of distant west African settlements which the Romans did not develop, but made Sala (Sallee) their limit; and after they established Egypt as a special province, India became the great goal in that direction, and Somali coasts were for a time the limits of actual east African exploration by sea. Strabo, who seems to have put all Africa north of the Equator, says that the southern part of the triangle formed by Africa could only be spoken of by guesswork[55]; that it could not be approached, or at least never had been, because of the great heat; 'all those who have sailed along the shores of Libya, starting either from the Arabian Gulf or from the Pillars, after going some distance, have been forced to turn back again because of various accidents; ... all the navigators called the last region they reached "Ethiopia" and described it as such'; he combats the idea that land blocked the circumnavigation.[56]

The great increase of trade between Egypt and the Indian Ocean under Augustus, as we described in Chapter 4, caused some merchants to wonder about the possible route round Africa from east to west, and it seems probable that the mission of young Gaius to the East in 1 B.C. was to include the circumnavigation not only of Arabia but also of Africa from east to west. Thus Juba, writing about Arabia in preparation for the mission, announced that to sail from Somali coasts to Mauretania and Gades you must use the wind Corus, that is the north-west wind which would be convenient as far as Guardafui.[57] In due course a fleet was put into the Red Sea to protect the proposed sailing round Arabia; the fleet found pieces of wreckage identified (mistakenly, of course) with parts of Spanish ships.[58] Maybe men had thought of Eudoxus and wanted to encourage the idea of trying again; but on the death of Gaius everything was given up. On the full development of the use of the monsoons by Nero's reign merchants thought still more about known India, and still less about unknown Africa, though their dealings in the Indian Ocean carried them far down the east coast, as we see in Chapter 4, thus causing them to realize that the coast went on and on southwards. But this very fact, the ever-lengthening out of the coast-line, caused an old mistake to revive and become permanent. Its growth was slow. Greeks reached Menuthias – that is Pemba, or more probably Zanzibar – by Nero's time, and others confused them both with Madagascar which no

Greek ever saw or named. Those who visited Pemba or Zanzibar reported that the unexplored coast beyond bent round westward to the waters of west Africa, as it indeed does after Madagascar.[59] No Roman subject ever went round, Dioscorus (early in second century?) reaching only Portuguese Cape Delgado, though it is possible that the Himyarites, trading with south Africa through Sofala held by Bantus at that time, went also round the Cape from east to west, bringing the coin of Antoninus Pius found far inland at Zimbabwe, that of Constantine found in Madagascar, that of Trajan found not far inland along the Congo of west Africa, and some others.[60] They tried to bar the Greeks; the coast still lengthened out; we may suppose that the great size of Madagascar was heard of; hence Ptolemy led the way in propounding not only that Africa could not be circumnavigated, but that the south-running east coast turned east after Delgado so as to face north and run south of the Erythraean Sea and join a westward-facing China and form the north coast of an unexplored southern tract. It is hopeless, men said; Africa cannot be sailed round. The farther they had explored, the more wrong had the old right idea of Africa seemed to become; all appearances seemed to prove that the east coast never ended at all. Thus did the 'Red' Sea become a lake once more, but much larger than before.

We can see why the presumably accomplished circumnavigation by Necho's Phoenicians was not repeated either way:

(i) *From East to West.* (*a*) The Arabian barrier and mist of falsehoods which, stretching originally uninterruptedly from the Red Sea to Mozambique, was never entirely removed from east Africa. Even if it had been pierced, the Carthaginians were for long another barrier in north-west Africa.

(*b*) Loss of the details of the voyage taken under Necho.

(*c*) Lack of any known commercial goal (as in the case of the North-west Passage in later times), any rich land, to be gained by sailing round, coupled with the vast distance, ever vaster as men sailed farther. Few would really want to reach Spain that way.

(ii) *From West to East.* (*a*) The Phoenician and Carthaginian barrier and mist of falsehoods in the western Mediterranean and on the north-west coast of Africa, and, if passed, the Arabian barrier east of the Cape.

(*b*) At a time when India was the goal, Eudoxus failed, apparently

THE CIRCUMNAVIGATION OF AFRICA

with disaster, as others did, though preserving their lives, and all the reported signs of ships from Gades were wrecks, apparently with never a sign of a survivor.

(*c*) The fact that during the early Roman Empire India was brought near by the full use of the monsoons; Rome held Egypt and crossed the Arabian barrier to the Far East. Therefore merchants said, 'Why sail round Africa for India?' The Roman Government said, 'Why let anyone try it? It would mean loss of revenue to Egypt and therefore to the Emperors.'

7. CONCLUSION

Subsequent ages followed Ptolemy's idea without testing it, so far as the learned were concerned. But strangely enough, in the backsliding of the Dark Ages the heads of the Church, opposing the learned, made Ocean flow round the inhabited world of their oblong-rectangular, circular, or oval schemes; but of course it was not enlightenment, but the contrary – geographical knowledge ceased to live in Christendom, though the Islamic Arabians by translating Ptolemy fared better; their limit was Cape Corrientes. Realization of the southern extension of Africa by Saracens, if it ever took place, did not reach the ears of Europeans. The translation of Ptolemy into Latin in 1410 made a wider acceptance easier, but of course the false idea about south Africa remained also. The final success was left to the Portuguese, inspired by Prince Henry the navigator, who himself was set thinking by Herodotus's story of the Phoenicians. They wanted to open a route from the Atlantic to India. They doubled Cape Bojador in 1434, Cape Verde in 1445, and crossed the Equator in 1471. The Congo mouth was found in 1482. Diaz rounded the Cape in 1488, calling it the Cape of Storms, but King João II, realizing the coming discovery of the route to India, named it the Cape of Good Hope. At last in 1498 Vasco da Gama sailed right round to Sofala and Malindi and went to India. Thus the Indian trade passed from Venetian into Portuguese hands; the goal was reached.

Readers will now, after due thought, see that the title of this chapter might well have been 'The Non-circumnavigation of Africa'. The title which it bears, however, is just as true, and offers much greater encouragement.[61]

CHAPTER 6

EUROPE

1. NATURAL FEATURES

OF all the continents Europe offers the most favourable conditions to the traveller and explorer. It has a pervasive system of rivers and is free from impassable desert tracts. Its climate does not move to murderous extremes, and its inhabitants have generally been quick to appreciate the benefits of foreign intercourse.[1] On the other hand, large areas of Europe which are now fully reclaimed were formerly covered with forest or swamp. The Continent's vast resources in minerals long remained unexploited, and in any case offered few of those glittering attractions which were the most potent stimulus to early commerce. Moreover the European mainland is comparatively inaccessible from its Mediterranean seaboard, the quarter from which its effective exploration was eventually accomplished. Its highest mountains are folded up against the Mediterranean coast-lands, towards which they present their steepest face, leaving only a few easy avenues into the interior, where a river valley or a sea-arm has made a break in the barrier. Besides, the natural difficulties of intercourse between the Mediterranean and its hinterland were reinforced by man-made bogies. The cooler winters and cloudier skies of central and northern Europe created an exaggerated impression of chill and gloom on the Mediterranean mind; and the native populations, many of whom had been organized since the Bronze and Iron Ages under military aristocracies, presented the appearance of being naturally fierce and intractable. The exploration of Europe, therefore, was only half begun before the Roman conquests: its completion was essentially the work of the Roman soldier.

2. THE SPANISH PENINSULA

In the present chapter Spain and Portugal require a separate section, albeit a brief one. The Iberian peninsula is so effectively cut off from the rest of Europe by the Pyrenees that its exploration may be accomplished quite independently of discovery in adjacent lands. Besides, its easiest lines of communication lie along the coastal plains, which are marked off sharply from the elevated inland; and its chief mineral resources are found close to the seaboard. Its internal penetration, therefore, was not achieved till long after the growth of maritime traffic along its shores.

The first indication of cross-country travel in the peninsula has been found in certain lines of Avienus, who measures the distance of some short cuts from Tartessus to the western and southern coasts in terms of days' journeys.[2] The mention of Tartessus is not quite conclusive as to the date of this traffic, for Avienus sometimes confuses Tartessus with Gades.[3] But the measurement by day trips instead of stades or miles indicates that these routes were established early, and presumably by the Tartessians. This conclusion is supported by finds of bronze statuettes, bearing unmistakable resemblance to early Ionic art, at the foot of the Sierra Morena.[4] The models from which these were copied were probably brought up the Guadalquivir valley by Phocaean traders of the sixth century or by Tartessian middlemen. But whatever knowledge Greek merchants might have obtained about inner Spain was obliterated by the Carthaginian conquerors of the Spanish coast. No acquaintance with the interior is revealed in the surviving fragments of Hecataeus or in Herodotus; and something worse than ordinary ignorance appears in Aristotle, who found the source of the river Tartessus (the Guadalquivir) in the Pyrenees.[5]

The Carthaginians remained content for three centuries to exploit the resources of the southern seaboard. Their interest in the hinterland was not awakened until 236 B.C., when Hamilcar Barca discerned in it an excellent recruiting ground for the Punic army. Of his progress into the interior of Spain nothing is known; but in 220 B.C. his son Hannibal made a foray from Andalusia as far as the middle Douro; and if he used the Peñarroya and the Valdepeñas passes through the Sierra Morena on his way out and home, he laid open the two best approaches from the southern seaboard.[6] In the following year, however, he turned aside to wage his war against Rome, and so the definitive exploration of Spain was left over to his Roman conquerors.

This task was accomplished in a series of small wars against the natives which lasted intermittently for two centuries. The Romans at first secured without much difficulty the valleys of the Ebro and of the southern rivers. Their real trial of strength came later, when they waged a protracted guerrilla across the mountain lands. Their severest struggles took place in the same difficult regions that witnessed the hardest fighting of the Peninsular War, on the desolate plateau between the lower Douro and Tagus, on the equally bare watershed between Douro and Ebro, and in the mountain fastnesses of the northwest. In Spain, as in every part of the Roman dominions, the road-builder completed the opening-up process begun by the soldier. Under the Roman emperors the country, though never an important thoroughfare of traffic, was at any rate permeable in all directions.

3. SCYTHIA (SOUTH-WEST RUSSIA)

In describing the exploration of the European mainland we shall in general proceed from east to west, beginning with Russia and the river Tanais (the Don), which was usually reckoned by ancient writers as the boundary between Europe and Asia.[7] This country, though once less bare in its western and central zones than at present,[8] has at all times offered easy communications between Europe and Asia, a fact of which Asiatic conquerors have again and again taken advantage. From the Mediterranean side, however, Russia was comparatively inaccessible (p. 37).

The first dim knowledge of the Russian plains peeps through in a reference by Homer to the 'Mare-milkers, the justest of men'.[9] Herein we may recognize a preliminary report by one of the earliest Greek travellers in the Black Sea, who found to his relief that not all the tribes of those shores were pirates. But a serious penetration of the hinterland was not attempted by the Greeks until they had established their colonies along the coast (after 650 B.C.). The Greek settlers found the inland under the control of another race of newcomers, the nomad Scythian tribes. These acquired a healthy taste for Greek decorative art, and in return for Greek metal-work and ceramics bartered the grain of the wheat-belt between the Dnieper and the Danube, and the gold of the eastern Carpathians. Greek merchants were in consequence admitted to their territories and travelled up the river valleys, which were the obvious avenues of trade. Evidence of this Greek penetration survives in the plentiful finds of pottery and bronze work along the Dniester and Bug, and more particularly along the Dnieper, up which the Greek ware went as far as Kiev.[10] But our knowledge of Greek exploration comes mainly from Herodotus, who visited the colony of Olbia (situated on the Bug, near the joint estuary of that river and the Dnieper), about 450 B.C., and either made personal expeditions up the rivers or, more probably, heard them described by other travellers. His informants knew the Bug for nine days' journey upstream, the Dnieper for forty days' travel; they spoke with discerning admiration of the clear waters of the latter river, of its wealth in sturgeon and other fish, of its rich meadows and tillage-lands.[11] They also gave Herodotus information about Scythian customs, to whose general accuracy modern archaeologists have paid high tribute. From the same source the historian likewise drew a more summary account of the peoples who dwelt round the borders of Scythia proper, including the effeminate Agathyrsi of the auriferous Carpathian region, the Neuri (proto-Slavs) of Poland, who were reputed to transform themselves into werewolves, and the cannibal proto-Finns of the Smolensk district.[12]

Yet it is evident from Herodotus that the Greek merchants in Russia did not wander (probably were not allowed to wander) far from the established river routes, and did not look far to right or left. Herodotus reckons the Donetz and other tributaries of Don or Dnieper as independent streams, and his hydrographic system cannot be plotted on a modern map.[13] Again, his information nowhere extended beyond the peoples adjacent to Scythia. He may be roughly right in deriving the Dnieper and other Scythian streams from four lakes (i.e., shallow meres, now reduced to marsh).[14] But he had to admit that of the upper reaches of the Dnieper nothing was known. To the north of these he has heard of nothing but deserts where the atmosphere is always thick with snow, and he disclaims all knowledge of a northern ocean beyond these.[15] It is true that some of his contemporaries were more confident of their Russian geography. They knew of 'Rhipaean mountains' of immense height, from which the Russian rivers were fed, and of 'Hyperborean' folk beyond (on which see also p. 237). But these brave conjectures merely throw into relief their authors' ignorance. The same range of knowledge and ignorance is also indicated by the archaeological record, which marks out Kiev and the Pripet marshes as the farthest north to which Greek merchandise was carried.[16] Furthermore, in later times the Greeks made no notable advance beyond the points reached by Herodotus's informants. From the fourth century B.C. the Scythian rulers of the hinterland were gradually replaced by a new nomad tribe, the Sarmatians, who were less receptive of Greek influence and clearly discouraged Greek travel. Hence Herodotus's successors could add little or nothing to his account. The fourth-century historians Ephorus and Theopompus, whose Scythian geography is excerpted in the verses of pseudo-Scymnus,[17] are but an echo of the older writer. Strabo, who knew the Black Sea coast well, is reticent concerning the hinterland; his statement that the Dnieper was navigable for 600 stades only (c. 70 miles)[18] may be taken as marking the limits of Greek penetration in the days of Augustus. Neither did the extension of Roman influence to the Black Sea encourage the Greek settlers to fresh exploration. Under the emperor Hadrian, who gave effective protection to the seaboard towns, the trade route along the Dnieper was reopened as far as Kiev. But the trickle of Hadrianic coins does not extend beyond this city.[19] A confused passage in Pausanias (c. A.D. 170), which describes how amber was forwarded from the 'Hyperboreans' along a chain of peoples to Scythia, and so to Sinope and the Aegean area,[20] might be taken as evidence of a belated opening up of trade relations between Olbia and the Baltic. But it is by no means certain that Pausanias is describing a line of traffic of his own or of any previous time. Moreover the absence of Roman finds in eastern Poland or Lithuania tells against this view,[21] and decisive evidence against it may be found in Ptolemy's geography. Ptolemy enumerates a number of stations along the Dnieper, and shows acquaintance with the Pripet and the Beresina, but he describes these as tributaries of the Dnieper. Worse still, he locates seven mountain ranges in Russia, of which the Carpathians alone are now on view; among the other six he finds room for the inevitable 'Rhipaeans', which he trails across the Russian lowland from lat. 57° 30' to 63°.[22] Neither in ancient times nor in the Middle Ages did Mediterranean travellers pursue their explorations

4. THE DANUBE

Of the waterways which lead into Europe from the Black Sea the Danube has played the largest part in the history of travel, both by reason of its magnitude and because of its diagonal course through the heart of the continent. In prehistoric times it was an avenue by which the civilization of the Near East passed westward;[24] and at the dawn of the historic period (c. 1000–700 B.C.) it was the main track for the advance of the Scythian vanguard from Russia into the Balkan lands.[25] In the sixth century it became a gateway of Greek exploration. The pioneers of Greek travel were the colonists of Istria on the river delta, who played the same part on the Danube as the merchants of Olbia on the Dnieper and Bug, and found equally good customers among the western Scythians. At the junction of the Danube and the Seret they established a depot for trade up the streams of Moldavia; the boldest of them, if we may judge by native imitations of Greek bronze-ware in Hungary, advanced beyond the Iron Gate. We may perhaps attribute to an isolated colony of Ionian Greeks in the lower Danube basin the mysterious gifts which in Herodotus's time were conveyed from some undefined northern land to the temple of Apollo at Delos.[26] After the sixth century the Istrians appear to have extended their explorations no farther. But about 300 B.C. the wine dealers of Rhodes and Thasos, presuming on the military prestige of Philip and Alexander of Macedon in the Balkan lands (see p. 140), and on the peculiar persuasiveness of their merchandise, passed on from Moldavia into Wallachia and worked their way up the tributaries of the Danube to the southern Carpathians.[27] The tracks of these wine-merchants are marked by the remains of their amphorae, and by the silver money of Thasos, which enjoyed such a vogue in the Danubian lands that imitations of it were struck beyond the limits of Greek travel in the fastnesses of Transylvania and Bohemia.[28]

The Danube was described in unusual detail by Herodotus, who proclaimed that it was the most voluminous of the world's rivers. His information, it is true, was plainly at second-hand, and got badly damaged in transmission. He transposed tributaries from the left bank to the right, and imagined that the last great bend of the main stream inclined to the south.[29] Yet the residue of his description implies a tolerably familiar acquaintance with the lower Danube, and a slight knowledge of its middle course. Herodotus correctly notes the equable flow of the river in its lower reaches, and he enumerates no less than fifteen of its feeders, among which one may recognize the Atlas (Aluta), the Tibisis (Theiss), and the Maros (Marosh, a tributary of the Theiss). Two other streams, Alpis and Carpis, which are defined as 'flowing northward from the country above the Umbrians', should probably be identified with the Save and the Drave.[30] It may be that Herodotus's information on the Theiss and the Marosh was derived from the Agathyrsi of the eastern Carpathians, and about the Save and Drave from travellers on the Adriatic–Baltic route (p. 144). But the more likely explanation is that it came ultimately from some Greek pioneer who had adventured himself up the middle reaches of the Danube. On the other hand, it is clear that neither Herodotus nor any Greek of the fifth, fourth, or third century knew anything about the upper Danube. Herodotus found its source in 'Pyrene', Ephorus in the 'Pillar of the north' (the Alps), Timaeus in the 'Hercynian Forest' (anywhere in central Europe), Timagetus (an obscure author of c. 350 B.C.) in a 'Celtic lake', Apollonius Rhodius in the universal watershed of the 'Rhipaean Mountains'.[31] Plainly the sum total of this is zero. Still more significant is the delusion which first appeared in Scylax and Theopompus (c. 350 B.C.), but soon became firmly established, that an arm of the Danube ran out into the Adriatic Sea, as if forsooth the Save flowed west and connected across the Carnic Alps with the Isonzo.[32]

The history of modern discovery shows that the exploration of large rivers has more often proceeded down-stream rather than against the current.[33] In ancient times the problem of the Danube was similarly solved from the fountain head. Under the emperor Augustus Roman armies, groping for a suitable northern frontier to the Roman dominions, felt their way to the upper and middle Danube and established themselves along its banks. The chief credit for these discoveries belongs to Augustus's stepson and eventual successor Tiberius. In 16 B.C. Tiberius, following up a campaign on the

EUROPE

upper Rhine (p. 146), turned northward and happened upon the actual source of the Danube. From 13 B.C. to A.D. 12 he fought a succession of wars which carried the Roman arms into the great bend of the middle Danube. The effect of these movements stands out clearly in the geography of Strabo, who gave a substantially correct account of the Danube,[34] and still more so in that of Ptolemy, who had mastered the problem of its bends and used it as his chief line of reference for plotting the map of Central Europe.

5. THE BALKAN PENINSULA

Despite its greater proximity to the Mediterranean, exploration in the Balkan peninsula began later than in the Danube basin. The Aegean and Adriatic sea fronts of the peninsula are shut off from the interior at most points by coastal ranges of mountains. From the Aegean Sea three rivers – the Hebrus (Maritza), the Strymon (Struma), and the Axius (Vardar) – offer an apparently easy avenue into the continent, but progress along any of these is obstructed by cross-ranges which confine their streams in difficult gorges.[35] The Adriatic sea-wall is indented with fjords, but is not penetrated by rivers of any length. The native Illyrians and Thracians of the Balkan lands, moreover, were not as good customers for Greek merchandise as the Scythians, and had less to offer in return.

In the second millennium successive bands of migratory peoples forced their way down from the Balkan interior to the Aegean area. But the doors of the hinterland closed after their passage, and in the early first millennium no counter-movement by Greeks into the continent was attempted. In Thrace a momentary opening up was effected by the Persian war-lord Darius (c. 515 B.C.). It is probably no mere accident that the contemporary geographer Hecataeus could draw up a detailed list of tribes from Dardanelles to Danube,[36] or that Herodotus was able to describe in some detail the peculiar customs of the Thracian group of peoples.[37] Presumably these authors drew upon some Asiatic Greek who had served under Darius and his generals. But after the withdrawal of the Persians from Europe nothing further is heard of Greek visits to Thrace except by occasional soldiers of fortune like Xenophon and the remnant of the Ten Thousand.[38]

Similarly there is as yet no evidence of Greek travel up the Vardar and Struma valleys, despite the philhellenism of the Macedonian kings.[39] Under Philip and Alexander of Macedon (355-335 B.C.) the Balkans were again traversed by military reconnaissances. Philip crossed Mt Rhodope and occupied the Maritza valley; Alexander threaded his way up the Struma valley and through the main Balkan range and made a dash across the Danube.[40] But the progress thus achieved was nullified by Alexander's greater explorations in Asia, which engaged all the resources of Macedonia in an opposite quarter. Indeed, after 280 B.C. irruptions of Celtic tribes from the region of the middle Danube threw the Macedonians on the defensive and thus once more sealed up the interior.

The incipient exploration of the Balkan lands after 350 B.C. is illustrated by finds of Greek coins. Money of Byzantium made its way along the Black Sea coast into Bulgaria and Rumania; gold pieces of Larissa in Thessaly travelled up the Vardar and Morava valleys as far as Hungary; the silver of Philip and Alexander penetrated into Transylvania and Austria, and was counterfeited there by local dynasts. But it is difficult to determine how much of this money was carried northward by Macedonian soldiers and Greek traders, and how much by Celtic looters after a successful raid.[41] In any case the temporary Macedonian conquests did nothing of consequence towards clearing up Balkan geography. Current Greek belief on this subject is shown up by a theory of the historian Theopompus (c. 350 B.C.), that the peninsula tapered into an isthmus between the Black Sea and the Adriatic, and a wild surmise of his that both these waters could be seen from a peak in the Balkan interior.[42] This theory was tested in 181 B.C. by Philip V of Macedon, who made a laborious ascent of a mountain (probably Mt Vitosh, south-west of Sofia)[43] in the fond hope of beholding both the seas, and the Danube and Alps into the bargain.[44] After his descent Philip maintained a diplomatic silence concerning his experience, and thus gave a reprieve to the prevailing illusion, which was not dispelled until the Roman conquest of the Balkans.[45]

The first century of Roman rule in Macedonia (from 148 B.C.) was marked by one notable foray on the part of Marcus Livius Drusus (112 B.C.), who made a counter-raid along the Morava valley and

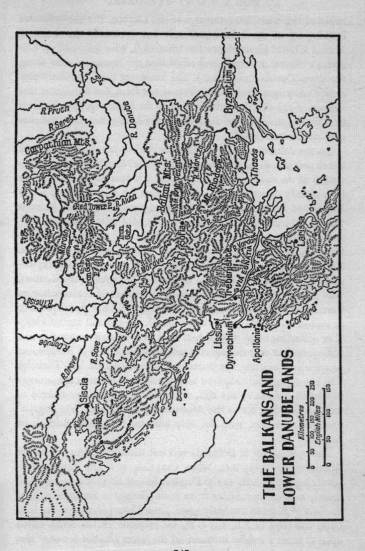

THE ANCIENT EXPLORERS

explored this main route through to the Danube. But the definitive opening up of the Balkan peninsula was essentially the work of Marcus Crassus (grandson of the triumvir?), who followed up Octavian's victory at Actium and established the Roman frontier along the lower Danube (29–28 B.C.). The history of these singularly successful campaigns survives only in skeleton form, but it indicates that Crassus re-opened Alexander's route up the Struma and past Sofia to the Danube, and explored the Balkan and Rhodope ranges by marches and counter-marches.[46]

On the Adriatic side of the Balkans there is no clear evidence of Greek penetration except up the valley of the Narenta. Here the winetraders of the third century B.C., whose activities in the Adriatic we have already noticed (p. 36), climbed over the watershed and across Bosnia and Serbia to some road-centre (presumably in the Morava valley), where they met brothers of the road who had come presumably by way of the Danube, the Save, and the Dun.[47] Where Greek joined Greek, there was the tug of corks, and the natives had a choice between vintages of Thasos and of Corcyra. But the exploration of Albania and Dalmatia was mostly left over to the Romans, who had a special interest in finding direct lines of communication from Italy to the Aegean and the lower Danube. Their first experience of Balkan travel was in the campaigns of 199–198 B.C. against Philip V of Macedon, when their armies were set the almost impossible task of forcing their way through the mountain wilderness of Albania, yet succeeded in crossing the watershed into Macedonia and Thessaly. The knowledge thus acquired was put to good use after the annexation of Macedonia in 148 B.C., when a short cut was constructed to the new province across the Albanian highlands. This 'Via Engatia' remained until A.D. 1916 the only tolerable road from Adriatic to Aegean.

The opening up of Dalmatia was not seriously attempted by the Romans until 35–34 B.C., when Octavian, as a first step towards advancing the frontier to the Danube, fought his way across the still densely wooded mountains from Senia (Zengg) to Siscia (Siszeg) on the Save.[48] A more southerly route, from Lissus (Alessio) to the Iron Gate, was tried in A.D. 105–6 by the emperor Trajan, when called upon to meet a sudden invasion on the lower Danube sector.[49] But these were only emergency tracks. In ancient as in later times the main approach to the Adriatic hinterland was from the valleys of the Save and Culpa, i.e., from the continental side. The Roman penetration of the Balkans did not lead to any extensive colonization or road-building in the peninsula. The coastal fringes and the Danube valley always remained the centres of traffic and of population.[50] But the framework of the land and its chief lines of cross-communication were definitely made known by the Roman soldier-explorers.

6. FROM ADRIATIC TO BALTIC

From the head of the Adriatic the lowest of all the Alpine passes leads to the valley of the Save, and thence to the lower Danube, or along the valley of the Raab towards the great Danube bend. From this latter point another easy path through the Moravian Gate connects with the north European plain. Finally, an almost direct route heads due north from the upper Oder valley by fords and land-bridges across the rivers and marshes of the German-Polish borderland to the Vistula and Baltic.[51] As the Danube and the Rhine form the 'cardo maximus' of Europe, so the route from Venice or Trieste to Danzig constitutes the 'cardo decumanus'.

The first section of this cross-road provided in prehistoric times one of the chief passages from the Mediterranean to Central Europe. From *c.* 1500 B.C. the bronze work of northern Italy was conveyed along it to the Danube, and thence into the Carpathian lands or Bohemia.[52] About the beginning of the Iron Age (*c.* 900 B.C.) the entire route was brought into use for the importation into Italy of Baltic amber, which henceforth supplemented, and even supplanted, the amber from Jutland (on which see pp. 52–3, 146).[53] In return Etruscan bronze-ware was traded as far as the Vistula.[54]

In the historical age traffic on the trans-European route, or at any rate on its southern sector, remained predominantly Italian. It has been supposed that the mysterious gifts transmitted to Apollo of Delos in the days of Herodotus (p. 137) consisted of amber. But the fact that they were packed in straw, and disappeared from view at Delos, indicates that they were perishables, e.g., the honey of his 'Hyperborean' bees, or wheat. Herodotus's tale therefore does not prove that the Greeks ever participated in the amber trade. Neither is it possible to demonstrate Greek travel to the Baltic on the evidence of two stray coins of the Hellenistic period which have been found in Silesia:[55] in all probability these were conveyed in native hands as mere curios. The somewhat more numerous pieces of Apollonia and Dyrrachium, belonging to the second and first centuries B.C., which have been unearthed in Transylvania together with Roman money,[56] may have been carried by Greek traders travelling in company with Italians. But at best the Greeks only gained a very slight acquaintance with the route from Adriatic to Danube.

From c. 700 B.C. the traffic between Italy and the Danube lands was interrupted by movements of Scythian or Celtic tribes,[57] and in the fourth century the importation of Baltic amber tailed off.[58] But in the last two centuries B.C. Italian adventurers opened up the lands of the middle and lower Danube afresh. At this time Transylvania and Wallachia were visited by merchants who paid in Roman denarii.[59] About 150 B.C. a discovery of placer gold in Styria caused a rush of Italian fortune-hunters into this temporary Eldorado. These undesirable pioneers, to be sure, were soon driven out by the natives; but other more legitimate traders established a regular traffic in Styrian iron.[60]

The decisive move in the exploration of the middle Danube regions was made under the early Roman emperors, who advanced the frontier to the Save and thence into the Danube bend. With these military movements went a resumption of Italian commerce beyond the Danube as far as Silesia, of which plentiful evidence survives in Roman money and Capuan bronze-ware, which now replaced Etruscan bronze as the chief article of export.[61] From a remark by a Greek writer named Philemon, who probably lived in the age of Augustus, concerning the 'fossile amber of Scythia', it may be inferred that imports of amber from the Baltic were resumed towards the beginning of the Christian era.[62] A new fillip was given to this traffic in the reign of Nero, when a Roman knight of unknown name set out from Carnuntum (near Vienna) for the amber coast, and reached it after a journey of some 500 miles.[63] From the length of his trip it is a fairly certain inference that he followed the ordinary route through the Moravian Gate and Silesia. The nameless knight returned safely to Rome and brought back so much amber that his employer, the 'procurator munerum' or Minister of Sports, was able to stud the safety-nets at the amphitheatre with the precious stones. Of all explorations by Roman civilians this was the most extensive and had the biggest results. Not only did it benefit the intermediate trade with Silesia and Posnania, where Roman coins of the late first and early second centuries are particularly frequent,[64] but it led to an expansion of commerce along the German coast, across the Baltic to Gothland, and in a lesser degree to Bornholm, to Oland, and even to the Swedish mainland.[65] Although the traffic beyond Carnuntum was left in the hands of German chapmen,[66] the route to the Baltic none the less became accurately known to Greek and Roman geographers, and most of their information about the Baltic reached them along this track. In his *Germania* (written in A.D. 98) Tacitus writes with unwonted precision concerning the tribes of eastern Germany and the Suiones of Sweden and the adjacent isles. Ptolemy, though somewhat confused about the Suiones, whom he transfers to the coast of Germany,[67] shows some knowledge of the east Baltic shore as far as Riga.[68]

An extension of Roman rule beyond the middle Danube was projected by the emperor Augustus, and in A.D. 6 his step-son Tiberius set out from Carnuntum for the invasion of Bohemia (see also pp. 155–6). But an insurrection in the cis-Danubian lands forced him to a speedy retreat, and the whole scheme of conquest was postponed indefinitely. Hence ancient geographers learnt very little about the mountain bastion of Bohemia, and ancient explorers made no use of the Elbe as a convenient line for the penetration of Germany.[69] On the other hand, the trans-Danubian plateau of Transylvania was conquered and annexed in A.D. 105–106 by the emperor Trajan. This country, though beyond the range of ordinary Greek travel on the Danube, had already come within the sphere of Italian traffic (p. 144). Yet Trajan had to conduct a search for the most suitable approach through the southern Carpathians, and his expeditions were veritable journeys of exploration.[70] By means of a superbly constructed road in the Iron Gates he overcame the natural difficulties of entry into Transylvania, and thus laid his new conquest wide open to traders and colonists. Subsequent transverse roads were driven across the eastern Carpathians, so as to connect Transylvania with the Black Sea coast.[71] In the third century A.D. the work of the Italian soldiers and traders in east-central Europe was mostly undone by the German migrations. But for the time being it had opened up cross-communications along the minor axis of Europe and had contributed more than any exploration since the cruise of Pytheas to a knowledge of its northern seaboard.

7. THE ALPS

It is a paradox in the history of European travel that one of the earliest trans-continental routes passed through the Alps. About the middle of the second millennium B.C., when a land route for the conveyance of Jutland amber to the Mediterranean was first opened, its course ran up the Elbe and its tributaries (Saale or Moldau) to the Danube, and thence by the Inn valley and across the Brenner Pass into Italy.[72] Conversely Italian bronze-ware was conveyed on a trans-Alpine route to northern Germany and Denmark.[73] After the discovery of the alternative amber route from the Baltic (p. 145) the traffic across the Alps declined in importance. In the first millennium Etruscan bronze-ware went as far as the Rhine (presumably by the Brenner and Splügen Passes).[74] But the rare finds of Greek imports into the Alpine lands and southern Germany do not indicate a lively intercourse,[75] and it is significant that the coins of Philip and Alexander, which penetrated both to the eastern and to the western forefeet of the Alps, did not circulate within the mountain area.[76] The Alpine regions, like most other difficult parts of Europe, were left over to the Romans to explore effectively. In 16 B.C., as we have seen (pp. 138–9), the emperor Augustus extended the Roman frontier to the upper Danube. To this end his stepsons Tiberius and Drusus made a converging march upon Lake Constance: while Tiberius ascended the Rhine from Basel, Drusus crossed the Tyrolese Alps by the Brenner or some adjacent pass. The gains of this brilliant campaign, which finally brought the Alpine *massif* under Roman rule, were consolidated with metalled roads across the Brenner, Splügen, and Maloja Passes, so as to connect Italy with the Inn and the upper Rhine.[77] Besides solving the problem of the upper Danube, Tiberius and Drusus cleared up the confusion between Alps and 'Hercynian

Forest' (the wooded heights of Thuringia and Bohemia) under which Greek writers had laboured.[78] Yet in the long run the natural obstacles to mountain travel outmatched the skill of the Roman road-builder: the main lines of communication from Italy to northern Europe had to be designed so as to circumvent the Swiss and Tyrolese highlands. Hence the Alpine countries did not serve as a base for exploring south Germany. What little was done to open up this region was accomplished from the middle, not from the upper Rhine.

In contrast with the Alpine passes that lead into Germany, those which open upon France were made known at a comparatively late period, but once explored remained in unbroken use. There is no clear evidence of travel across the western Alps before the Celtic migrations from Gaul to Italy (c. 400 B.C.). But from this time one of the western passes, the Great St Bernard, became a regular connecting link. It is possible that this was one of the routes by which the gold coinage of Macedonia and the money of Tarentum entered France.

Although the Greeks gained their earliest information about the Alps by way of the Rhône valley (p. 150), it is probable that this knowledge was supplemented by travellers in north Italy, where Greek traffic is attested by plentiful finds of red-figure pottery and Ionic bronze-ware.[79] It is hazardous to see any allusion to the Alps in Herodotus's remark about a 'river Alpis that flowed northward from the land above the Umbrians';[80] but beyond doubt the fourth-century historian Ephorus (excerpted by pseudo-Symnus) had them in mind when he spoke of 'the pillar of the north that dips down to the sea' (the Gulf of Lions).[81] This expression suggests a point of view in Italy. An observer on the Superga by Turin, surveying the chain of the western Alps, with the gleaming bastion of Monte Rosa towering at their northern edge, might describe his panorama in similar language. On the other hand, the continuous high altitude of the western Alps was not properly appreciated by Apollonius Rhodius (c. 225 B.C.), who invented a waterway by which the Argonauts could pass directly from Po to Rhône.[82] In any case, it would seem that the Gauls brought only the two St Bernard transits and the Mt Cenis Pass into regular use.[83] When Hannibal in 218 B.C. selected a more southerly defile for the crossing of the Alps, he was in effect making a voyage of exploration, and his severest trial lay in the unforeseen abruptness of the descent into Italy. Thanks to the ambiguities and ellipses in the narratives of Polybius and of Livy, it remains a matter of controversy whether Hannibal used the Col du Clapier, the Mt Genèvre, or the Col d'Argentière.[84] From the geographer's point of view the chief consequence of Hannibal's exploit was that it stimulated Polybius to reconnoitre the western Alps in person and to make the first definite attempt to place them on the map. The Greek historian oriented them wrongly,[85] but he duly emphasized their enormous massiveness and gave a good description of the chamois.[86]

But the opening up of the western as of the eastern Alps was more particularly the work of the Romans, whose acquisitions in France in the second and first centuries B.C. required them to establish direct communications across the mountains. Whereas Polybius only knew of the coastal road, of Hannibal's pass, and of the Little St Bernard, Varro (c. 50 B.C.) enumerated three passes intermediate between the coast road and the Little St Bernard.[87] One of these roads (probably the Mt Genèvre) was made more commodious for the passage of troops by Pompey (c. 75 B.C.);[88] Augustus paved the Little St Bernard,[89] and one of his early successors did as much for the Great St Bernard. So well was this last road constructed that in A.D. 69 a large Roman army crossed it at the season of the melting snows and avalanches without suffering any serious mishap.[90]

8. FRANCE

Of all the gateways that open into Europe from the Mediterranean, none offer a more easy passage than those of France. From Narbonne the gap formed by the valleys of the Aude and the Garonne provides a short cut to the Atlantic seaboard. From Marseille or Arles the most attractive of all river-entrances into Europe leads past Lyon to east, north, and west. From Lyon the Rhône may be followed to the base of the western Alps, or it may be ascended as far as Lake Leman, whence an easy track passes over into the valleys of the Aar and the upper Rhine. The Rhine may equally be reached by the Saône and the Burgundian Gate, the Atlantic coastlands by way of the Seine or the Loire; and on none of these routes do the rivers or their watersheds offer serious difficulties to travel. Yet despite these natural advantages

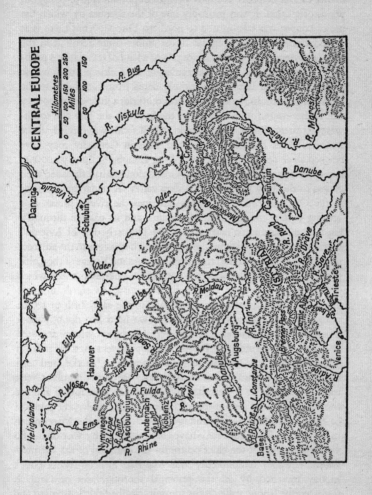

France cannot be proved to have been an important land of passage in prehistoric times. It was probably one of the avenues by which the use of copper was transmitted to northern Europe;[91] but there is as yet no evidence of any regular tin or amber route across it in the dawn or forenoon of ancient history.[92] As a land of travel it was probably of little account until the coming of the Celtic peoples, whose arrival in western Europe is variously dated from the eighth to the sixth century B.C. In the sixth and fifth centuries a little Greek pottery and bronze-ware found its way up the Saône to the Jura and southwest Germany. But it is not certain whether this was carried all the way by Greek traders; indeed the finds in Germany may have reached their destination by way of the Danube.[93] But by 500 B.C. it may be assumed that occasional Greek pioneers had threaded their course to the very head of the Rhône valley. A remarkable passage in the Ora Maritima of Avienus describes the river's source in a 'gaping cavern' by the 'Sun Mountain', and its passage through a big lake and through a narrow gorge.[94] The early date of Avienus's source may be inferred from the names which it assigns to the adjacent populations. Instead of the familiar Allobroges and Volcae of the Celtic period we are here introduced to otherwise unknown tribes which the Celts had displaced by the fifth century. Avienus's ultimate informant must therefore be dated back to the early fifth or to the sixth century. The writer in question, we need not doubt, was a Massiliote. It should be added, however that the knowledge acquired at Massilia concerning the Rhône did not prevent Aeschylus from presenting it to an Athenian audience as a river of 'Iberia', and confusing it with the amber stream Eridanus.[95] Proof of regular Greek traffic up the Rhône is furnished by coins of the later fourth and the third centuries, the gold pieces of Philip and Alexander, the mintage of Massilia, and the Gallic imitations of either. Of the Macedonian money a certain amount may have entered France by way of the Po valley (p. 147); but the chief centre for the diffusion of the Mediterranean coinages undoubtedly was Massilia. Henceforward Greek money travelled by all the principal thoroughfares of Gaul. It followed the Rhône valley to its summit, climbed the St Bernard and descended into north Italy, thus forming a counter-current to such traffic as may have passed from the Adriatic into France.[96] From the

Lake of Geneva it also threaded the corridor between the Alps and Jura into the Aar valley, where finds of alluvial gold were a source of attraction to foreign traders.[97]

On the Atlantic side of the watershed, to which the early Greek bronzes and pottery-ware only penetrated in a sporadic way, the coins of Massilia, Macedon, and Tarentum passed down the valleys of the Seine and Loire, those of Emporiae and Rhoda (in northern Spain) down the Garonne.[98] It is true that the Greek money in general did not approach within 100 miles of the Atlantic coast, which perhaps means that Greek traders were not yet pushing their way as far as the seaboard. But it is clear that by 135 B.C. at the latest through communications between the Mediterranean ports and the Loire estuary had been established. At that time Scipio Aemilianus met at Massilia or Narbo a number of merchants from Corbilo (on the lower Loire).[99] It is not clear whether these traders were Greeks or Gauls, but undoubtedly they were dealers in Cornish tin. The method of overland transportation has been described by Diodorus. The metal was cast into concave ingots and slung on the backs of horses at the French port. The duration of the journey, which probably extended from Corbilo to Arles or Marseille, and therefore measured some 500 miles, was of thirty days.[100] Though no direct proof exists, it seems likely that the overland cargoes went in part by the Seine or the Garonne, though the main route lay along the valley of the Loire.[101]

It cannot be said that the opening up of France by Greek traders produced a corresponding increase in general knowledge of French geography. In extending the cruise of the Argonauts to Switzerland and France, Apollonius Rhodius adventured himself far out of his own depth. In his story the Argonauts, having passed from the Po to the Rhône (p. 147), sail across 'wintry lakes in the Celtic land'; re-entering the Rhône, they reach another fork (the third on their fissiparous course), and taking the wrong limb are about to stray towards Ocean, when a traffic policeman in the shape of Hera drops from the skies on to the 'Hercynian Rock' (the Alps) and signals a turn to the left.[102] From this passage it appears that Apollonius might have heard of other lakes in western Switzerland beside Lake Leman, and of the 'bise' which sweeps over them from the Bernese Oberland; but obviously he knew nothing of the river-system of France, though it was precisely by these rivers that the Greek pioneers had explored the country. The reason for his ignorance may be found in the deliberate reticence of the Greek merchants engaged in the up-river traffic. When Scipio Aemilianus questioned the traders of Massilia and Narbo, and visitors from Corbilo, on the subject of Britain, they with one accord disclaimed all knowledge thereof,[103] and no doubt would have given a similar reply if Scipio had asked about the way to Britain.

In France, as in the Danube lands, the Italian trader preceded the Roman soldier. Italian money-lenders made themselves indispensable in many parts,[104] and vendors of wine penetrated everywhere, except only into a few 'dry' areas.[105] But to go by the finds of Roman coins, these traders were not numerous and did not begin their activities before 100 B.C.[106] To the Roman world in general Gaul remained a strange country until Julius Caesar explored it by fighting his way along its length and breadth. From the geographical no less than from the political standpoint Caesar's chief discovery was that of the Rhine,[107] which in conjunction with the Danube gave the Roman emperors their best frontier and Greek geographers their best lines of reference in marking out the map of Europe. Caesar also searched out the two most secluded fastnesses of Gaul, the area of swamp and forest between the Rhine and the Somme, and the 'massif central' between the Loire and the Garonne. By 50 B.C. no considerable district of Gaul had been left unvisited by him and his lieutenants, and it only remained for Marcus Agrippa (*c.* 20 B.C.) to round off Caesar's work with his surveyors and road-makers. The effect of Caesar's campaigns, moreover, was as rapid as it was far-reaching. In conquering Gaul Caesar simultaneously made it known to the general Roman public: as primers of geography his commentaries will compare with Wellesley's dispatches from India.[108] The results of his conquest show up well in Strabo. While this writer retained a very inaccurate idea of the Atlantic seaboard of France, which he imagined as running in an unbroken line from west to east,[109] he gave a competent description of the inland regions.

9. WESTERN GERMANY

We have already seen that the ocean seaboard of Germany was made known by Pytheas and the Roman navy, and its eastern margin by the Roman amber traders, but that neither the Elbe nor the upper Danube was used as a base for the opening up of the southern and central regions (pp. 145–6). In the present section we shall consider the tentative explorations which proceeded from the line of the Rhine.[110]

In the middle of the first millennium B.C. the upper Rhine and upper Danube became the channels of a brisk commercial movement between the Jura districts and Bohemia, which originated in the export of the fine iron-ware of La Tène (near Lake Neuchâtel), and anticipated the yet greater traffic from Gaul to the Danube lands under the Roman emperors.[111] At the end of the second century B.C. the entire stretch of the middle and upper Rhine was traversed up and down by those unwilling explorers, the Cimbri and Teutones; their line of march is revealed by hoards of miscellaneous coins which evidently were droppings from their variegated war-loot.[112]

The definitive opening up of western Germany appeared to be in store when the Romans established themselves along the Rhine. Julius Caesar, it is true, contented himself with two brief raids across the middle Rhine,[113] but Augustus planned the advance of the Roman frontier from that river to the Elbe. The expeditions that were undertaken from 12 B.C. to A.D. 6 were attended with difficulties similar to those that beset the British and French armies operating in North America during the eighteenth century, for the Romans had to blaze their trails through regions of dense forest. The armies, as we have seen (p. 60), were assisted by naval explorations up the Ems, Weser, and Elbe; but the principal lines of advance ascended the Lippe and the Main. Two of these campaigns were of special promise for the opening up of western and southern Germany. In 9 B.C. Augustus's stepson Drusus forced his way as far as the river Elbe. On his outward march he apparently ascended the Main or Lahn, and skirted the Harz to north or south; on the return journey he tried out the more southerly route along the Saale. In A.D. 6 his brother Tiberius arranged a converging movement of two armies, one of

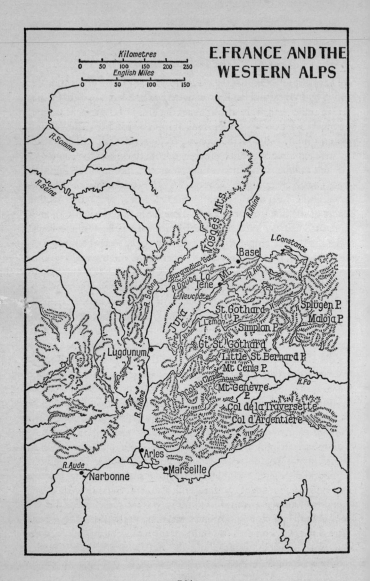

EUROPE

which, as we have seen (p. 145), was to invade Bohemia from the Danube, while the other made for the same objective by threading its way up the Main; by these operations southern Germany was to be encircled as Switzerland had been ringed off in 16 B.C. (p. 146). But the insurrection which broke out in his rear necessitated the retreat of both his divisions.[114] Furthermore in A.D. 9 a less skilful strategist, Quinctilius Varus, lost his bearings and his army to boot by straying from the line of the Lippe into the forests of north-western Germany, and this set-back induced Augustus to abandon his scheme of conquest. Henceforth the usual Roman policy was to establish a glacis beyond the Rhine and upper Danube, and to prevent a free flow of traffic across those rivers. A partial exception to this rule was allowed at some selected points. On the northern sector trading centres were established at Noviomagus (Nymwegen) and Asciburgium (opposite the junction of the Rhine and the Ruhr), whence Capuan bronzeware and Roman denarii travelled across north-western Germany into Denmark.[115] This traffic, although in native hands, gave the Romans a somewhat closer acquaintance with the region between Hanover, the Elbe, and the Atlantic seaboard. But in general the Roman embargo on cross-river traffic was strictly enforced, and Central Germany, in particular, remained unexplored by official order.

Because of the patchy and incomplete manner in which Germany was opened up by the Romans, the geography of the country was never properly coordinated by ancient writers. Tacitus, who described in some detail the tribes on the eastern and western fringes of Germany, left a wide gap between them. Ptolemy endeavoured to fill this lacuna, but knowing little of the Elbe and even less of the Oder he was unable to arrange his data of peoples, towns, and mountain systems on any consistent plan.[116] For similar reasons he failed to correlate his information about Scandinavia. Having collected one set of details from the standpoint of the Atlantic seaboard, another set from that of the eastern Baltic, he was unable to combine them in a correct stereoscopic view, but described 'Scandinavia' (or 'Scadinavia') on the western side, and the 'Suiones' on the eastern, as if they had no connexion.[117] The failure of the Romans to conquer Germany has often been discussed from a political angle. From the standpoint

155

of trade and travel it involved the result that the easiest lines of communication across Europe from east to west were never opened up in antiquity, and that the European continent remained essentially a mere hinterland to the Mediterranean seaboard.

CHAPTER 7

ASIA

1. GEOGRAPHICAL. PREHISTORIC MOVEMENTS. MESOPOTAMIA AND EGYPT

THE nature of man's development in mind and spirit during the era of ancient history makes this chapter almost wholly an account of slow penetration into, and even conquest of a part of, the vast tracts of Asia by the civilized peoples of the enterprising West. It is the story of the exploration of one huge plateau by men from Mediterranean lands. The western borders given to Asia are not and have not been based on real division of races, and it is not untrue to say that the countries of the eastern half of the Mediterranean interacted on one another. We shall take as the western boundaries of ancient Asia the same, with one exception, as those which generally held good in the eyes of the Greeks. Thus from north to south the line shall be the River Don (Tanais); the eastern and southern coasts of the Black Sea; the western and southern coasts of Asia Minor; and Palestine as far as the Gulf of Suez. Unlike the Greeks, we shall exclude Egypt as belonging to Africa.

Asia shelves up from the lowlands of Siberia to a great divide stretching from Asia Minor through the Elburz and Hindu Kush to Tian Shan and away to Behring Strait; south of the eastern half of this comes the vast depression of Gobi, and then another divide consisting of a mighty mass of mountains leaving the northern divide in the Pamir region and passing from the Himalayas and Kuen Lun Mountains into the mountains of north China as far as the river Amur. South of this come the low plains of India, farther India and China. This colossal system of formations was never explored by any single Eastern people nor by the ancient Greeks and Romans, except that they sailed in the Indian Ocean, exploring the southern coasts, and penetrated (as far as the Pamirs) the extensive plateau which consists of all Asia west of India, except Arabia.

This plateau, a wide expanse of about 2,500 miles from west to east, they did come to explore because of the nearness of Asia Minor to the Greek world and the fact that one can travel from the Caspian to Karachi without crossing heights greater than 5,000 feet; but since even Asia Minor rises from west to east, with mountains enclosing a central salt desert and falling into narrow coast plains and a few inland ones, even here, penetration by the seafaring Greeks was slow. Arabia stood apart, and was never explored in ancient times. It may be said truly that Asia possesses barriers which Europe has not. Such features as the Armenian heights, the Iran deserts, and the terrifying mountains of central Asia are obvious instances. There are also extremes of climate, of cold and heat, associated with each of the very regions here named. And further it can be truthfully said that the greatest civilizations which developed in it, except the Arabians and Phoenicians and the Greek invaders, were not quick to see the advantages of intercourse.

With the prehistoric movements of the peoples of Asia we cannot deal; indeed, it is difficult even to guess at the conditions of mankind in Asia before about 2000 to 1500 B.C. In the dim mists of antiquity there descended on the aboriginal races, such as the Veddahs of Ceylon, the Ainus of Japan, the Kols and Santhals of India, and so on, invaders from other regions; thus the Chinese came from the west, the Aryan Hindu and Persians from the north and north-west, the Burmese and Siamese from the north. Asia Minor was held chiefly by mysterious Hittites who were invaded by Aryans before the eleventh century; Semite races, with local ideas, lived on the east coast of the Mediterranean and Arabia; the Iranians were Aryans perhaps from central Asia. India had had two waves of invading peoples from the north and north-west; China was a shifting league of feudal states; all else in inner Asia was a gently surging mass of nomads, except in Babylonia and Assyria, which were the centre of culture and activity until the Persian Empire swept over them. Yet even this source for the spreading of civilization was fixed and local, bursting forth however at times in outbreaks of a military kind and conducted by the Assyrians when they were great, before which time one or two invasions of Elam, Kurdistan, and 'the West' are all we know of. But Shalmaneser I (thirteenth century) attacked the Hittites and other people, and explored Cappadocia, while Tiglath-Pileser I (eleventh

ASIA

century) pushed towards Armenia and the upper Euphrates. Tiglath-Pileser III after 745 harassed Armenia and Media and shook the Hittites, and Sargon II reached the Elburz heights and the place (today Hamadan) where Ecbatana rose later, but even this was going a very little way, and in Arabia little was known beyond Teima; after the attacks on Egypt in 674 and 670 the empire began to break up.[1]

Egypt, too, had done something by way of real exploration of Asia from without. Expeditions to South Palestine started under the sixth dynasty, but it was the glorious eighteenth dynasty which from 1580 onwards did most to keep Syria and Phoenicia secure against the advancing Hittites. Thothmes I, and perhaps his predecessor Amenophis I, reached the Euphrates, which Thothmes III crossed and followed southwards to Niy, where he hunted the elephants which still lived there. Says Amenemhab: in Niy he 'hunted 120 elephants, for the sake of their tusks. I engaged the largest which was among them which fought against his majesty; I cut off his hand' (trunk, of course) 'while he was alive before his majesty, while I stood in the water between two rocks. Then my lord rewarded me with gold; he gave me ... 3 changes of clothing.' Other kings of the dynasty were also strong in this direction, and by the twenty-eighth dynasty a part of Hittite Asia Minor was peacefully explored. But the Israelites were becoming a barrier, and from the twenty-first dynasty (about 1100) onwards, Egyptian power in Levant varied until the all-conquering Persian in 525 overran the land of Egypt.[2]

2. THE EARLY GREEK ACTIVITIES

We find that the Minoans or Aegeans, Mediterranean in their ideas, did hardly anything, though under Achaean leadership they may have sailed the Black Sea to Colchis,[3] and certainly overthrew the Asiatic city of Troy, gaining some knowledge of a few features of Asia Minor[4] as the Homeric poems show. Of Assyria and Babylonia and Arabia these poems have no sign. The Hittites for a long time were a strong barrier against the penetration of even Asia Minor. After these ceased to be a power the Hellenic Greeks began to develop their culture and became seafarers to an extent not dreamt of by their 'Aegean' predecessors, so soon as the Dorian invasion had spent its force. This invasion and other influences had pushed the Greeks about 1130–1000 over the Aegean to the west coast of Asia Minor on the first part of their great colonizing activities. The colonization of this coast by Greeks settled in cities which prospered, and later colonies on the southern and eastern shores of the Black Sea,[5] brought a knowledge by exploration of all the coasts of Asia Minor from Issus (on the Gulf of Scanderoon) to the Phasis (Rion, mentioned in a late passage of Hesiod)[6] and a good deal more knowledge obtained by hearsay (particularly from the centres Sinope and Trapezus), of inland regions hardly explored at all by the seamen Greeks, except perhaps western Phrygia, Maeonia, and Lydia (by the early half of the seventh century) and the vicinity of Phasis (a town based on the Rion trade) and of Dioscurias (Sukhum Kaleh, receiving much Caucasian trade) and of Panticapaeum (Kertch, prospering on trade down the Don). Before the end of the century the Greeks had heard of tribes such as the Issedones[7] east of the Caspian between Merv and Balkh, along the upper Oxus valley; Aristeas especially, of Proconnesus, learnt a good deal about wild tribes of the Asiatic Steppes of the north and east of this sea, and travelled some way among them himself, though his credentials are not certain.[8] The tales told by Phoenicians and the spread of Assyrian power even to Cyprus made the Greeks interested in Assyrian kings, so that Assurbanipal (668–626) with his easy luxury became the prototype of Sardanapalus.[9] When, after the fall of Assyria, Nebuchadnezzar (604–562) made Babylon great, a few Greeks travelled to the place in Phoenician and Arabian caravans from Syria, and others served the king as paid soldiers.[10] These were in a sense the first Greek explorers of Asia.

3. THE PERSIAN CONQUEST AND ITS RESULTS

The fall of everything before the Persians as far as the Aegean by 545 brought the Greeks even of Europe face to face with a power of a new kind, and their clash with this power both hindered and helped western knowledge of Asia. The Empire of Cyrus stretched from the Aegean over the plateau of Iran to inner Afghanistan and Bactria (Afghan, Turkestan, and Badakshan) and Sogdiana (Bokhara) to the Iaxartes (Syr Darya) and the new town of Maracanda (Samarkand).[11]

ASIA

Darius who, as Herodotus puts it, discovered the greater part of Asia, extended the dominion to the Caucasus,[12] annexed the valley of the Indus and the gold-producing lands of Kafiristan, Kashmir, and Dardistan, making perhaps the Beas the boundary, warred against the Sacae of the Pamir plateau and regions north of it, and caused Scylax and probably Ionian Greeks to explore the Kabul and Indus rivers to the ocean, to report details of the lands on either side of the Indus, and then to sail to Suez as we describe elsewhere[13] (see p. 78). Arabia was neither conquered nor explored, but only touched along the desert fringe between Suez and Euphrates. If the Scythian expedition (about 512) really took place, Darius explored a little east of the Tanais. This king made better roads, particularly the secure and well-provided Royal Road beginning at the satrapy capital Sardis, reached from the Aegean coast in three days by Greeks with guides,[14] and ending at Susa after a long journey through Lydia, places later called Acmonia, Ipsus, Laodicea the Burnt, and Cybistra, and then the Cilician Gates and across the Euphrates at Zeugma to the Tigris at the ruins of Nineveh, turning south after Arbela across the Greater and the Lesser Zab to the Choaspes (Kherkah) and Susa. This road was now travelled on by exploring Greeks, and mapped by them as a straight line. One other route the Greeks had probably trodden, leaving the Royal Road at later Ipsus, and taking a great loop round the salt Anatolian desert through what were later Pessinus and Ancyra (Angora), across the Halys (Kizil Irmak) and then south-east across the Halys again, to the later Mazaca, and then south to the strong Cilician Gates in Mount Taurus. There can be no doubt that many Greeks, including Herodotus, confused these two roads.[15] Witness what he makes Aristagoras of Miletus say when he came to Sparta in 499 to ask for help in the revolt against Persia. Aristagoras showed to Cleomenes, the king of Sparta, a bronze tablet 'on which was engraved the circuit of the whole earth, and all the sea, and all the rivers'. It was probably a map based, as many others were, on the first ever made by Greeks, that of Anaximander of Miletus in the first half of the sixth century. Pointing out with his finger on the map Aristagoras said: 'Next to the Ionians here the Lydians dwell in a lovely land, and are very rich in silver. ... Here we have Phrygians next to them on the east, the richest in sheep and fruits of all men known to me. ... Next to the

Phrygians are the Cappadocians, whom we call Syrians; bounding these are the Cilicians stretching down to the sea, where, look you, lies Cyprus Island. ... Here, next to the Cilicians, we have the Armenians, and these too are very rich in sheep. Next are the Matienoi, holding this Cissian land in which, look you, here is Susa lying beside this river Choaspes. ...' Cleomenes said he would put off his reply till the third day. When it came, Cleomenes asked Aristagoras how long a journey it was from the Ionian Sea to the Persian king's court; Aristagoras, forgetting for the moment the dislike of Spartans of travelling far from Sparta, especially away from the sea, blurted out: 'Three months. ...' 'Go away before sunset,' interrupted Cleomenes.[16]

Further knowledge of utterly unexplored regions of inner Asia came with the Persian invasions in 490 and 480 when European Greeks saw in Xerxes's army the varied races from all Asia between the Aegean and Bactria and north-west India, so that not only could Hecataeus of Miletus, who had travelled in Asia, add to the Asiatic part of his work a list of the Persian possessions as far as India, mention of which and of the Indus he is the first to make,[17] but also Aeschylus, who fought at Marathon and Salamis, in his best plays at Athens was able to bring in various people and towns of the Persian Empire, journeys in Asia Minor, and so on,[18] though since he was not an explorer his geographical ideas were utterly wrong and even extraordinary.[19] It is clear that the Persians, who, as Herodotus said, looked on Asia as their own,[20] tended to exclude the Greeks from actually visiting the regions of inner Asia, so that even Herodotus does not know much beyond the Persian Empire except round the Caspian; but a certain number were admitted to the court of the kings at Susa for special reasons – Telephanes under Darius and Xerxes because he was a sculptor;[21] Democedes as a captive, and later at a very high fee under Darius, because he was a good doctor of medicine;[22] the Greek vassal-tyrants of the coast of Asia Minor, for imperial reasons until 499; and Diotimus and other diplomats of Athens, especially in order to arrange the peace of 448 between Athens and Persia.[23]

After that date travel in Persian lands was easier; Xanthus not only visited Armenia, but studied the volcanic wonders of western Phrygia and Lydia – the craters, earthquakes, and vapours; Democritus of

ASIA

Abdera travelled widely in Asia, and may have visited the Persian satrapy in north-west India,[24] while Herodotus is supposed to have travelled extensively in Asia both before and after 448. At any rate he visited Colchis, and in another direction, Tyre, from which he went to Mesopotamia, so that he knew the Euphrates and the Tigris tolerably well, and coasted from Tyre to Gaza, but it is probable that his information about Persia and its satrapies was got largely from Hecateus and the satrap at Sardis, and in writing about Asia Minor he confused the main roads and did not know the Taurus Mountains or the Anatolian Desert, though of course he and men of his time knew mere names in plenty. Lack of exploration combined with interest in the ruling power caused Herodotus to be in a sense ethnographical rather than geographical even in dealing with Asia Minor, which he, like writers also, believed had a waist which made crossing it a short matter. He also confused the Araxes and the Oxus or the Iaxartes. The same dimness applies to the plateau of Iran, where names and some details and relative positions were heard of, but no Greek had explored it; not even the great mountain ranges of the plateau were known. To the Greeks India was merely the district of the Indus, stretching away to the east, and very populous. No one knew what lay beyond it except uninhabited sandy deserts (that is the Thar), nor whether Asia was bounded by sea or land; India was the easternmost part of the inhabited world. Of its northern mountains nothing, of its products barely anything was heard of; all that they knew came from the Persians by hearsay.[25] At the end of the fifth century Ctesias of Cnidus, physician to the court, made Asia better known in three ways: he made special note of trade-routes between Ephesus and Bactria and between Ephesus and India and lengths of daily travel on them; he knew that the north-west of India was mountainous; and he also learnt and reported some details about the precious products and the size, much exaggerated, of India, but unfortunately he gave way to a love of the marvellous[26] (see pp. 238–9).

4. EXPLORATION OF THE CASPIAN AND ITS STEPPES

In one direction, however, the Greeks had made remarkable explorations in search of furs and gold from the Urals and the Altais, namely

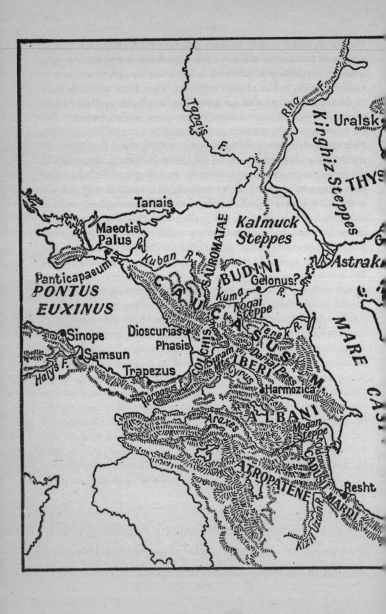

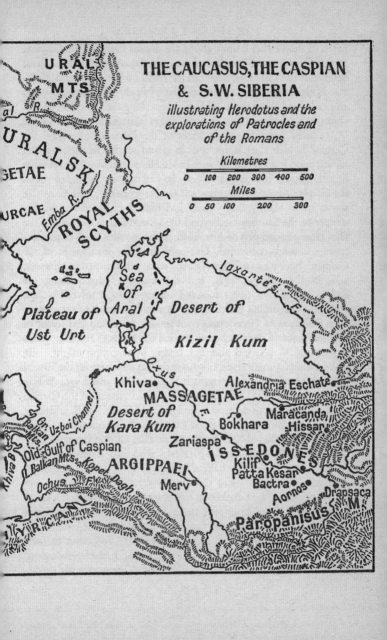

north of the Persian Empire round the distant shores of the Caspian. In this direction the Greeks had become acquainted with the great heights of the Caucasus, and men of towns on the north and northeast of the Black Sea had pushed inland (we do not know how far) and had heard about the marshes of Russia along the unexplored Don, about tribes between this and the Volga (though not much of the Volga itself), and about the races of the lowlands all round the Caspian, which they learnt was an inland sea by frequenting and sailing upon it apparently in row-boats. Herodotus's sources for what he tells us were traders and sailors who had travelled across or round the Caspian – namely Scythians and Greeks from the Black Sea.[27] He shows quite plainly that Greeks had indeed explored the treeless steppes of the Sauromatae beyond the Don,[28] that is the land of the Don Cossacks and part of Astrakhan, to the wooded lands of the lice-eating, nomad, red-haired, blue-eyed Budini[29] north-west of the Caspian but west of the Volga, who hunted otters, beavers, and, in the Caspian, seals, and that fugitive Greeks along the Kuban and Kuma rivers founded among them a town Gelonus, which was becoming barbarized.[30] One of these explorers may have been Damastes of Herodotus's own time, a man who had learnt of tribes apparently near and even beyond the Urals.[31] But Herodotus ignores the Volga, and although traders in reporting might easily make little of a mere river, this suggests that here among the Budini lay the limit of Greek penetration in this direction, and that the reason why he says that the lands even beyond round the east of the Caspian were well known to the west (through Greeks as well as Scythians of south Russia west of the Don) is either that he attributes to Greeks what really applies only to Scythians; or else, all that lay east of the Budini had been explored by Greeks who had sailed across the Caspian near the southern shore and explored from the south-east coast inland through a vast plain in several directions; in these we may well include the Greeks of the mart on the Borysthenes (Dnieper) who reached [32] the Argippaei, that is Turcomans east of the Kopet Dagh heights, and Aristeas (*c.* 550), who reached the Issedones, as we saw, and learnt fables about regions beyond.[33] We cannot be sure. Anyhow, much had been found out beyond the Budini. Seven days' desert (between Astrakhan and Guriev) led to Thyssagetae and Iurcae,

ASIA

who were hunters dwelling near Uralsk, and Royal Scyths north of the plateau of Ust Urt.[34]

A rugged and rough tract (deserts of Kizil Kum and Kara Kum) then stretched to the bald, flat-nosed, long-chinned Argippaei at the foot of the high mountains (Kopet Dagh range); with these the Scythian traders talked through seven interpreters. They fed upon bird-cherries[35] like the Calmucks today, whose sacred men are bald. The Issedones came next, but beyond them on the east nothing was known but fables[36] except the Massagetae of Khiva and Bokhara,[37] while in the north the Argippaei reported goat-footed men (mountaineers) of high mountains (Urals) which blocked the way to people who slept for half the year, that is, who experienced long Arctic winter nights.[38] All beyond was uninhabitable through cold.[39] Subsequent influence and presence of the Greek merchants inland on the north-east shores of the Black Sea are shown by excavations in the Kuban valley and so on,[40] but it is curious that the exploration of the Caspian shores and regions not only ceased but was forgotten, so that for centuries the Caspian was regarded as a water connected with the northern Ocean.

5. THE TEN THOUSAND

At the very end of the fifth century, when Greek soldiers were beginning to serve barbarian authorities on a scale unknown before, there occurred a remarkable piece of exploration which may be said to have begun the opening up of Asia Minor and even Armenia. Cyrus, satrap at Sardis, who intended to overthrow his brother Artaxerxes, king of Persia, took into his service, besides his orientals, about 13,000 Greeks of various states under the pretext of reducing the hillsmen of Pisidia.[41] The terrible adventures of these Greeks, calling forth all their bravery, discipline, and staying power, are told to us in the lively and authentic pages of Xenophon.

In 401 B.C. Cyrus led his army out in springtime from Sardis and went along a known route across the Maeander to Celaenae near its sources. Having picked up further Greek forces, he went north to the Potters' Mart (Iskam Keui), east through several rising towns and through rough and wild Lycaonia and Cappadocia to Tyana, and

167

then south to the strong Cilician Gates crossing the Taurus,[42] meeting at Tarsus Menon, who had been guided along a shorter route near the coast. Here the Greeks, no longer deceived, and almost terrified at the idea of marching for three months into inner Asia, refused for three weeks to advance, until they were won over by the unpopular martinet Clearchus and by Cyrus's alleged intention not to go beyond the Euphrates. The Sarus and Pyramus rivers were crossed to Issus and then the army went through the Jonah pass to Myriandrus on the Gulf of Scanderoon.[43] After a week's rest here they crossed Mount Amanus by the pass of Beilan, and in twelve days reached the Euphrates[44] at what was then its normal crossing, Thapsacus, after which the distances along the roads were unkown. Here Cyrus had to confess what the Greeks were guessing – that the expedition was aimed at the great king, but his promise of a handsome reward and the example of Menon carried the Greeks across the river by fording early in August, and a journey along the river began. A pleasant march to the Khabur river was followed by a fortnight passing through the smooth and treeless 'Arabian desert' full of strange beasts and birds, to Pylae, where Babylonia, which was not a desert in those days, began. At Cunaxa (near Fellujah and the mount of Kunish) the forces of Artaxerxes met them, and the death of Cyrus left the uncrushed Greeks leaderless in an unfriendly land. Artaxerxes let them alone, but how were they to return without provisions? Since they did not know the course of the Tigris they accepted the offer of Tissaphernes to guide them by another road.[45] Passing the 'Median Wall' near Babylon, they reached and crossed the Tigris by a bridge somewhere south of Baghdad. Following the left bank into Media, they crossed the Lesser Zab. On the banks of the Greater Zab, the treacherous abduction of the Greek generals by Tissaphernes caused the election of new ones, including the able Xenophon, and the army went on, using local guides when they could get them.[46] The Greeks' one wish was to go home to the sea and the Greek world. Having crossed the Greater Zab they travelled for about ten days on the left bank of the Tigris, harassed by Tissaphernes, through a hostile hilly tract, past ruined Calah and Nineveh to the mountains of Kurdistan just above Jexireh. They found the river too deep and the opposition too strong to cross on the way west of which they had been told, and so

ASIA

they went up into the Kurdistan Mountains, [47] which Persia had never really conquered, and into Armenia where a route westward might be chosen.

What followed was the exploration without permanent guides and compass, and in mid-winter, of rugged, difficult and inhospitable regions which were hardly explored again by travellers (let alone by women such as those who must have marched too with these Greeks) till modern times, and never have been by an army, namely the mountains and table-land of Armenia which had no good roads. The Greeks were strongly opposed by the Carduchi or Kurds, and there was seven days' fighting. Pushing through the mountains the Greeks reached in December the river Centrites (Buhtan Chai) with hillsmen on their rear. They came to an agreement with the satrap of part of Armenia, and pushed ahead towards the north-west. They crossed the Bitlis Chai, and by way of Mush, the Teleboas (Murad Su) on the high tableland of Armenia.[48]

We cannot be sure of the route actually taken thence across intricate and rugged regions. From the Teleboas it appears the Gunek Su was reached. To get away from the satrap of Armenia they then went (helped by local guides) across trackless hills amidst snowstorms, crossed, wet to the waist, the western branch of the Euphrates west of Erzeroum, moved east along the river in the face of blasting bitter winds, snow-blizzards, snow-blindness, frost-bite, and hunger, to some almost underground villages on the tableland of Erzeroum, where there was a month's rest. A good many of the soldiers had been left behind – 'those whose eyes had been blinded by the snow, and those whose toes had been rotted off by the cold. But it was a guard for the eyes against the snow if a man walked along with something black in front of them, and a guard for the feet if a man was moving and never took rest, and took off his footwear for the night. But whenever any took their sleep with shoes tied on, the straps sank into their feet, and the shoes froze round them; for, since their old shoes had worn out, their footwear was brogues made of newly-flayed cattle-hides.' Then apparently they were misled to, and eastwards along the Araxes, called locally Phasis. Thinking it was the Phasis (Rion) which flows along the Black Sea, they followed it for seven days, returned along it for two days, and then pushed

northwards away from it and on amidst great privations by way of the Kars plain, across the Kur and the Olti and westwards down the Harpasus (Chorokh), and then again northwards.

The city of Gymnias was the first sign of civilization, and they were welcomed there, being told that the Greek city Trapezus (Trepizond) on the Black Sea was not far off. A guide promised to show them the sea, and indeed on the fifth day, with a scrambling tumult of joyful running and shouting, they saw it from the top of Mount Theches. Then, with tear-dimmed eyes, they kept embracing each other, and their generals and captains too. A few more days brought them early in 400 to Trapezus, where they rested a month.[49] Numbers had died from exhaustion; others had lost toes, others their eyesight. Had not the half-underground dwellings of the Armenians, holding both men and cattle and good supplies, given some shelter and relief, far fewer would have won through.[50] Ultimately they marched along the known Black Sea Coast to Cerasus and Cotyora, sailed to Sinope and Heraclea, and marched again to Chrysopolis (Scutari), whence they crossed out of Asia to Byzantium.[51]

6. THE FOURTH CENTURY TO ALEXANDER

Henceforth the Greeks felt more strongly than before that the Persian Empire was weak and accessible, and in the early part of the fourth century two things helped to further their travels in Asia: diplomatic business carried on between the states of European Greece and the Persian king, and the decline of the old ideal of the city-state, a decline which made Greeks serve for pay oriental, and especially Persian, rulers, to whom they proved their great worth. For instance, when Darius III had to face Alexander (as we shall relate), he relied partly upon Memnon of Rhodes, and he had thirty thousand Greek mercenaries with him at the battle of Issus. A good many Greeks of this kind must have known personally many parts of Asia Minor. Traders too and artists, cultured men of all kinds, and even harlots, played their part in obtaining knowledge, and Greek influence was felt strongly in Asia Minor (especially Caria and Lydia) and Phoenicia. For a time the Persians were even arbiters of Greek affairs, but this caused a revulsion in which men began to demand war against them. Isocrates was the most prominent spokesman of this feeling, and it is

ASIA

probable that he, like Philip of Macedon when he became supreme in Greece, had a vision of a Greek conquest of Asia Minor as far as the Taurus.[52] Ephorus the historian made an attempt to deal with the ethnography of the people of Asia Minor,[53] but inner Asia remained almost, though not quite, as dim as before. Writers on Persian history made the lands between Armenia and Bactria a little better known;[54] the Hindu Kush and Himalaya Mountains were dimly heard of (as Parnasus), and the rivers Balkh, Kabul (as Choaspes), Indus, and Iaxartes or else Oxus (as Araxes) flowing from them were known by name, but even Aristotle, the wisest of thinkers, not being a traveller, was quite confused in his ideas. India was the farthest east, Scythia the farthest north of the inhabited world, and the sea lay north of Parnasus.[55]

To the north of the Persian Empire the tribes east of the Black Sea were somewhat better known, but otherwise there was no further exploration made. Indeed, men ceased to penetrate as far as they did in Herodotus's time, since they had found two great discouragements: dreary steppes and an unusual climate which was fearfully cold in winter. The Caucasus was wrongly credited with many lakes and large rivers, and the Don was thought to be a branch of the Iaxartes, so that the Volga, the Kirghiz Steppes, Uralsk, the Sea of Aral, and so on, were still unknown. Aristotle could take the 'Hyrcanian' and Caspian as two inland seas, and men began to believe that this Caspian sea was connected with the northern Ocean.[56]

7. ALEXANDER THE GREAT

(a) Asia Minor

However, the great opening-up of Asia Minor was at hand. Alexander III the Great, king of Macedon and supreme in Greece, achieved the greatest of all military exploits, and did what no other man has done: he advanced with a conquering army through Asia from west to east. It is right to call his advance after the death of Darius not only campaigning but exploration of new lands, followed by further exploration in retreat. This penetration therefore by the Greek West of the vast unknown Persian East, forming a new era in geographical discovery, and almost doubling the geographic knowledge of the

THE ANCIENT EXPLORERS

Greeks, and at the same time increasing their knowledge of known lands, was a military one, with commercial principles ever growing out of it. It was also a promise, curiously unfulfilled for some centuries, of further exploration of unknown regions outside the limits of the Persian Empire. Even in Strabo's time little of inner Asia was known by really western peoples, and even he could say that it was Alexander who had revealed to the Greek and Roman world the 'greater part' of Asia,[57] and the writers of Alexander's time, and also Megasthenes, were always the chief sources of knowledge for the farther parts of the continent.

Alexander, inheriting his father Philip's power and ambition, had as his first object the dethroning of the Persian king Darius. Having prepared an army of Greeks of Macedon and of other states, he crossed the Hellespont early in 334 B.C. and won his first victory at the Granicus (Khodja Su). With Sardis and all Ionia, Caria and Lydia subdued, he went along the Pamphylian coast to Side, and then inland past Lake Buldur and Celaenae, forcing his way through the fierce Pisidians in unexplored lands and subduing Phrygia. Early in 333 Gordium on the river Sangarius was reached and there he met Parmenio, who had come from Sardis through Lydia into Phrygia.[58] By way of Ancyra, Cappadocia, and the Cilician Gates, Alexander reached Tarsus, and thence, after an illness and the reduction of Cilicia, went to Myriandrus, Parmenio having been sent from Tarsus to hold the passes (Kara-Kapu and the 'Pillar of Jonah') between Cilicia and Syria. The battle of Issus and the flight of Darius followed (November 333). Parmenio was sent to Damascus and found other Greeks there intriguing with Persia.[59] An Alexandria (near Alexandretta) was founded at Issus. The sieges of Tyre and Gaza (after a land march) and the entry into Egypt followed,[60] so that Alexander secured his rear successfully.

(b) Through Iran to Central Asia

Early in 331 Alexander went to Tyre and crossed the Euphrates by a boats-bridge at Thapsacus, and from here the hot dryness of the route down the Euphrates, about which his excellent intelligence department was able to warn him, made him go north and then east (avoiding the Armenian heights) to the Tigris, which he forded at

ASIA

Bezabde. Four days took him through the plains, and in September Darius was put to flight at the battle of Gaugamela close to Mosul between the Tigris and the Greater Zab.[61]

Now began the crossing of the lands between the Tigris and the Beas in India in several directions by a Greek army – a thing which has never been paralleled; the soldiers saw great wildernesses, vast steppes, rich fields, amazing plants, awesome snowy mountains of unheard-of heights, and other things which intelligent men with Alexander tried to describe in writing afterwards. The battle of Gaugamela laid open to Alexander the route to Babylon and treasure-filled Susa, both of which, after chasing Darius as far as Arbela, he entered without a battle, Susa having been secured in advance by Philoxenus sent on from Arbela. Alexander crossed the Pasitigris (Kuren) into lands unmapped by the Greeks; the Uxian hillsmen (Persian 'Huzha') were quelled, and Persepolis in the valley of Mervdasht, and Pasargadae up the Murghab valley fell, with much treasure, into his hands after travel along a route little known until recent times.[62] His really new discoveries had begun. After an interval the conquering explorer went by forced marches north-west to Ecbatana (Hamadan), where he found much treasure, and then followed the retreat of Darius to Rhagae and through the Caspian Gates. It was an exciting chase, king hunting king, the chased over known, the chaser over unknown ground, both making unheard-of efforts and leaving behind all laggards. Alexander caught Darius too late to save his life,[63] snuffed out near Damghan by the murderer Bessus. He now made a halt at Hecatompylus near Damghan, from which he won control of the route to the Caspian Sea by the submission of the Tapuri in the forests of Mount Elburz, after sending his baggage by way of Shahrud, himself going to Bander i Gez by the Caspian. Parmenio had been sent from Ecbatana to occupy the Cadusian regions on the south-west of the Caspian. Having subdued the Mardi, perhaps reaching Amol, Alexander went to Zadracarta (Astrabad), capital of Hyrcania (Tabaristan with Mazenderan).[64] Then hearing that Bessus in Bactria had taken the title of Great King, he left Zadracarta[65] and went towards Bactria into lands where he soon felt the new towns were an imperial necessity. By way of the Gurgan river, Bujnurd, the Kashaf Rud valley, and Meshed on a well-used route he may have reached the Murghab river, but the disturbed state of Aria made him turn south to Artacoana, near which he founded 'Alexandria of the Arians' (Herat).[66] Drangiana (roughly Seistan) came next, where he wintered, founding a city at Phrada (Furah? or Faranj?), later called Prophthasia, receiving the submission of Carmania (Kerman and part of Persian Baluchistan) and Gedrosia (Makran), and dealing with discontent in his army.[67]

In the spring of 329 he followed the Helmund and the Argandab into Arachosia, founded 'Alexandria of the Arochosians' (Kandahar), made his way up the Tarnak, and founded another Alexandria, perhaps at Ghazni.[68] Then he climbed with his army and crossed the mountains, in spite of cold and snow-blindness, to the Kabul valley, and founded an Alexandria (probably Opian, near Charokar) called 'of the Caucasus' because some of his officers thought that they had reached this range.[69] In reality it was the Hindu Kush, which Alexander, after wintering at the foot, crossed early in 328 through the Panjshir-Khawak Pass by an astounding effort in the face of cold and lack of food except raw meat and asafetida, to find that Bessus had wasted Bactria. He occupied Drapsaca (Kunduz), Aornos (Tashkurgan), and Bactra (Balkh),[70] being in lands which until modern times travellers were not able to visit.

(c) Central Asia

But the great exploration was not yet ended. In pursuit of Bessus, Alexander advanced from Balkh to the Oxus (Amu Darya) opposite Kilif, in the midst of much thirst and heat, and here the men crossed[71] into Sogdiana lying flat on rush-stuffed skins (just as today inflated skins have to be used), and paddling. Bessus was taken by Ptolemy.[72] Having occupied Maracanda (Samarkand), Alexander pushed on to Ferghana and the Iaxartes at its southern bend, the end of the known world. Some thought that he had reached the Don. The revolt of Sogdiana and preparations for attacks by men north of the Iaxartes caused him to finish in a hurry 'Alexandria the Farthest' (Chodjend), a city begun not with any reference to a silk route from China – about which he knew nothing – but as a boundary of the empire and a military fort to control the pass over the Tian Shan to Kashgar, of which his scouts must have told him;[73] and he crossed the Iaxartes

ASIA

as he had the Oxus and scattered the mounted Turcomans in order to create a wholesome fear of him north of the river. The troubles in Sogdiana now led to the exploration of part of the river Sogd to Sogdiana town (Bokhara?) and Zariaspa (Charjui) on the Oxus, where he wintered, 328–7.[74] Having crossed the Oxus again in 327, Alexander found a well of petroleum by the river, but this discovery was not developed.[75] The district of Paraetacene (Hissar) was subdued. The Sogdian Rock of which we hear controlling the entrance pass was perhaps near Derbent and the River Vakhash south of Raisabad. Returning to Bactra, Alexander re-founded this as another Alexandria, while another one was founded at Merv.[76]

In these campaigns the navigability of the Oxus (which is today navigable up to Patta Kesar above Kilif), and of the Iaxartes, and the Indian trade carried towards the Caspian, were perhaps discovered: undoubtedly vague reports about the Sea or Lake of Aral reached the ears of the Greeks,[77] and Alexander meditated an expedition against the 'Scythians' between high Asia and the Caspian;[78] indeed, Pharasmanes, a ruler in Khiva, offered to guide him by a northern route apparently round the north end of the Caspian to the Black Sea (on which Alexander desired to sail and explore the north coast) and so disprove the statement that the Caspian had an outlet into the northern Ocean.[79] But Alexander was bent upon going to India so as to reach the one-time limit of the Persians, and to find a sea for his Greeks. It was indeed high time to move somewhere, for the dryness of the climate and the badness of the water led to deep wine-drinking and beastly behaviour after it. Fearful was the drunkenness among westerners in Turkestan until Constantine Kaufmann, made Russian governor of this region in 1867, established in the army the use of the best of all man-made drinks – tea.[80]

(d) *India*

Alexander therefore made preparations for a campaign into India.[81] Knowing nothing about the vast stretches of northern and eastern Asia, about farther India, about the Indian desert (Thar), and about the peninsula, he pictured India as the land of the Indus, joined possibly to the Nile and Egypt, jutting east into a sea flowing right round so as to come a little north of the Iaxartes near the foot of the Himalayas (called Imaus) and the Hindu Kush (called Caucasus or else Paropanisus), with an inlet into the Caspian.[82] On this expedition Alexander took with him, besides his army, shipbuilders and oarsmen from Phoenicia and elsewhere, and expert scientists also, because he wished to hold what Persia had held, to secure the frontier and trade of the Indus, to test the supposed connexion between India and Egypt, and perhaps to sail round east Asia to the Caspian, which he thought would be easy. His camp was a kind of moving capital city of his empire, full of all sorts of men.

Leaving Bactria in 327 and crossing the Hindu Kush by the Kaoshan Pass, 14,300 feet high, to Alexandria of the Caucasus, he made further preparations at Nicaea (Kabul?, Bagram?), where also he made a friend of Taxiles, the king of distant Taxila, and sent him with Perdiccas and Hephaestion on through the Khyber Pass to build a bridge of boats across the Indus,[83] while he himself went by way of Laghman and the Kunar river across to Bajour of the Aspasii, which he quelled, and then across the Landai river to the Assaceni of Swat.[84] Passing through the Shahkot Pass to the Yusafzai (the capital being Peucelaotis), which he pacified, he halted at Mount Aornos, which is now regarded as Pir-sar Ridge rather than Mahaban, and took it with difficulty. All this part of his journey, across great ridges at all seasons, and necessitating the reduction of mountain forts, was a colossal feat. Whether Alexander entered Chitral or not, he found the Indus successfully bridged to Ohind close above Attock, and rich and pleasant lands to march through.[85]

Crossing into the Punjab early in 326, he reached Taxila (Shahderi), crossed the Hydaspes (Jhelum) above Jhelum, defeated on the Karri plain the Paurara or Puru king Porus of the regions between the Hydaspes and the Acesines (Chenab), reconciled him with Taxiles, and founded two cities, perhaps at the modern Jalalpur and Mong.[86] Having conquered the Glausae on the borders of Kashmir, he crossed with some loss the Chenab, so swollen as to be more than a mile across, and then the Hydraotis (Ravi), defeated the Cathaeans (who practised suttee) and took their city Sangala (near Amritsar?), and reached the Hyphasis (Beas) probably near Gurdaspur.[87]

Here the army, tired of battles and rains, miserable in their rags, and fearful at the prospect of crossing the Thar desert, refused to advance

ASIA

to the Ganges and its surrounding lands, which the exile Sandrocottus (that is Chandragupta, afterwards founder of the Maurya Empire of India) stated to Alexander to be conquerable, telling him how it flowed from the Himalayas past Palibothra (Pataliputra, now Patna) to the Ocean. Sandrocottus also told Alexander of the Sutlej, and perhaps of the now lost Hakra, and of the Gandaridae near it, and of the Ocean beyond.[88] Little, if at all, had Alexander gone beyond the limits of the old Persian Empire of Darius I; still, he intended to make the Indus the boundary of his direct rule, with only protectorates beyond, and he wished to find a sea for which his followers longed. Now, before this, upon noticing crocodiles in the Indus and Egyptian beans in the Acesines, he thought he had found the Nile's sources; but natives of India told him that the Indus river flowed into the 'big sea' with two mouths.[89] So Alexander completed a fleet, put Nearchus in command of it, and set sail down the Jhelum.[90] Below its join with the Chenab, Alexander attacked the Malli (Mahlava) on the lower Ravi, and was wounded in the siege and capture of a place near Multan. Then he went on down the Chenab and the Indus,[91] which like other rivers of north-west India ran a quite different course in those days, joining the Hakra and flowing into the Rann of Cutch, which was then an estuary. Having founded two more Alexandrias and overcome the resistance of a group of three rajas, without, however, reconciling the Brahmans, he reached Patala (Haiderabad?) in July 325, and began to make it a great trade-centre.[92] He explored both arms of the Indus, sailed on the Indian Ocean, thus proving that the Indus and the Nile were not connected, and sent Nearchus to explore the coasting route to the Persian Gulf, as we describe elsewhere (pp. 80ff.).[93]

(e) The Return through South Iran. Further Plans

In September Alexander began his march through South Gedrosia,[94] a dreadfully dry and lonely country, with about 12,000–15,000 soldiers besides women, children, and camp-followers, the army being intended to support with provisions Nearchus's fleet. Alexander knew something of the difficulties, but hoped that the result of the summer rains would be more water. In the first stage the low mountains west of the Indus were crossed, the Oritae of Las Bela submitted, and their Rhambacia became an Alexandria.[95] Returning to the coast, Alexander then made a depot at Cocala, but the revolt of the Oritae caused at once a lack of food. Real trouble began at the river Hingol, for the Taloi Mountains forced him to go inland and the guides lost themselves. The result was 200 miles of suffering in hot, lonely, waterless wastes of rocks and sinking sand, where marching was done at night and sometimes in haste in order to reach water. The baggage animals were killed for food, the carts broken up for fire. Alexander himself walked and refused water when others lacked it. Many poor wretches died before the marchers reached the sea at Pasni, where there was enough fresh water. From Gwadar they followed a trodden route across the river Dasht to Pura (Fahraj, opposite Bampur), where, tired and worn-looking, they rested.[96] Gulashkird was then reached by way of the rivers Bampur and Halil. Here Alexander was joined by Craterus, who, sent previously to subdue a revolt in Arachosia, seems to have come by way of the Mula Pass (rather than the Bolan Pass), Kandahar, the Argandab and Helmund valleys, the Seistan Lake, and south-west through Nazretabad to Gulashkird.[97] When Nearchus came too, there was much rejoicing and holiday, and the founding of another Alexandria.[98] The army then went through Persepolis to Susa, and rested there in the spring of 324, while the fleet went by sea (see p. 86).

Alexander now planned to make Babylon his capital. He explored the Pasitigris, and the Tigris up to Opis, and then went to Ecbatana. In winter 324–323 the Cossaeans of Luristan[99] were quelled, and Alexander was able to return and explore the Babylonian marshes, founded on the Persian Gulf and Alexandria, later Charax,[100] began preparations for the Arabian scheme (see pp. 86–7), and sent Heracleides to explore the Caspian Sea and find out if it was really connected with the northern Ocean and even, as some said, with the Black Sea.[101] This project, like sailing round Arabia, was dropped on the death of Alexander. Arabia was not explored by land, nor did he intend to conquer it[102] any more than he intended to conquer all the known world.

(f) Results

The great explorer and conqueror died (having lived too hard) in

ASIA

June 323, not yet thirty-three years old. Rarely has Asia felt a man like Alexander, in spite of the imperfect state of his own work, which his successors did not finish. The results of his campaign[103] sprang as much from his exploration as from his conquering. Firstly, Alexander having founded about seventy new cities with the object of creating a mixed Greek and Asiatic empire, there were now little bunches of Greeks in many parts of Asia, though there was no common ideal created. Secondly, Asia was opened up to the west; the outlook of the civilized westerners was enormously enlarged; they could think of the world as a common possession of civilized men, trading with each other in international commerce along new routes. Provided that they were able and willing, Greeks living both in and out of Asia could travel on a certain number of surveyed roads in Iran. These had been measured for Alexander by an official staff[104] of 'bematistae' – 'steppers' who measured distances by pacing – in particular Diognetus, Baeton, and Philonides. The first two had surveyed the road from the Caspian Gates through Herat, Phrada, Kandahar, and Kabul, and across the Indus and the Hydaspes to the Beas.[105] Baeton apparently had also surveyed the road from the Caspian Gates to Bactria and the Iaxartes,[106] and even sent out preliminary explorers among the cannibals and valleys of the Himalayas.[107] The roads became great traffic routes later. Others surveyed the road from Thapsacus to Arbela, Ecbatana, and the Caspian Gates; roads from Thapsacus to the Persian Gulf; one from Susa to Carmana and one from Amisus on the Black Sea through North Persia.[108] One Amyntas appears to have edited all these surveys before the rise of Parthia.[109] Routes like the one through Makran were shunned by Greeks and Parthians alike. Thirdly, great geographical features of Asia – the ranges of Iran and central Asia – became known to western Greeks, so that before 285 Dicaearchus could draw a parallel through the Taurus and the Himalayas, though the north and south offshoots of the Iran Mountains had not been explored. Fourthly, the products of Iran and north-west India and even of southern India, became known and were gradually prized, but though Ceylon was heard of, nothing about the Indian peninsula came through except that it was large.[110] All the details collected and put into the care of Xenocles by Alexander[111] cannot have gone far outside north and north-west India.

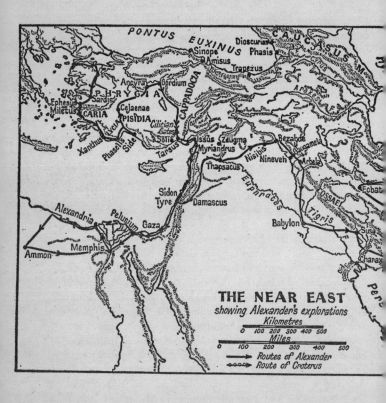

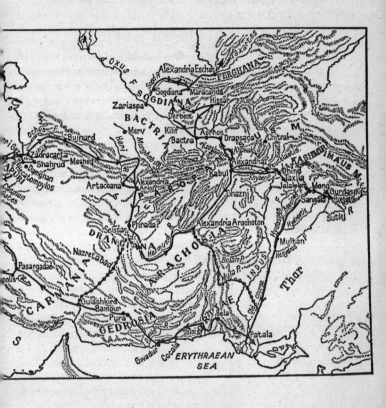

THE ANCIENT EXPLORERS

Lastly, we can say that Greek culture in western Asia, caused by Alexander, led to the growth there of the idea of spiritual unity, as pointed out by Christianity, and that by inspiring Chandragupta to found the Maurya Empire Alexander caused the spread of Buddhism and perhaps the union of China under the first Han Dynasty.[112]

8. SELEUCUS AND HIS SUCCESSORS

The impression created by Alexander lasted for some time, but great drawbacks against further exploration were found in the several decades' struggle of Alexander's generals for his imperial inheritance, and also in the gradual breaking up of much of the Asiatic part of it into separate and primarily oriental states. Seleucus (murdered 281), who had served under Alexander, founded in 312 in Asia the 'Seleucid' dynasty which formed one of the three chief monarchies into which Alexander's empire crystallized by 276, the 'Seleucid' capitals being the new Antioch in Syria and the new Seleucia on the Tigris. But Pergamum, Bithynia, Pontus, Cappadocia, Armenia, and north Media were separate kingdoms, all (except at times Armenia) independent, and all except Pergamum and Bithynia more eastern than Greek. About 250–240 eastern Iran fell away when Bactria and a so far small Parthia became free. Armenia was lost permanently after 190; Pergamum seized all west of the Taurus in 189; Persia was really free after 164; 143 saw the free Jewish state; and Parthia soon absorbed Media and (by 129) Babylonia; and (by 88) Mesopotamia. Troubles increased, and in 64 Syria became a Roman province.[113]

What can we say of penetration and exploration during this period? The earlier part, when the Seleucids were strong, brought a greater spread of Greek culture and new discoveries which find a place here. Parts of Asia Minor became full of Greek cities which flourished greatly upon newly opening trade-routes,[114] except during periods like that when the Celt (beginning in 278) checked Greek culture in the north. Greeks now knew the mountains of Asia Minor well, and climbed high ones like Argaeus (13,150 feet) in Cappadocia to see the view.[115] But inner Asia was not neglected, especially by Seleucus I. The chief work was the founding of Greek cities and settlements in order to hellenize parts of Asia and improve[116] the

ASIA

newly surveyed routes. Says Mr Tarn: 'Roughly, city organization in northern Syria and Babylonia-Susiana was primarily due to Seleucus, in Iran to Antiochus I and Antiochus II, with a noteworthy extension everywhere later due to Antiochus Epiphanes.'[117] Even in Iran, Punjab, and Bactria, Greek was for some time the chief spoken language.

Between 311 and 302 Seleucus I conquered and doubtless re-explored all Alexander's eastern empire to the Iaxartes, across which he sent Demodamas to investigate the regions beyond towards the Caspian,[118] and to the Indus, beyond which Seleucus may have penetrated to the Jumna, but was blocked by, and was forced in 302 to give up India and much of Afghanistan to, Chandragupta Maurya, who had established his rule over all northern India.[119] Antiochus I (281-262) and Antiochus II (261-246) were also active in farther Iran, the former re-founding, for instance, Merv as Antiochia Margiana and exploring the Iaxartes again, re-founding Alexandria Eschate (Chodjend) as an Antioch.[120] The relations between all these three kings and the Mauryas deserves a special section (10), but meanwhile Seleucus had carried out an important exploration in another direction.

9. PATROCLES AND THE CASPIAN

We have seen (p. 180) how the intent of Alexander to explore the Caspian was not fulfilled in his life. Long after his death the exploration was indeed begun, but curiously enough, being incomplete, it confirmed the wrong belief that the sea entered the northern Ocean. Pliny says [121] that fleets of Seleucus and Antiochus sailed from India round into the Caspian! A wonderful voyage had it been possible and true. What really happened was this: Patrocles was a trusted official of Seleucus and Antiochus.[122] At a time when from the Caspian a gulf apparently ran in from Khiva Bay between the Balkan Mountains over the Kara Kum desert and along the Uzboi Channel towards the Aral, he was sent about 285, at Seleucus's command, by Antiochus (after the latter had become eastern governor at Seleucia about 296) to explore the Caspian by sailing on it in order to find new trade routes, especially from northern India. He started from the Kizil Uzain river, or else from the bay at Resht, and took a voyage northwards, on which he noted the Cadusians, the Albanians, and probably the mouth of the Kur. He learnt that the sea stretched on towards the north. Coming back again to his starting-point he went on another voyage, eastwards this time, passing the Mardi and Hyrcani and pushing up the eastward side past the Ochus (Attrek), which he found was navigable, to the Khiva and Balkan Bays, which he entered and then sailed on the now dried-up gulf leading towards the Oxus, the Aral, and the Iaxartes, all of which he learnt of by hearsay. He discovered or confirmed the Indian trade coming down the Oxus,[123] but learnt little about the nomads east of the Caspian.[124] The fact that he had explored no more of the Caspian than the southern end; the presence of the Sea of Aral, receiving by separate mouths the rivers Oxus and Iaxartes, and brought near to the Caspian by that mysterious gulf; the Volga flowing into the Caspian from the north; and vague misunderstood native reports about all these – caused Patrocles to assert that besides the Ochus, the Oxus and Iaxartes also flowed into the Caspian, and that this sea was about as big as the Euxine, and that one could sail from India to the Caspian by water, though few admitted it.[125] Reports like this caused the Greeks to go on believing that the Caspian opened into the northern Ocean. We may compare with this Champlain's belief that Lakes Ontario and Huron were connected, and Speke's partial exploration of Victoria Nyanza.

Under the influence of Patrocles's reports, Seleucus planned to cut a canal from the Black Sea north of the Caucasus range to the Caspian, perhaps in order to sail round Asia close by the Himalayas. But this was not tried. After the death of Seleucus, the kingdom was cut off from the Black Sea and lost interest in the northern trade and the problem (supposed to be solved) of the Caspian. A few Greeks continued to sail in the southern part of the sea, but, north and east of the Black Sea, Dioscurias, the Phasis, and the Tanais remained the boundaries of really known regions. On the latter river a mart called Tanais was rising, but Strabo says that only a small tract above the mouth was explored because of the cold and desolation and unfriendly tribes, who excluded travellers from otherwise accessible parts, and even from the navigable river. Hence mistakes rose about the Tanais.[126]

10. MEGASTHENES; THE MAURYAN EMPIRE

When Seleucus had made peace (in 302) with Chandragupta (321–296), several embassies passed between the two kingdoms. Seleucus in 302 sent Megasthenes to reside at Palibothra (Patna) on the Ganges,[127] where resident he collected and transmitted to posterity much new detail about northern India, being what we may call an ambassadorial explorer with a thirst for knowledge, but a limited range. He travelled on and organized a survey of the Mauryan 'Royal Road' from Alexander's limit on the Beas across the Sutlej and the Jumna to the Ganges, through Rhodopha and Calinapaxa, past the confluence of the Jumna and the Ganges and through Palibothra to the Hughli mouth of the Ganges.[128] He made note of the tributaries of that river,[129] but did not explore them himself, and heard vaguely of nomad tribes of Sikkim and Morang and a hint of their trade in Chinese silk with India, and of Tibetan and Himalayan gold-miners.[130] Of the organization and management of Chandragupta's empire he naturally learnt much, taking particular interest in the Brahmans and in the several castes, and perhaps the Buddhists and their veneration for Buddha.[131] He learnt of the large number of Indian cities and peoples, vaguely realized the shape and jutting coasts, but not the size of India, learning really nothing of the Deccan and southern India, though he heard naturally new things about Ceylon and its pearls and shipping and the trade up the east coast of India and perhaps from Malay, and vaguely knew of the absence of slavery among the Tamils.[132] He may have heard of the Chola, Pandya, and Chera Tamils dwelling round the southern point of India.[133] Under Chandragupta's successor Vindusara (296–264), Deimachus[134] was sent as Seleucid ambassador, but did not explore; there must have been a representative sent to the great Asoka (264–*c.* 227) also, who not only received Dionysius as ambassador sent by Ptolemy Philadelphus through Palestine and then (secretly, perhaps) through Iran or by way of the Persian Gulf, but also sent Buddhist missionaries and ambassadors to Greek kings, including Antiochus II.[135] After Asoka's time, and perhaps before, the Seleucid connexion dissolved, and soon came the decline of Mauryan power. Moreover, under Seleucus II

THE ANCIENT EXPLORERS

(246–227), among other troubles, the provinces of the Seleucids in East Iran and beyond were allowed to slip away and a small Parthia struggled towards independent power.

11. ANTIOCHUS III AND ANTIOCHUS IV. DECLINE. THE GRAECO-BACTRIANS AND INDIA. PARTHIA

Antiochus the Great (233–187) tried to recover the east; in 209 he occupied Parthia and took the road from Hyrcania to Bactria, and having made peace with king Euthydemus,[136] crossed into the Kabul valley, received the respects of the Indian king Sophagasenus (Subhagasena), and came back through Arachosia and Drangiana by way of the Helmund valley to Carmania and Persis – a little-known route.[137] His clash with Roman interests early in the second century brought the Romans into Asia Minor (190), and by the peace of 189–188 Rome divided his lands north of the Taurus among her friends. The distant provinces again rebelled and Antiochus met his fate in Luristan (187). The Seleucids did not now rule east of Media and Persis or west of Cilicia. Antiochus IV Epiphanes (176–164) went east in 166 and temporarily won back something, but from now onwards the history of the Seleucids is a wretched story of struggle and dissolution. Pontus, Armenia, Parthia – these were the powers, all except Parthia yielding before the advance of Rome. In spite of this, however, commerce crowded more and more on the roads of Asia Minor, especially between Ephesus and the Euphrates.[138]

In far Asia an isolated part of the Greek world still flourished. Diodotus, who revolted[139] from the Seleucids before 240, bequeathed to his son Bactria and Sogdiana. After 225, when the Mauryan Empire was declining, the usurper Euthydemus and his son Demetrius crossed the Hindu Kush and won control over the Indus to its mouths, over Kandahar, and the limits of Gujarat.[140] Various usurpers arose; one of them, Eucratides (*c.* 175), seems also to have reached as far as Gujarat and Cutch if we judge by the finds of his coins; but the power was weakened, Parthia pressed on from the west, and native ways came in. Under Euthydemus's son, murderer, and successor, Heliocles, the lands north of the Hindu Kush were lost to the migrating Sakas and Yueh-chi,[141] who by 130 were struggling desperately with Parthia.

ASIA

Only men like the Greeks Apollodotus and Menander, both of a new Kabul-Punjab dynasty, held out. Both tradition[142] and coin-finds tell us that Menander (*c.* 155-150) reached the Jumna, and possibly the Ganges (for Indian tradition takes Greek sway to Patna) as well as the Indus mouth, Gujarat, and even Broach where coins of Apollodotus and Menander were current in Nero's time.[143] This was a great achievement. But besides this they came into contact with Chinese Turkestan, the Phrynoi or Phounoi, that is, Huns, and the 'Seres', that is, not the Chinese really, but their intermediaries in High Asia, being nomads with a regular trading-beat both north and east of Bactria; the eastern Greeks also colonized Ferghana, and perhaps developed the silk-trade.[144] To them probably are due the seal-impressions found on documents in cities of Khotan.[145] How active these Graeco-Indians were is shown by Heliodorus, who was sent from Taxila to Bhilsa in Central India in the name of Antialcidas (*c.* 140-130), and by Hyspasines, a Bactrian who turned up at Delos in the Aegean.[146] But almost from the beginning these Greeks had been cut off from the west by the Parthians, through whom alone a part of their history came to the west, and then chiefly though writers on Parthian history like Apollodorus. Hence Strabo says that the Parthians increased the knowledge of the Roman West.[147] In reality they were a barrier against exploration by all Greeks coming from west of the Euphrates, and the Greeks who read the *Histories* of Polybius could be tantalized by his account of Seleucid activities without being able to take advantage of them.

12. THE FIRST ROMAN EXPLORATION OF ARMENIA AND THE CAUCASUS

The Romans had entered Asia against Antiochus in 190, but after the settlement did not stay there. But by 126 the province of 'Asia' was fully formed out of western Asia Minor. Rome's interferences became more and more widespread. The Taurus was crossed in 78-76 by her troops, and before long their military activities in the farther part of Asia Minor and beyond formed a most important piece of exploration, occurring in the third war with Mithradates, king of Pontus, whose rule extended over Lesser Armenia and part of Cappadocia and north

THE ANCIENT EXPLORERS

over Colchis, Tauric Chersonese, and even lands between the Dniester and the Don, and who was allied with the strong king Tigranes of Armenia. Of the regions from the north of the Black Sea round to Caucasus and Colchis his generals had made reports.[148] Rome clashed with him over making a province out of Bithynia. The first two campaigns were in known parts of Asia Minor, but the flight of Mithradates into Armenia and the power of this kingdom led to a fresh military exploration by Lucullus of a land almost unvisited since Xenophon's time. The Romans now for the first time passed through Anti-Taurus on to the terrible highlands where were the sources of the Tigris and the Euphrates, suffering much from the hardness of the climate, and also crossed the Taurus and the Euphrates into Sophene, where the western branch of the Tigris rises.

In 69 Lucullus crossed the Tigris and took Tigranocerta (somewhere near Diarbekr?), defeating his enemies on the eastern branch of the Euphrates.[149] Having spent the winter in South Armenia, he then moved north in the spring of 68 on to the Armenian tableland through the Pass of Nardjiki and across the Murad Su towards Artaxata, again defeating his enemies. But the winter snows of Armenia were too much for the Romans. On the mutiny of his soldiers he retreated round the eastern shore of Lake Van across the Tigris into Mesopotamia, where he captured Nisibis.[150] He was soon (66) superseded by Pompey, who explored the lands between the Black and the Caspian Seas. He had to drive Mithradates out of Asia Minor and receive the submission of Armenia. He then followed Mithradates, who had gone with an army to Colchis and thence along the Black Sea shore to Kertch by a most difficult way through hostile tribes whose languages Mithradates had probably learnt.[151] Pompey subdued the Iberi on the southern hills of the Caucasus and round the upper Kur, and the Albani living round the lower Kur as far as the Caspian (65). He was now in lands known to no Roman and but few Greeks, and hence Theophanes, who was with him, recorded with care all the discoveries made, being helped perhaps by others, including Metrodorus and Hypsicrates, who also now travelled in the Caucasus. Pompey crossed the Kur, but while still three days' travel from the Caspian was put off by exaggerated stories about real poisonous snakes of the plain of Mogan.[152] The barrier of the Caucasus

ASIA

was fully realized and partly explored, the Darial Pass noticed, and the Suram range duly observed. The Kur was carefully explored, perhaps even to its mouth, and the habits of the Iberi and the Albani noted in detail. Seventy tribes using one hundred and forty interpreters were found trading at the Greek Dioscurias, and some were pirates on the Black Sea.[153] Pompey learnt fresh details about the Indian trade coming through Balkh and down the Oxus to the Caspian and then by the Kur and the Rion to the Black Sea. The most interesting thing which the explorers came across was the practical use to which the natives of cold Caucasian and Armenian heights put the pleasant pastimes of tobogganing and roller-skating, the one of which they used for conveying wares to market, the other partly to prevent the mishap of sliding against one's wish upon parts of the body most usefully employed in sitting, lying, or digesting. Says Strabo: 'The summits are untreadable in winter, but in summer the people go up them by binding under their feet shoes of raw ox-hide, flat out like timbrels, because of the snows and ice-covered stretches. And they go down, each lying with his load on a skin and sliding down, which is the custom also in Atropatene and on Mount Masius in Armenia; but there little wooden wheels with spikes on are put under the people's shoes.'[154]

Pompey did not pursue beyond the Rion (Phasis) since he felt the task to be dangerous, but turned back, wintered in Pontus, subjugated Syria (64), settled Judaea, and startled the Arabians of Petra. The Euphrates became the Roman boundary, west of which everywhere urban life and commerce were encouraged, partly in new cities founded by Pompey.[155] The Romans were now badly bitten by visions of conquering all Alexander's former empire to the Indus. Even Lucullus had dreamed a little, but the money-magnate Crassus, put in command of an army intended for the invasion of Parthia, passed all bounds by wishing to conquer Parthia, Bactria, and India to the outside Ocean.[156] He crossed the Euphrates at Zeugma and instead of a longer but safer route through part of the Armenian Mountains, or one down the Euphrates, took a dangerous one through the trackless wastes of the Mesopotamian plains, guided by the treacherous sheikh Abgarus, with much suffering from heat and drought, reaching the river Nahr Belik. He was defeated at Carrhae

(Haran) in 53 B.C. and met his death. Prisoners captured in this battle were taken by Orodes to Merv, and the result was greater knowledge of that district.[157]

Caesar meditated the conquest of Parthia and 'Scythia', and the exploration of the Caspian, but attempted none of these.[158] Antony, advancing against the Parthians in 36, was induced to push instead through Armenia by a long route into Atropatene (Azerbaijan), which he thus first made known to the west. Forced to retreat, his harassed army suffered terribly from cold and hunger as it was guided for a month (by a Roman soldier captured at Carrhae) to the boundary of Armenia, through the highlands of which Antony pushed to Syria. He subdued Armenia again in 34.[159] When Augustus became emperor, all hoped that he would conquer as far as the Indians and Seres,[160] but instead he held to the Euphrates as the Roman boundary and was recognized as overlord of Armenia in 20 B.C.

13. THE ROMAN EMPIRE AND ARABIA

One piece of exploration came so near the beginning of the Empire as to belong almost to the previous period, and it stands alone of itself. Throughout our story Arabia had remained unexplored. Eratosthenes shows that by the end of the third century B.C. the caravan-routes between South Arabia and Akaba and between Hadramut and the Euphrates were known by hearsay, and the central desert was known as such,[161] but even Juba's[162] information about inland towns and routes (never of the inland plateau, within which nothing was known even by hearsay) comes from Greek coasting voyagers. The expedition of Gallus sent by Augustus in 25 B.C. was therefore exceptional.

It was a military exploration directed in a blundering fashion against the commercial monopoly of the Himyarites in South Arabia.[163] A fleet with Egyptian troops, Jews, and doubtful Nabataeans of Petra under the faithless Syllaeus, assembled at Cleopatris in the Gulf of Suez, crossed to El Haura with difficulty, and was held up there for a summer and a winter by sickness due to bad food and water. Then Gallus set out over desert tracks, carrying water on camels. Thirty days through lands of friendly Aretas and fifty more through the

ASIA

roadless desert of the Bedouin king Sabos led them to the watered and fertile lands round Negrana, today Nejran (?), where 'cities' were taken and supplies obtained. The army then went to Marsiaba (Mar'ib?) of the Rhammanitae round Rhadman (?) under King Ilisaros (Elisar or Iliazzu), but was unable to keep up a siege through want of water. Gallus heard that he was two days' march from the Aromatic region, that is of Hadramut with Yemen. Six months were then wasted in marches through the treachery of his guides, as Strabo seeks to impress on us, after which the suspicious Gallus went back to the city Negrana, took eleven days to reach 'Seven Wells', and pushed on through deserts and villages to Egra on the Red Sea, this return taking only sixty days, whereas the outward journey had taken six months. The army then crossed to Myos Hormos (Abu Scha'ar) in Egypt, sadly reduced in numbers through disease, fatigue, and hunger, nor was much information gained about Arabia by or through the expedition except, as Pliny shows, some points about the Sabaeans, Himyarites, and Minaeans. Difficult as are the identifications of the names in our account, we can safely say that Gallus was led through Nejd and Hajr to the borders of Hadramut and Yemen and was for a period hopelessly lost. 'The Arabian deserts may be passed by armies strong enough to disperse the resistance of the frenetic but unwarlike inhabitants; but they should not be soldiers that cannot endure much and live of a little ... Europeans, deceived by the Arabs' loquacity, have in every age a fantastic opinion of this unknown calamitous country.'[164] Some impression, however, was made upon the Himyarites, and the Greeks learnt something about their monuments.

Augustus's policy was never revived; all that came through about Arabia during the imperial era was the result of (i) merchants and ambassadors passing between Rome and Charax by way of Palmyra or Petra along desert routes, after a more detailed exploration under Claudius of the river Jordan, of the vague frontier of Palestine, Syria, and Arabia, of Lebanon and Anti-Lebanon, of the desert town of Palmyra and of routes from that town to Mesopotamia; (ii) the creation of a Roman province permanently out of the Nabataean territory by Trajan in 106 and the making of new roads along the whole Arabian frontier and to the Persian Gulf;[165] and (iii) the enormous increase of coasting voyages along Arabia as we describe on pp. 95–8, 101. Hence without exploration appear full details of inner Arabian routes, wells, villages, and wadis in the Geography of Ptolemy.[166] In other parts of Asia things were quite different, as we shall now see.

14. THE ROMAN EMPIRE AND IRAN. EXPLORATION TO INDIA AND CHINA

Rome ruled over Asia Minor to the Euphrates and Colchis, and over Syria and Palestine, and these were the limits of real knowledge. The west was a peaceful whole, and in it the wealthy demanded rich oriental wares; the Greeks could get these; Roman capital could help them; most travellers were inspired by gain, not a desire for knowledge of conquest; so that we expect an outburst of new but peaceful exploration in the east. So there was, but it took place by sea as the commercial channel, the exceptions being such explorations of inner Asia east of Mesopotamia as we deal with in this section. Since the Romans were not good travellers this exploration was done by Greeks for them. After 20 B.C., when Rome and Parthia were more friendly, men had better chances of personal observation, as Strabo says,[167] and before 1 B.C. a good deal of surveying, and even exploration, was done for Augustus beyond the Euphrates, in Parthian regions only, by special treaty with Parthia in connexion with Gaius's mission to the East and under the influence of Indian embassies. The Greeks were allowed to go from Antioch down the Euphrates to Greek Seleucia and Parthian Ctesiphon, then on what was by that time a silk-route through Iran, by way of Jebel Tak, Hamadan, and the Caspian Gates to Merv, and then south along the Murghab through Herat round to the Parthian frontier at Kandahar.[168] At this period the Greeks apparently preferred to include Merv on their travels rather than to leave out that place by turning south along the Tejend (instead of the Murghab) on their way to Herat. Beyond Merv eastwards they were not allowed or else were not willing to go, though they learnt vaguely of the Sacae (Sakas) and of the silk-producing Seres in Central Asia,[169] nor did they go beyond Kandahar on the branch towards India. To both limits Isidore of Charax (in Media?), who could influence Parthia, made a survey. But no other routes in

ASIA

Iran were systematically examined for Rome,[170] though it is probable that Isidore's work was continued for Augustus, over not yet re-explored routes, by king Archelaus of Cappadocia, for he was a χωρογράφος of the lands trodden by Alexander, and traded with India, and so probably re-edited the old 'bematists'[171] (see p. 183). Local maps and large maps were made and Agrippa compiled a survey and wall-map of all the Empire and its neighbouring regions. Paved roads became gradually a feature of Rome's rule to the Euphrates, and material for the later 'Itineraries' was slowly collected.[172] We should bear in mind the statement of the old Chinese chronicles, that the Parthians barred the Roman emperors from access to China, because they wanted the Romans to do their trading in Chinese silk through them.

Under succeeding emperors of the first two centuries the progress of exploration by land in disturbed Asia was slow and in only one direction sure. In Claudius's reign the Tigris and Lake Van became much better known.[173] Nero's settlement with Parthia was lasting and by his reign, though the Chinese were unknown, as Seneca shows, in direct commerce, yet the Romans were learning about the silent trade of the Seres in the Himalayas, and these Seres (or Chinese as known along the land routes) were definitely placed north of India. Even a few dim geographical details of China itself were filtering through, and other Chinese wares besides silk were reaching the west. Trajan in his Parthian wars made Mesopotamia a province (A.D. 115), explored down to the Persian Gulf and drew close Rome's bonds with the Kushans of north-west India. Men expected him to add Arabia, Bactria, and the Chinese to his conquests. Hadrian's new peace with Parthia and his reversion to the Euphrates frontier proved also a great help, as we shall see.[174] The story of Apollonius's travel through Iran and Kabul to the Beas and return from the Indus to the Persian Gulf must be taken as an imaginary repetition of Alexander's advance and Nearchus's voyage,[175] but at any rate by Domitian's time there were prospects of travel from Zeugma by the Parthian silk-route through Merv east to Balkh. Says Statius: 'Thou shalt tell of the swift Euphrates and royal Bactra and the sacred riches of Babylon and of Zeugma, a journey of Latin peace.'[176] Later on, Dionysius Periegetes alludes to travel of Hadrian's time along the Parthian route to the Hindu Kush – that is, through Merv, Herat, and Kandahar.[177] In 97 Kan Ying the Chinaman visited Antioch as ambassador soon after the China conquests in central Asia had made safe for a time both silk-routes – the southern one through Cherchen, Khotan, and Tashkurgan (the one in Sarikol) to Bactra, and the northern one through Kusha, Kashgar, and Samarkand to Merv. Hearing about this, one Maes, a merchant of the Roman Empire, started about 120 to send out agents to meet the Chinese along the routes, the southern being favoured by the Chinese and thought to be more civilized. We may plausibly reconstruct this interesting penetration (recorded only by Ptolemy) as follows: Possibly by way of north Parthia, but probably by way of the Persian Gulf, the Indus and Kushan dominions (which were penetrated and mapped by Ptolemy's time),[178] the agents reached Merv and began their explorations which broke new ground. From here some went along the northern route across the Jaxartes to Samarkand and Kashgar, but not beyond. Most, however, went to Bactra, finding it to be a great trade centre controlled by the Kushans, and on to the Stone Tower (Tashkurgan) where they found not only Parthians, Indians, and Kushans, but also the Chinese and their middle-men (both classed as Seres) with silk. They then went on and, joining the northern route at Kashgar, came shortly to a place which, like the Stone Tower, they found to be a resting centre where the Chinese met foreigners. Here they stopped. We can see a reflection of these activities in sources other than Ptolemy's Geography. Bactrians as well as Indians, and perhaps even Chinese, began to visit Asia Minor, as Lucian shows; this writer also alludes to travels made to Bactria; Bactrian (Kushan?) embassies visited Hadrian and Antoninus Pius probably as results of or in connexion with Maes's activities. Beyond Tashkurgan and Kashgar his agents had merely heard of the seven months' journey from Kashgar (or from Tashkurgan along the Yarkand valley) along the northern route through Kusha and Miran, or along the southern route through Khotan, Cherchen, and Miran, to Daxata (Singanfu, where Roman coins have been found) and Sera (Loh-yang, the new Han capital). They gained a rough idea of the Pamir, Tian Shan, and Altai ranges, and vague details came through about towns, rivers (Hwang-ho and Yang-tsze-Kiang) and mountains of China and of routes towards the Altais, though nothing

ASIA

was heard of the sea east of China or north of Siberia except as a theory, nor of the vast extension of Asia both north and east. Ancient writers give fear of deserts, cannibals, cold, and wild beasts as the reason for this. Maes's agents rightly reported to the west that silk was not a plant-product, but a thread spun by a caterpillar.[179]

The travels of Maes and his like may well have some connexion with articles of western origin which have been found in Afghanistan and in Punjab, though the source of ingress may well have been the sea-trade such as is described in the fourth chapter of this book. Be that as it may, at Begram (about forty-five miles north of Kabul) have been found much glass made in Italy, Egypt, and Syria (mostly of the second century A.D.), bronze bowls, bronze feet or stands, and steel-yard-weights, all of western origin, Graeco-Roman bronze statuettes, jugs and other vessels of alabaster, and also medallions, made of plaster and furnished with reliefs showing Mediterranean subjects, apparently to be used as models by workers in metal. At Faxila (Sirkap) have been found fine examples of Graeco-Roman sculpture in stucco and bronze, a silver ornament, some pots made of bronze, a wine jar, two or more gems, a coin of Tiberius, but nothing such as Arretine pottery. Places like this were not end-points of trade in the more popular articles of the western world. The district of Peshawar has produced a bead in terracotta which is modelled on the Apollo Belvedere, a stone relief, in which western and eastern things are mingled, and a stucco head of a soldier which looks Roman in inspiration.[180]

Other Roman subjects followed in the track of Maes and beyond, after the sea-trade of the Roman Empire declined and the more commercial Sassanids had replaced the unbusinesslike Arsacids. They reached Lop Nor and Miran where Sir Aurel Stein found Egyptian, Greek, Roman, Christian and Byzantine, and especially Syrian, influence in art. Christianity spread by way of Kashgar even to Peking; early in the sixth century all the stations on the route were known, and two monks who made the journey to China smuggled eggs of the silk-moth from that land to Constantinople for Justinian. By the seventh century a Syrian mission was established in China.[181]

With these exceptions all travels after the third century were in already explored lands. During the imperial period the unstable relations between Rome and Parthia (and then Sassanid Persia), the uncertain power of China in disturbed Central Asia, and later the activities of the White Huns, hampered 'Roman' exploration into distant Asia, though the Sassanids from the third century onwards were a better medium for exploration than the Arsacids had been; but these Sassanids rose at a time when the Roman West declined.

15. THE ROMAN EMPIRE AND THE CAUCASUS, THE CASPIAN, AND ARMENIA

In the regions north of Armenia and Parthia the object for penetration was mainly military, the motive being the empire's 'security' (in that wide sense in which good patriots often use the word); but the trade of the Steppes also counted. Not only were Greeks not exploring round the Caspian, but not even the Parthians had explored or navigated it, and as for the Scythians east of it, hardly anyone but Alexander had even thought of attacking them. Nearly everyone believed that the Caspian entered the northern Ocean, and in fact Nepos adduced Indians found by Metellus Cleer to dwell among the Suevi in Germany as evidence that Indians had sailed from India round the east and north of Asia through this Ocean; Pliny believes firmly that the voyage from India to the Caspian had been made.[182] Of Armenia, and in effect also of the Iberi and the Albani, the Romans were overlords during the reigns of the first two emperors,[183] but after Tiberius's death the Parthians sometimes controlled Armenia. Nero's government kept the Caucasian tribes friendly, and that emperor contemplated an expedition against the Alans through the Darial Pass, and an exploration to the Caspian,[184] and further, the campaigns of Corbulo in Armenia were a fresh exploration of that difficult land and his staff saw fit to send to Rome a map of it including the Caspian regions.[185] Mela shows that by this reign the water north of the Caspian was known by some to be a river (the Volga), but none yet saw it or named it. None explored even the Don, though men had heard that it did not flow also into the northern Ocean.[186] The raids of the Alans[187] now greatly hindered exploration beyond the Cau-

ASIA

casus and the north-west of the Caspian, so that Vespasian put a garrison at Harmozica in Iberia.

For a short time only (under Trajan) Armenia was a Roman province and the Caucasians were definitely clients; Hadrian renounced Armenia, but he made the Black Sea a Roman lake, had the coasts carefully surveyed, and kept the Alans and others well in check, and under his rule Phasis and other towns became starting-points for explorations on which even the Caspian was reached, though not often at first.[188] But these efforts led to further exploration along the southern coasts even to the south-east shores, so that Dionysius Periegetes could speak a good deal about Scythian tribes round the coasts and even mention the Huns to the east, though he shows that none had yet gone round the northern end.[189] Then, under Pius or Marcus Aurelius, the exploration was completed, though we cannot estimate it fully, since Ptolemy failed to collect all the available material, unless indeed the gain-seeking informants ignored, as usual, much of the geography of the regions they visited. But we do know that the Greeks from the Black Sea once more established the truth that the Caspian Sea was an inland lake, and nowhere near the northern Ocean. They discovered the Rha (Volga) flowing in at the northern end, learning something of the bend by which it approaches the bend in the Don. Ptolemy's wrong idea about the Palus Maeotis and the shape of the Caspian shows that the discoverers of the river had found it not by travelling by land, but by isolated voyaging up the middle or along the west coast only. Thus the Aral was not explored, unless indeed it was joined to the Caspian in those days. No one knew of the sources of the Volga and reports of what lay east and north of the Iaxartes (which like the Oxus was not explored in any part at all) were chaotic.[190]

The era of the decline of the Roman Empire and the early Byzantine Empire saw no further exploration here except the opening up of a silk-route round the north of the Caspian by the Turks for Justinian in the sixth century A.D.[191] But the story of exploration generally after the rise of Constantinople finds no place in our account of ancient discoverers; the achievements of the new Arab civilization after the seventh century belong to a new era.

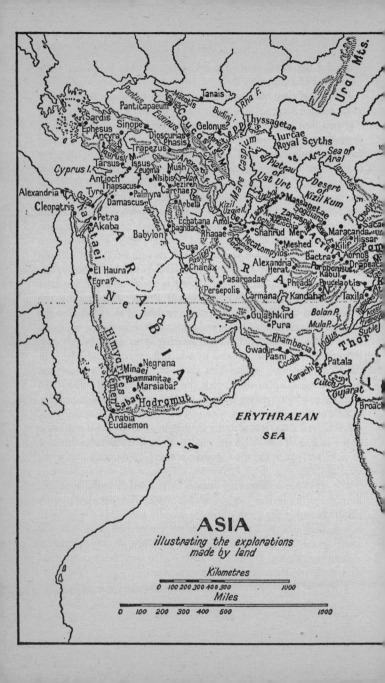

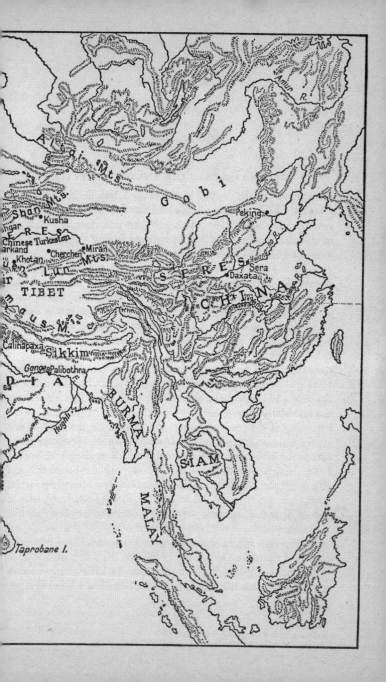

CHAPTER 8

AFRICA

(a) The Nile Basin

1. GEOGRAPHICAL. THE ANCIENT EGYPTIANS

EASTWARDS of the dreaded barrier of the Sahara lies the Nile – the great river-highway between the Mediterranean and inner Africa. It gave from the earliest times a passage by ship or on foot into what are now Nubia and Egyptian Sudan; eastward from it desert tracks led to the Red Sea, and from the West paths perhaps came in from sandy wastes of the Libyan desert through the oases. But in effect it was and is only the river which makes Egypt differ from the rest of the Sahara. Throughout Egypt it was always navigable, and is now navigable at 'high Nile' as far as Fort Berkeley. But this is a place beyond the farthest limits reached by civilized man in ancient times, nor was it sailed on by African natives above the Sobat, because nobody took up the task of clearing away the 'sadd' or 'sudd', as the Arabians call the masses of decayed plants which choke the water, making a barrier soon felt after the Bahr el-Ghazal confluence with the White Nile, especially along the Bahr el-Jebel. This barrier was not sufficiently cleared until it was dealt with by Major Peake at the beginning of the present century. In ancient times the limit of uninterrupted navigation was Syene (Assuan) and the first cataract. From here to the Blue Nile, through a stretch which represents the land of Nubia, navigation was partial only, because of the constant falls and rapids; but the Blue Nile and the White Nile were always navigable. From their junctions the Atbara and the Blue Nile led towards the hills of Abyssinia and their complicated water-system; but this, like the systems of the upper Nile southward of 10° N., lies outside our story, for the ancient explorers never knew more than the main streams of the basin. On the lower Nile men sailed in boats and ships which in their varied kinds do not seem to have changed much through long ages. We find that there were canoes made out of single

AFRICA

tree-trunks and shallops fashioned out of papyrus, and we hear even of earthenware boats; but the vessels chiefly in use on the lower Nile were the high-prowed and high-sterned square-rigged ships and, for heavy loads, the large 'baris'.[1]

The ancient Egyptian monarchy, lasting about three thousand years, ruled direct as far as the first cataract and the central mart at Assuan, and for long periods kept a loose control to the second cataract. In ancient times Ethiopia, though often applying to Nubia in particular, meant all that part of Africa south of Egypt and extending to the Red Sea and the Gulf of Aden. The stretch from Syene to Meroe (Bakarawiya, Begerawiyeh) was in the stricter sense Nubia, famed for its gold of the desert eastward of the Nile, and holding loosely as subjects the people round the Atbara, the White Nile, and the Blue Nile. Far distant Sudan also sent much commerce, but under the earliest dynasties the wares, like those of 'Punt' and 'God's Land' (see p. 75), were fetched to the upper Nile on the southern frontier of Egypt not by Egyptians but by tribes foreign to them.

At the end of the third dynasty (c. 3100) we find that Snefru raided the Ethiopians of Nubia and the process of subjection went on slowly from his time onwards, and distant explorations were carried out, so that by the sixth dynasty Egyptians were pushing far southwards by land as well as by sea. Thus under Mernere a gentleman of Assuan called Harkhuf, who was sent to Yam (southern Nubia) several times, tells us that on the first journey he was sent 'to explore a road to this country. I did it in only seven months, and I brought all kinds of gifts from it. I was very greatly praised for it'. He came back 'with 300 asses laden with incense, ebony, heknu, grain, panthers, ivory, throw-sticks, and every good product. I was more excellent and vigilant than any count, companion, or caravan conductor, who had been sent to Yam before'. Under Pepi II one Sebni was sent by land even as far as the Tigre highlands and obtained samples of trading wares. The 'Theban' or 'Middle' Kingdom of Egypt controlled fully to the second cataract, worked directly the Nubian gold-mines from the time of Senwosri I onwards, and learnt to despise the 'wretched Cush' who were centred in the Nile valley between the third and the sixth cataracts and spread over large tracts of the highlands also. Under the twelfth dynasty (2212-2000) settlements of Egyptians were extended

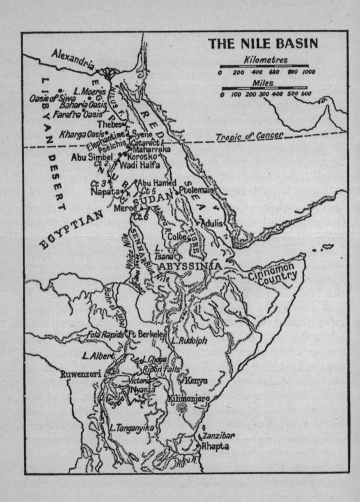

AFRICA

as far as Wadi Halfa and Batn el-Hagar, while the eighteenth dynasty (*c.* 1580–1322) pushed Egyptian conquests and culture to the fourth cataract, and perhaps beyond, for the great expeditions of this dynasty towards the south to the regions of incense were conducted by land as well as by sea. The 'wretched Cush' were often subdued and made to pay tribute. The collapse of this empire after the death of Ramses III left Nubia still Egyptian, but before long Ethiopia, in this case a part of Nubia with Napata (old Merawi near Jebel Barkhal) as its capital (transferred later to Meroe), broke away, and even invaded Egypt. But Egyptian exploration went on, for, dating from about 700 B.C., the Egyptians penetrated frequently to the Blue Nile, as is shown by discoveries made not long ago at Jebel Moya near that river. The Assyrian invasions were the beginning of the end, though there was a revival until the conquest by the Persians.[2]

The Egyptians had penetrated southwards to the sources of the Blue Nile and perhaps of the Atbara, and perhaps also to the Bahr el-Ghazal confluence with the White Nile; but beyond this the sudd barred their way. We now turn to the Greeks and to the Persian conquerors, by whom subsequent explorations were done.

2. THE GREEKS AND PERSIANS TO ALEXANDER

The long-lasting penetration by the Egyptians caused the Greeks of the Homeric poems, who, as the story of Menelaus shows, visited Egypt, to receive dull echoes of races very far to the south of Egypt, of which they knew by hearsay the Nile (as 'Aegyptus river') as far as Thebes.[3] The Egyptians had told of eastern and western Burnt-faced men (Aethiopes) on the borders of 'Ocean river'; as we show in another chapter (p. 62), both these branches of Ethiopians may be African peoples partly of Somali lands, partly of inner Africa south of Egypt.[4] The Pygmies too who fought with migrating cranes were dwarf races such as the Akka tribe of inner Africa.[5] Both these and the Ethiopians reappear in Hesiod, who at last mentions the Nile by its name which is of unknown origin.[6] Strabo notes that even in his day, at the beginning of the Roman Empire, visitors to Egypt always asked first about the Nile, of which Hecataeus said Egypt is the gift.[7] But Egypt was almost a closed land until Psammetichus I after 665 let the

Greeks and others in. Various Greek states soon had factories at Naucratis in the delta, and when Psammetichus II (594–589) sent a mercenary force into Ethiopia as far as Abu Simbel, Greeks were included, whose names are still scratched on the temple-statues there. But generally the Greeks were confined to the delta until Amasis widened their activities.[8] After that they visited Egypt with some frequency in order to see the wonders of the land and to learn wisdom from its priests. We will name Solon, Alcaeus, Pythagoras (c. 550), Hecataeus (who visited Thebes c. 500), and Democritus of Abdera,[9] who is said to have gone to Ethiopia – probably the southern border of Egypt.

Cambyses the Persian, when he conquered Egypt in 525, subdued Ethiopians beyond that frontier, but met with disaster in attacking the 'long-living' Ethiopians in the south, that is, the kingdom or kingdoms of Napata and Meroe in Nubia. We learn about it from Herodotus:[10] 'Cambyses planned three campaigns, against the Carthaginians, against the Ammonians' (at the oasis of Siwah), 'and against the long-living Ethiopians dwelling in a part of Libya upon the Southern Sea. And when he had taken counsel he decided to send his fleet against the Carthaginians, chosen foot-soldiers against the Ammonians, and, against the Ethiopians, spies first of all to take a look at what is called the table of the sun among these folk, to see if it were truly there, and besides this to spy out other things, taking as a pretext gifts for their king.' He chose certain Fish-eaters for the task and sent them as ambassadors with gifts – a purple garment, a linked golden necklace, armlets, an alabaster box of ointment, and a cask of date-wine. These Ethiopians, to whom Cambyses sent embassy, are said to be the tallest and handsomest of men. Herodotus goes on to describe how the Ethiopian king was not deceived, and made mockery at the gifts, except the 'wine', which he enjoyed. He showed them everything (we are told), including the table. Cambyses in a rage set out, like the madman he was, without proper supplies, passed through Thebes, where he separated part of his army to attack the Ammonians, and pushed southwards. 'Before the army had travelled over one-fifth of the journey, all the grain-food which they had with them ran out, and after that the yoked beasts were eaten up and failed them; ... the soldiers, so long as they could get anything from the earth, lived on by eating grassy plants. But when they came to the sand, some of them wrought a fearful thing: in tens they chose by lot from themselves one man and ate him.' When Cambyses heard of this he gave up his plan and returned to Thebes. From Herodotus's geographically useless account it is difficult to tell what route Cambyses had really taken, but it is probable that the desert which he entered was the horrible one between Korosko and Abu Hamed, and that his object was Nubian gold. Famine may well have made retreat inevitable because of mutiny.[11]

The men sent against the Ammonians suffered an unknown fate. They were known to have reached a 'city' called Oasis (the Kharga Oasis), seven days' journey from Thebes. The Ammonians reported that when they were about half-way across the sand between Oasis and their goal, which was the Oasis of Siwah, 'a great and violent south wind blew upon them while they were breaking their fast, and bringing heaps of sand buried them underneath and in this way made them vanish'.[12]

This is all the Persians could do. Greeks had gone with Cambyses towards southern Ethiopia, but after this for a long time all they knew about the Ethiopians came from the Persian government in Egypt, and what that authority had to say was not much. Some men thought that the Ethiopians bordered upon the Indians (see pp. 79–80, 177, 179). In the early part of the fifth century most Greeks who went to Egypt did not go beyond the delta of the Nile,[13] but after 448, when Athens had made a peace with Persia, penetration up the Nile was easier, and men like Herodotus in thirst for knowledge reached Elephantine near Syene without difficulty, though of Egypt itself there was very little exploration except near the Nile, for even in Egypt the camel was not yet in use[14] (see pp. 216, 218). Nor was even the Nile explored any longer beyond Elephantine and the first cataract.[15] Herodotus did not know what caused the flooding of the river;[16] he states that it flowed out of Libya through Ethiopia to Egypt, but 'about the sources of the Nile no one, whether Egyptian or Libyan or Greek, who has talked with me, has admitted that he knew anything, except the clerk of the holy pots-and-pans of Athene at Sais in Egypt; but I think he was joking when he said he knew accurately'. For, resident in the delta, he told an impossible story of the Nile rising from two deep holes between Syene and Elephantine and flowing thence north and south.[17] Herodotus then goes on to tell how he himself went as far as Elephantine and there heard vaguely of details of distant waters and regions of the Nile.[18] Thus he learnt of the gradual rise of the land, of the need of towing the boats, of the windings of the river, of Derar near Dakka, and of long weeks of travel by river and (after Wadi Halfa) by land and again by river to Meroe, south of which he could speak only of the 'Deserters' or Sembritae who had left the service of Psammetichus long before and had made their homes in Sennar. Of the affluents and tributaries of the Nile he knew nothing. Great heat, it was reported, prevented men from living in lands beyond the deserters. Naturally men tried to guess the cause of the summer flooding; Thales said it was lack of counter-winds blowing from the north; Hecataeus thought it was a connexion of the river with the earth-encircling Ocean; Aristagoras's guess that the cause was melting of snows on Libyan heights was a good one; but Herodotus looked to the sun – its withdrawal in winter caused a shrinkage of waters.[19]

The fourth century saw Egypt free for about sixty years and helped by Greeks against Persia until the conquest once more by the Persians and then by Alexander the Great. During this period there was no further penetration to the south of Egypt, and men still had false ideas about the flooding of the Nile,[20] though others like Aristotle guessed the real cause, and it is just possible that the dimmest reports possible of the sudd above the Sobat, of rivers rising in the Abyssinian hills, and even of Alpine heights like Ruwenzori and Kenya and Kilimanjaro, reached the Greeks[21] through Egypt. Alexander's visit to the famous oracle of Ammon at Siwah was no new exploration except that he took an army across the desert to it and inspired men to describe in much detail this curious region, while the tradition that he sent a force up the Nile to find out the cause of the yearly flooding would appear to be a false one.[22] It remained therefore for the lasting dynasty of the Ptolemies in Egypt to add to Greek knowledge of inner East Africa.

3. THE NEW EXPLORATIONS UNDER THE PTOLEMIES

But the Ptolemies found that the most natural channel for reaching the desirable parts of Africa stretching from the Nile eastwards was the Red Sea, and the elephant-hunts of the first kings caused the races of the west coast of that sea, and stretching back inland, to become well known, but this finds its place in the story of the exploration of the Red Sea (p. 209 ff.). Still, we should note here especially the trade-route opened for Philadelphus's new station 'Ptolemais of the Hunts' inland to Meroe and the Nile; and the penetration from the early Ptolemaic foundation Adulis (near Massowa) inland into Abyssinia at least as far as Coloe (Kohaito) on the way to the Atbara and the highlands. This caused a Greek knowledge of the upper Atbara and its several sources in the Abyssinian hills, and verified the rumours about the flooding of the Nile being caused by water from these highlands, which could be seen from the Red Sea; it also verified the reported existence of Lake Psebo, called today Lake Tsana.[23] These had been only vaguely heard of before through the brisk overland trade and penetration up the Nile, which, as we shall now show, opened up to the knowledge of the Greeks regions not reached from the Red Sea.

This inland exploration was made in spite of the barrier of the second cataract and other obstacles which the Greeks now discovered. Early in the reign of Ptolemy Philadelphus one Dalion went farther than Greeks had hitherto explored, namely beyond Meroe, which was now the capital of Ethiopia, and wrote an account of his exploration. Soon afterwards Aristocreon did the same. This may have prompted Philadelphus to increase his geographical knowledge and to obtain at closer hand strange animals and costly wares of Ethiopia. He sent an expedition into Ethiopia, probably one of pure exploration, for no annexation was made.[24] This seems to have led Greeks to visit Meroe in greater numbers; one Simonides, a cultured man, resided there for five years and wrote a book about Ethiopia,[25] and the court at Meroe began to be 'hellenized' in a very superficial way. Somewhere about 230, or perhaps earlier, Ergamenes king of Meroe made all Ethiopia his own[26] and was closely joined in diplomacy and trade with the fourth Ptolemy, the meeting-ground covering the stretch between Philae and Derar.[27] The advisers of the fifth Ptolemy contemplated subjecting the Ethiopians; what happened is not known,[28] but the penetration which went on caused Greeks by the end of the third century to know well the Nile and its great bend (a feature not verified until last century) between Syene and Meroe, and to be pretty well acquainted with Νοῦβαι and the 'island of Meroe' formed by the Nile and the Atbara. They learnt too, but less accurately, the existence of the Blue Nile (Bahr el-Azrek), the White Nile (Bahr el-Abiad), and possibly of the marshes of the White Nile above the junction with the Sobat. But of the upper Atbara and of Lake Psebo and of the Abyssinian tableland, bringing the waters which swelled the Nile, only vague rumours were known until, as we saw just now, the voyagers down the Red Sea made them known a little. That something about these distant regions should have reached the ears of Ptolemaic Greeks is not to be wondered at when we realize that the Meroitic civilization spread at least a part of the way up the Blue Nile, and besides, men like the Egyptian annalist Manetho living under the first Ptolemies may well have found notices in the old Egyptian records.[29]

During the second century more Greeks visited the distant kingdom and wrote about it, for instance, Bion and Basilis, the former especially becoming a well-known writer about Ethiopia.[30] The gold-mines of Nubia were also extensively worked under a system of horrible cruelty.[31] Under the sixth Ptolemy an attempt was made to put the frontier of Egypt at Wadi Halfa,[32] but we cannot trace any penetration into new regions before the time of the Roman Empire. As Strabo says, the Greeks could not define the boundaries of Ethiopia or of Africa, or even of the lands close to Egypt, still less of parts by the Ocean, although Syene, Meroe, Ethiopia (here meaning Abyssinia), and the Cinnamon country (of the Somali) – looked on as the southernmost part of known Africa – could be and had been reached by land. Diodorus says that the voyage up the Nile to Meroe was costly and dangerous.[33] But it is worth while mentioning on Strabo's authority that three oases of Libya, dependent on Egypt, were well known.[34] The first, the Kharga Oasis, was reached from Abydos in seven days across the desert; the second, that of Baharia, was opposite to Lake Moeris; while the third, that of Siwah, was still reached in five days from the coast, though the once celebrated oracle of Ammon was now neglected. But of the upper Nile neither Strabo nor Juba[35] could add anything to the knowledge gained by Dalion and his successors.

AFRICA

4. THE ROMAN PENETRATION

(a) *Petronius and Ethiopia*

The interests of the Romans were not different from those of the later Ptolemies; they were bent on developing Egyptian trade because the emperor had special rights in Egypt. But here again the Red Sea was the natural channel for commerce even with inland districts of Africa. At the beginning of her occupation of Egypt, Rome had to make terms with the kingdom of Ethiopia, which was now ruled by energetic queens. The Ethiopians in 29 B.C. were given freedom to control under Roman lordship a strip of the Nile leading down to the first cataract, but in 25 it was found necessary to send Petronius against queen Candace, and he carried out a military re-exploration which gave much new information to the Roman West. We learn about it from Strabo and from Pliny, who followed different sources. He drove the invading host back to Pselchis, which he took. He then went by a direct journey across the desert from Korosko to Abu Hamed in order to avoid the bend in the Nile, and took Premnis (unknown) and the capital Napata. After about two years he was forced to come back from Alexandria again and relieve his garrison in Premnis. By a treaty made in 21 the Romans were able to organize a military frontier district between Syene and Hierasykaminos (Maharrakah), beyond which their power ceased. But they continued to survey the Nile beyond the 'frontier', for a Greek inscription found at Pselchis, and dating from A.D. 33 during the reign of Tiberius, tells us that 'Titus Servillus, a soldier of the Legion III Cyrenaic, having made a district-map, was mindful of his parents in the temple of lord Hermes in the twentieth year of Tiberius Caesar Augustus on the twenty-sixth of July'.[36] The Ethiopians were now quiet, with Rome on one side and the new kingdom of Axum on the other, and desert tribes began to harass and to ruin the realm.

(b) *Nero's Explorers. The White Nile and the Blue Nile*

This brings us to another of those rare instances in ancient times when an expedition was sent out with exploration and geographical discovery as its final object. With motives partly commercial and warlike, partly coming from a love of inquiry, Nero planned a military expedition against the Ethiopians, particularly perhaps the Axumites in Abyssinia, and actually sent in advance a small party to see if they could find the sources of the Nile. Says Pliny, speaking of Meroe, 'soldiers of the praetorian guard recently sent to explore by the emperor Nero (who was thinking about an Ethiopian war among his others) brought back a report that many old towns on the way to Meroe had disappeared and there were wildernesses' in the kingdom, which was thus already fallen low. They reported fully (with a map) details as far as the 'island' and city of Meroe, noting things like parrakeets and dog-faced baboons. Since Pliny gives us nothing between Pselchis and 'Primis' (that is, Premnis), these explorers must have avoided the bend in the Nile in the usual way – by crossing the desert. They told how Napata was small; grass was green round Meroe; there were woods too and traces of rhinoceroses and elephants; but the city, though ruled by queens still, had only a few buildings, and there were forty-five other monarchs in Ethiopia; beyond Meroe there were no more towns on either bank. The explorers went on and on much farther until an obstacle blocked their way, as we learn from Seneca who had spoken with some of them. Having mentioned an opinion that the Nile, rising out of the earth, received its flood-water, not from rains, but by outbursts of more water from the bowels of the earth, he says, not without flattery of the emperor:

I have heard two centurions, whom Emperor Nero (strong lover of truth above all, as he is of other virtues) had sent to investigate the source of the Nile, tell how they travelled a long journey by the time when furnished with military help by the Ethiopian King, and recommended by him to neighbouring Kings, they had penetrated to regions beyond his realm. 'And indeed,' they said, 'we came to immense marshes, the outcome of which neither the inhabitants knew nor can anyone hope to know, in such a way are the plants entangled with the waters, not to be struggled through on foot or in a boat, because the marsh, muddy and blocked up, does not admit any unless it is small and holding one person.' 'There,' said one of them, 'we saw two rocks, from which a great force of river-water came falling.' But whether that was the source of or merely an addition [of water from below] to the Nile ... do you not believe, dear reader, that whatever it is, it rises up from a great lake of the earth?[37]

These men therefore did not claim to be certain that they had found the sources, and indeed they had not; Pliny rightly says the Nile rose 'from uncertain sources, flowing as it does through desert and heat and for an immeasurable length of distance and sought out by peaceful hearsay only without wars which have found out all the other parts of the earth'.[38] It was these men perhaps who told Seneca how small boats, each holding two boatmen of whom one steered and the other baled for dear life, were accustomed to shoot the rapids south of Egypt. The explorers brought back valuable samples of African commerce, for instance, ebony,[39] having pushed beyond the limits known to any previous Greeks and Romans and probably ancient Egyptians. We cannot explain the two rocks unless it be a fact that the explorers heard of the Fola Rapids – a most improbable thing, since they reported that they had seen this wonder, which if we consider the distance implied, and the sudd, we know cannot have been those rapids. But it is clear that the men had reached far beyond Meroe, even as far as the great marshes formed partly by sudd on the White Nile above the point where the Sobat comes in.[40] They reached regions not visited again by Europeans for more than seventeen centuries – regions discovered again in 1839 and 1840.

Nero intended to supervise his coming 'Ethiopian war' in person, but in the end no war was made at all. Still, his peaceful exploration which we have described was not the last which took place along the waters of the Nile basin. At some time between the death of Nero and the time when Ptolemy wrote his geography a few more Greeks or Romans pushed beyond Meroe; they did not explore extensively, and their reports were confusing, for even Ptolemy makes the 'island of Meroe' a true island by joining streams which do not join. But new ground was broken, namely along the Blue Nile; for along this stream men pushed right up to the heights of Abyssinia, where Ptolemy says lay the hills of Maste round Lake Tsana, a city Maste, a race of men, and a high ridge Garbatus south-east of the lake. A few villages along the river were noted. Traces of this exploration by merchants may be recognized in the many Graeco-Roman bronze vessels which have been found at Makwardam on the Blue Nile.[41]

(c) Diogenes and the Sources of the Nile

We have now described the last piece of extensive exploration made by land in ancient times southwards from Egypt, except one very interesting though but dimly seen achievement of a Greek explorer whose activity lay chiefly in the Indian Ocean, but seems to have had a place within Africa also. It had been found that the obstacles against exploration up the White Nile beyond Meroe were so great that no one could hope to reach along that way the goal – namely the source of the Nile. In much the same way as it happened in the last century, this mysterious unknown was suddenly learnt about and probably seen by a man who sailed down East Africa and gained his knowledge by an approach from the coast.

Of those who from Claudius's time onwards, as we describe elsewhere, sailed down the east coast of Africa, one Diogenes, blown by accident for twenty-five days as far as Rhapta (Dar-es-Salaam near Bagamoyo), seems to have landed and to have travelled inland until he reached the neighbourhood of two large lakes and a range of snowy mountains in a region where the Nile, he found, took its rising; or else he heard of these things from the Arabs who governed Rhapta. He wrote an account and perhaps drew a map which were used first by Marinus of Tyre and then by Marinus's follower Ptolemy, whose results we have – meagre evidence of a noteworthy increase of knowledge. It appears that Diogenes reported the existence in the very far south of a range called the 'Mountains of the Moon', stretching eastwards and westwards for five hundred miles and capped with snow, which in melting fed by streams two lakes towards the north. From each of these lakes flowed northwards two streams which in uniting formed the Nile; and this in its turn, as men had found before, received a river flowing from Lake Coloe (Psebo, Lake Tsana) in Abyssinia. Thus Ptolemy; that he places the sources too far southwards we may ignore.

Diogenes, or his Arabian informants, apparently discovered and passed not far from the snow-topped Mounts Kilimanjaro and Kenya, but falsely assumed (unless the mistake is Ptolemy's) that from these peaks were fed the waters which he saw or heard of. We may take it that he or they saw Lake Victoria (Victoria Nyanza) – that huge

AFRICA

reservoir which, receiving from the west the Kagera, the remotest headstream of the Nile, emits that great river at the northern end; and perhaps viewed the grand issue of its waters over the Ripon Falls between high cliffs in Napoleon Bay. But from this point onwards we may surmise that hearsay from African natives was all that came to hand, for the other lake, named Lake Albert (Albert Nyanza, not nearly so large as the Victoria), receives and emits the Nile as one stream after it has passed through the marshy Lake Kioga (Choga). Likewise Diogenes may have heard of the snow-capped Ruwenzori range, but we may conclude that he did not see it; being often cloud-hidden it could easily be missed, as it was by Stanley in 1875. It is, however, to be noted that the abrupt and isolated peaks of Kilimanjaro and Kenya could not be called a range, whereas the Ruwenzori, possessing as it does thirty miles of glacier, might suggest a landscape like that of the moon. In any case, there is no range here stretching five hundred miles eastwards and westwards.[42]

No further discoveries were made in Africa by ancient explorers for centuries. Ptolemy's Mountains of the Moon appeared in various shapes on maps until at least 1834, and the idea of a connexion between the Nile and the Niger (due to Juba) was still entertained after 1800. Early in the seventeenth century two Portuguese divines saw the source of the Blue Nile, which was not explored successfully until 1771, when Bruce explored it, but thought it was the main Nile. After this interest and efforts increased until Mehemet Ali had the White Nile explored between 1820 and 1842, though effective penetration took place only later in the century by Sir S. Baker (1862–4) and Gessi Pasha (1886). In 1850 Knoblecher reported hearsay rumours of the lakes, and about the same time the missionaries Krapf and Rebmann on the Zanzibar coast reported a huge inland sea where the Nile was believed to rise. Men now thought more of Ptolemy's geography. The Bahr el-Ghazal was explored, and at last Speke discovered, on a journey from Zanzibar, a lake which he named Victoria Nyanza, and in 1862 found the outlet of the Nile from it. Even this and other results of his exploration left much to be learnt about the Nile basin.[43]

(b) The Sahara
1. CONDITIONS OF TRAVEL

The problem of exploration in Central North Africa presented itself to ancient travellers in much the same light as to those of any age preceding the invention of the aeroplane and the motor-caterpillar. In one respect the pioneers of the Greek and Roman era were probably better off. Although the Sahara region was certainly an extensive desert since the end of the Ice Age, there is reason to believe that it was less completely desiccated two or three thousand years ago, and that its oases and water-points lay closer together.[44] Within the last hundred years scientific observers have noted a further encroachment of the sand dunes upon the cultivable land; and remains of neolithic settlements in the now uninhabitable river valleys of the northern Sahara suggest that pools of water may still have been found there in antiquity.[45] It is not certain whether ancient travellers in the African desert made use of camels. But these animals had been introduced into Tunisia not later than the first century B.C., and in any case they are not far superior to horses or donkeys for purposes of light travel.[46] On the other hand, the scientific interest that now attaches to desert exploration made little appeal to the ancient world. Besides, the commercial profits of trade across the Sahara were not sufficient to induce adventurers to search it out. The gold of the upper Niger, the ivory and slaves of the Sudan could be more easily transported by way of the west African coast, and alternative supplies of ivory were available from Somaliland, from Morocco and southern Algeria.[47]

2. EXPLORATIONS FROM CYRENE AND FROM THE BAY OF TRIPOLI

Westward of the Nile valley the first possible base for Sahara travel lies in the region of Cyrene. The immediate hinterland of this district is a steppe, infested of old with dangerous beasts, yet inhabited by nomadic tribes, and a neighbouring oasis (at Aujila) facilitates the first stage of travel in the actual desert. But the next oasis, that of

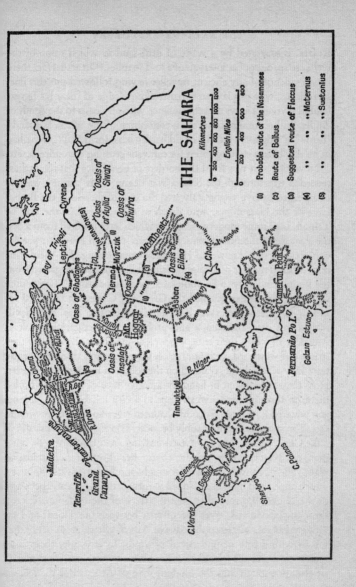

Kufara, is separated by a waste of drift-sand in which some of the wells are spaced out at intervals of 150–180 miles.[48] In actual fact there is no evidence of any ancient traveller having followed out this line. Herodotus, who displays an extensive knowledge of 'beastful Libya' behind Cyrene, shows no acquaintance with any oasis to the south of Aujila. Still more significant is his conception of the Sahara as a sandy 'brow' or ridge,[49] which he could hardly have formed if Cyrene had in his day been a starting-point for caravans crossing the entire desert.

Nevertheless it was from the borders of Cyrenaica that the earliest recorded trip through the Sahara was begun. This journey was accomplished by five young stalwarts, the sons of sheikhs in the Berber tribe of the Nasamones, who dwelt at the east end of the Bay of Tripoli. Like some modern escapades which have resulted from a bet, the venture of the Nasamones was a mere piece of bravado; yet the explorers at least knew the risks they were taking, for their tribe remembered the fate of a neighbouring folk who had emigrated from the steppe-land only to be obliterated in a sand-storm.[50] The outline of their story, which reached the Greeks of Cyrene at the third remove, has been preserved by Herodotus.[51] The five dare-devils started out over the steppe and probably made their first halt at Aujila, with which their tribe was well acquainted.[52] Eventually they struck out westward, and pushing on for many days through a sandy tract they reached the savannah on the southern border of the desert. Here they were taken in hand by a native race of dwarfs, who led them on through extensive swamps to a city inhabited by Negroes and watered by a large crocodile-haunted river flowing from west to east. The swamps might possibly be located on the western margin of Lake Chad, but the large eastward-flowing river can hardly be other than the Niger, whose middle course lies through an inundation plain. The city must therefore be sought somewhere near Timbuctu, which appears to have been lapped in former times by a northern arm of the Niger. We may then plot the track of the Nasamones across the oasis of Murzuk, over the plateau between the Tibesti and the Hoggar ranges, and so to the oasis of Asben, where presumably they fell in with a folk of undersized negroids. In any case, there is no reason to doubt the veracity of Herodotus's anecdote, for its details fit readily enough into the modern map of Africa. But the *tour-de-*

AFRICA

force of the Nasamones was an isolated performance, and it created no regular trans-Saharan traffic.

From the Bay of Tripoli several routes lead to the oasis of Murzuk, which is a good starting-point either for Lake Chad or for the more westerly track towards Nigeria.[53] This oasis was inhabited by a populous tribe, the Garamantes, whose chief town Garama is no doubt the modern Jerma (60 miles west-north-west of Murzuk). Herodotus relates that the Garamantes made raids on four-horse chariots into the Sahara, where they kidnapped fleet-footed Troglodytes (cave-dwellers), who 'squeaked like bats'.[54] The caves in which these dwelt were no doubt in the Tibesti or the Hoggar Mountains. It is not unlikely that the Garamantes also worked any traffic that plied between the Carthaginian seaboard and the Sudan: the rock-carvings of 'feathered Libyans' probably refer to this tribe.[55] The occasional presence of Carthaginian super-cargoes on the caravans is attested by a tall story in Athenaeus concerning one Mago who had thrice crossed and re-crossed the desert on a diet of dry barley without water.[56] This yarn is of the kind that only its own hero could take neat, but it shows that Carthaginians were believed to traverse the Sahara on business. This belief finds apparent support in an inscription between Murzuk and Asben whose characters have some affinity with Phoenician script.[57] But whatever the Carthaginians learnt about the Central Sahara they did not divulge to the Greeks, whether traders or geographers. It was left to Roman soldiers to lift the veil here and there.

3. MILITARY EXPLORATIONS OF THE ROMANS. RESULTS

In Africa the Roman armies had no such need or scope for exploration as in Europe, yet here too the exigencies of frontier defence imposed upon them occasional ventures into the unknown. The cultivable seam of north Africa was exposed to raids by tribes living on the brink of the desert or in its oases, and their incursions sometimes drew a Roman counter-attack which was also a voyage of discovery. Unfortunately we have only a few brief allusions to these operations. In 19 B.C. Cornelius Balbus the governor of Tunisia occupied the Garamantian capital of Jerma.[58] A handsome monument of carved stone,

THE ANCIENT EXPLORERS

said to be of Augustan date, perhaps betokens the presence of a Roman resident there.[59] About A.D. 70 or 80 another proconsul, Septimius Flaccus, replied to a Garamantian foray on the Tripoli seaboard by an inland march of three months' duration.[60] Since neither the rate of Flaccus's march nor the length of his rests in the oases is known, it is difficult to agree or to disagree with modern geographers who suggest that he advanced as far as Bilma (half-way from Murzuk to Lake Chad).[61] In any case there is no reason to suppose that he won through to Sudan. This distinction was reserved for another officer, Julius Maternus, at a date unknown. Maternus first visited Jerma, presumably on some diplomatic mission, but was here enlisted by the king of the Garamantes for a joint raid upon the 'Ethiopians'. From Jerma the allies proceeded southward for four months, until they reached a district named Agisymba, 'where the rhinoceroses forgather'.[62] This account leaves it uncertain whether Maternus followed the route to Chad or to Asben:[63] in any case he emerged from the Sahara on to the Sudanese steppe. This expedition, however, like that of Flaccus, caused no stir at Rome; we may assume that whatever caravan trade survived at that period remained in the hands of the Garamantes.

In the western Sahara, despite its comparative humidity, no ancient explorations are on record.[64] Behind the Roman province of Numidia (Algeria) the Gaetulians were as ready for mischief as their Tuareg descendants remain. But the Roman governors seem to have rendered them sufficiently innocuous by enlisting them as auxiliaries in their defence forces, and so had no need of retaliatory raids upon them. A slight acquaintance with the western Sahara is reflected in King Juba's theory that the Nile had its source in Mt Atlas, and after two disappearances underground emerged in the eastern Sudan.[65] Juba's fallacy presumably rested on a vague knowledge of the river Ger or the Jedi, but he might have culled this much information from the Gaetulians without exploration.

The only other journey that requires mention here is a march which Suetonius Paulinus, the future victor over Boadicea, achieved across Mt Atlas in A.D. 42. In quelling a revolt among the Moors, Suetonius pressed the pursuit right over the mountains and plunged on for ten further days across a torrid belt of black dust to the river Ger.[66] In traversing Atlas (it is not known by which pass) he noted the dense growths of scented timber and the snow that capped it in midsummer. Modern travellers on Suetonius's track have observed forests of larch and juniper under its summit, and have reported heavy going over loose grit to south of it.[67]

The results of these explorations were hardly commensurate with the exertions entailed by them. They did not create any considerable movement of trade or contribute much to elucidate African geography. The sally of the Nasamones misled Herodotus into identifying the 'crocodile river' of west Africa with the upper Nile.[68] The expeditions of Flaccus and Maternus deceived Ptolemy into extending the Sahara as far as 8° south. The same geographer rightly thought of the desert as mountainous; but he came to grief in tracing across it two rivers, the Gir and the Nigir, which he supposed to flow from west to east on the latitude of Timbuctu.[69] These streams were not wholly imaginary. Ignoring Ptolemy's latitudes, which are consistently too low for inner Africa, we may recognize in them the Jedi of south Algeria and the Igharghar of the north-west Sahara,[70] which still run in spates after rare rainstorms. But as described by Ptolemy the two desert rivers were as misleading as mirages, and they have caused much confusion among modern geographers.[71] In fine, the first effective voyages of discovery in the Sahara belong to the Middle Ages rather than to antiquity.

CHAPTER 9

RESULTS OF ANCIENT EXPLORATIONS

1. EXTENT OF GROUND WON

THE extent of the earth's surface explored by ancient travellers is indicated on map 15. It may be summed up as follows:

(1) The entire Mediterranean area.

(2) The seaboards of western Europe; of Asia from Suez to Tongking or Canton; of west and east Africa as far as Sierra Leone and Porto Delgado respectively (or to Cape Bay, if we accept the story of the Phoenician circumnavigation of Africa).

(3) All Europe south of the Rhine and Danube, together with parts of Germany and southern Russia.

(4) Asia Minor; the Caspian Lands; the plateau of Iran; the Indus system; and a silk-route to Singanfu in China.

(5) The Nile basin to the Sobat and the region of the sudd; a route from the east African coast to Lake Victoria Nyanza; two or three tracks across the Sahara to Sudan and Nigeria.[1]

2. ANCIENT GEOGRAPHIC LITERATURE

The advance in geographic knowledge achieved by ancient explorers is reflected (if not altogether accurately) in the geographic literature of the Greeks and Romans. This literature included in the first instance the reports of the actual discoverers. Surviving specimens of these are the narratives of the cruises of Hanno and Nearchus (pp. 63–8 and 80–6), and also the *Periplus of the Erythraean Sea* (pp. 96, 98–9 ff.). In this class we might further include Caesar's *Commentaries* on the Gallic War.

A considerable number of ancient geographic works consisted of practical manuals for travellers. One extensive series of guide-books was composed of sailing-directions by Greek seamen for coastal voyages. The earliest of these were probably restricted to particular sections of the Mediterranean seaboard; but a large proportion of

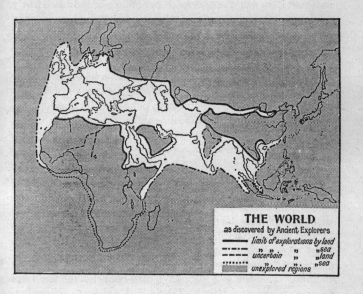

them comprised the entire circuit of the Mediterranean or the Black Sea. Thus the term περίπλους or 'sailing-round', which strictly speaking only applied to descriptions of circular tours, came to be used for sailing instructions along linear routes as well. As the range of naval exploration increased, guide-books for the 'Outer Seas', and especially of the 'Red Sea', were also published.[2] Of the extant examples of these naval manuals the earliest is probably the book of the Massiliote mariner of c. 500 B.C., whose description of the coast of Spain is partly preserved in the poem of Avienus (pp. 44–6). The oldest surviving guide-book for the entire Mediterranean is the work of Scylax, composed c. 350 B.C.[3] A report on a circular tour of inspection round the Black Sea, which the historian Arrian carried out by order of the emperor Hadrian, has also come down to us. A very competent work of this class survives in the *Stadiasmus Maris Magni*, presumably of the early third century A.D. This provides a point-to-point description of the Mediterranean coast, with distances uniformly expressed in stadia instead of the older 'day's journey', and additional information about harbourages and water-points in the style of our Admiralty Pilots. The *Periplus of the Erythraean Sea*, mentioned above, is also drawn up in the form of a guide-book.[4]

Of analogous manuals for journeys by land the Greeks significantly enough produced nothing previous to the continental conquests of Alexander. Of these ὁδοιπορίαι or σταθμοί, which mostly covered the long and difficult routes across the Asiatic mainland and were careful to mark the none-too-frequent halting-places, the only survivor is the *Parthian Stations* by Isidore of Charax, a contemporary of Augustus.[5] This text is a descriptive road-book for the trip from Zeugma on the Euphrates to Arachosia (Kandahar). The general opening up of all the three continents which followed upon the Roman conquests called forth a copious production of Itineraries. The largest remaining work of this class is the *Itinerarium Antonini*, a road-book of c. A.D. 200, which comprises many of the main routes in every province of the Roman empire. Other interesting examples are the *Jerusalem Itinerary* of A.D. 338, which sets forth in detail a route from Bordeaux through Italy and Constantinople to Antioch, and a record of stages and distances between Gades and Rome graven upon four vases found in northern Italy.[6]

RESULTS OF ANCIENT EXPLORATIONS

The earliest of the practical manuals enumerated above were scarcely older than the first χωρογραφίαι or general treatises on geography. The pioneer among these was the Γῆς Περιήγησις by Hecataeus of Miletus, written c. 520–500 B.C.[7] In outer shape Hecataeus's work resembled the περίπλοι of the Mediterranean, but to judge by the brief surviving fragments its object was rather to furnish information about the towns and peoples of the Mediterranean seaboard than to provide aids to navigation. The geographical excursuses of Herodotus were of still wider scope, in that they described continental regions as well as coastal districts, and gave as much attention to the customs of the peoples as to the physical features of the countries. The same kind of 'human geography' was written by the historians Polybius and Poseidonius,[8] who made known to the Greeks the results of the Roman military explorations in western Europe in the third and second centuries B.C.; by Agatharchides, a writer of the late second century B.C., who similarly followed up the discoveries of Greek traders in the 'Erythraean Sea';[9] and above all by Strabo, whose panoramic review of the lands of the Roman empire is the completest surviving treatise of ancient geography. The general *Descriptio orbis* by Augustus's colleague Agrippa, which was based on a systematic survey of the Roman provinces,[10] has unfortunately not survived. In addition to the above, numerous second-hand compilations for schoolchildren or the more hasty general reader were composed. Among the works of this class we have often had occasion to refer to Diodorus, to the fragments of Juba, to Mela and Pliny.

The remaining branch of ancient geographical literature consisted of mathematical geographies, such as the treatises of Eratosthenes, Hipparchus,[11] Marinus, and Ptolemy,[12] of which the last alone has come down to us. These works were not so much the outcome of explorations as of mathematical discoveries by which the shape and the perimeter of the earth had been determined, and their essential object was to plot its map by the mathematical method of reference to latitude and longitude.[13] Ever since the days of Hecataeus and Herodotus Greek geographers had been inclined to group the lands known to them in a symmetrical arrangement round the Mediterranean Sea. Shortly before the time of Eratosthenes, Dicaearchus had definitely adopted a line through the Mediterranean from the Straits to Mt Imaus (the Himalayas) as a median.[14] Improving upon this device, Eratosthenes measured off his distances from two main axes intersecting at the island of Rhodes, or from a small number of subsidiary base lines drawn parallel to the major axis. Hipparchus proposed to cover the earth's surface with a complete grid of meridians and parallels at, say, half an hour's distance from each other. In Ptolemy's Geography six books out of eight consist of lists of towns, mountain ranges, estuaries, and so on, located by latitudes north of the equator and longitude east of the 'Fortunate Isles'. Yet none of these writers could collect their data from astronomical observations alone, without recourse to the reports of travellers. In particular, Ptolemy's work was almost wholly the result of dead-reckoning from Peripli, Itineraries, and material supplied by actual explorers.

Lastly, the growth of geographic literature brought with it the development of cartography. The earliest known charts, it is true, antedate by many centuries the first treatises of geography. From Babylon there still remain two tablets of the third millennium B.C., showing land-surveys;[15] and from Egypt two papyri with plans of gold-mines in Nubia, and a third, dating back to the reign of Seti I (c. 1320 B.C.), which depicts a small part of Lower Egypt.[16] But these, together with the Forma Urbis, a street plan of Rome from the reign of Septimius Severus (c. A.D. 200),[17] were of little geographic value. The earliest geographic maps were constructed by two Greeks of Miletus, Anaximander (c. 575 B.C.),[18] and Hecataeus, both of whom plotted out the entire world so far as known to the Milesians of that period. Hecataeus's map illustrated his general geographical treatise, and henceforth it was a common practice of geographers to accompany their texts with maps.[19] The *Descriptio Orbis* of Agrippa was a commentary on a revised map of the world, which reproduced his survey of the Roman Empire on a reduced scale; and Books Two to Seven of Ptolemy's Geography might be described as an index to his general map. Independently of the text-books, maps for general reference had become common by the fifth century B.C. Since the geographic knowledge of the ancients applied only to the Old Hemisphere, and not to the whole of this, spherical maps could be of little use to them. But one huge globe, apparently ten feet in diameter, was set up at Pergamum by Crates of Mallus (c. 150 B.C.).[20] It is not certain whether

RESULTS OF ANCIENT EXPLORATIONS

the naval guide-books were furnished with coastal maps like the modern Admiralty Charts or the portolani of the later Middle Ages.[21] Of these maps nothing survives save two or three authentic copies of Ptolemy's chart, which are appended to some extant manuscripts of his text,[22] and the *Peutinger Table*, a roadmap of the Roman Empire and Iran (as far as India), arranged in strips to accommodate the pockets of travellers.[23]

3. THE DEFECTS OF ANCIENT GEOGRAPHICAL LITERATURE

It must be admitted that the extent and accuracy of the knowledge displayed in ancient geographic literature is hardly in keeping with the results of actual exploration. Ancient geographic writers are strangely reticent regarding some of the greatest and most fruitful voyages of discovery. The alleged circumnavigation of Africa by Pharaoh Necho's Phoenicians, and the Indian Ocean cruise by Darius's admiral Scylax, are known to us from nothing more than two brief allusions by Herodotus.[24] Pytheas's story of his voyage was appreciated at its proper worth by Eratosthenes, but was dismissed as a bad joke by Polybius and Strabo. The discoveries of Hippalus, again, were half forgotten. The merchants who benefited by his discovery of the law of the monsoons did indeed give him credit and appear to have spoken of a 'Hippalus' wind, a 'Hippalus' cape and sea; but the writers on geography do not mention him as a man, and seem unaware of his discovery that India was a peninsula.[25]

Not content with neglecting good information, geographic authors were prone to cherish false reports long after these had been exploded by fresh exploration. The Caspian Sea, proved before Herodotus's time to be an enclosed inland lake, nevertheless continued to be thought of as a great inlet of the northern Ocean until the time of Marinus and Ptolemy. Again, the 'Rhipaean Mountains' and Hyperboreans, which Herodotus tacitly or expressly ignored, were confidently accepted by a long series of Greek and Latin authors. The belief that a branch of the Danube flowed into the Adriatic was not disposed of until the time of Augustus, and then not completely. Atlantic bugbears, once created, died hard, and acted as a continuous deterrent to further exploration of those waters.[26] These are but the most glaring examples of a common form of error.

The general lack of accuracy in ancient geographers may best be gauged by the great standard work of Ptolemy, which was based on the fullest collection of data and aimed at the highest level of precision. The apparent exactitude with which his geographic entries are located by latitude and longitude has been shown up as wholly delusive. Not only, as we have seen, were his bearings more often derived from computations based upon travel records than from direct observation, but the computations frequently erred by a big margin.[27] We have noticed his errors in regard to Scotland and Scandinavia, to the Sahara, the west African seaboard, and east African highlands; we must observe too how, despite the opening up of the Indian Ocean by traders, he flattened India and swelled out Ceylon, and prolonged the Chinese coast by a returning curve so as to meet the African continent. Even the fully explored regions of the earth were faultily plotted out by him. The Mediterranean Sea was extended by him over sixty-three degrees of longitude, i.e., fifty per cent too far; the north African coast was levelled into an almost unbroken straight line. Italy was made to point in an almost easterly direction, and Sicily was twisted round correspondingly; and the Sea of Azov was prolonged into the heart of Russia. And if errors like these could beset Ptolemy, it is easy to imagine what might happen to Mela or Pliny.

Finally, after the second century A.D. not only was no further advance of knowledge achieved by further explorations, but much of the ground previously made good was lost again. The general regression of learning and of scientific interest which marked the third and succeeding centuries affected geography like all other branches of study, and in the writings of the Church Fathers preconceived ideas of the earth's outlines tend to replace the knowledge derived from actual discovery.[28]

The reasons for the imperfections of ancient geographic writers are partly to be sought in the nature of their sources. Not infrequently the records of explorations were left unpublished. It is most unlikely that the reports of Necho's Phoenicians or of Darius's admiral Scylax were committed to writing and survived to the age of Herodotus.

RESULTS OF ANCIENT EXPLORATIONS

But there is no doubt that many discoverers whose ventures had a commercial object kept the result of their investigations secret for fear of losing a profitable monopoly. Others issued deliberately misleading statements, in which fictitious terrors (sea-monsters, griffins, ferocious ants, etc.) played a nicely calculated part. Others again, though not intent on deceiving, contrived to do so by the narrowness or perfunctoriness of their observations, which did not extend beyond their immediate mercantile objective.[29] Again, the primitive character of ancient travellers' equipment impaired the accuracy of their records. In particular, the lack of fully adequate instruments for measuring distances and taking bearings could play havoc even with a scientifically minded explorer, and might seriously distort the map of his discoveries.

A further obstacle to the growth of geographical knowledge in proportion to the progress of discovery lay in the absence of a proper organization for testing the reports of explorations. For want of a body like the Royal Geographical Society, or the various modern academies, it was equally possible for a swindler to be taken at face value or for a genuine traveller to be refused credence.[30] Under such conditions of uncertainty it was unlikely that geographic authors should always sift the grain from the chaff of travel records with exact discernment.

4. INFLUENCE OF ANCIENT EXPLORATIONS ON POSTERITY

The influence of ancient exploration in medieval and modern times has been of slight importance, except in the Great Age of Discovery, but at this period its effect for good or bad was considerable. In medieval Christendom little was read of ancient geographic works except *mirabilia* like those of Pliny and Solinus; the more solid treatises, like Ptolemy's, were left over to the scholars of Islam. But in the fifteenth and sixteenth centuries the revival of classical learning, and especially the renewed study of Greek authors, brought with it a wider acquaintance with the more accurate and scientific writers on geography. The appearance of a Latin translation of Ptolemy in A.D. 1410 was an event of special importance, in that Ptolemy was studied more carefully by the scholars of the Renaissance than in his own day, and influenced both the existing conception of geography and the plans for future exploration.[31]

A curious feature of the influence which ancient geographers exerted upon those of early modern times is that the mistaken ideas of the ancients were sometimes more potent than their exact knowledge. The 'Rhipaean Mountains' reappeared on modern maps, and were not finally expunged until the eighteenth century; the 'Mountains of the Moon' were again marked across east Africa and remained a goal of discovery to the days of Speke, if not of Stanley.

Three ancient fallacies in particular had an effect upon the course of later discovery. The persistent belief in Atlantic terrors, and especially of a 'coagulated sea' in which ships remained fast, was one of the factors which retarded exploration down the west African coast.[32] Another preconceived notion which long misled modern pioneers was the doctrine of a Terra Australis which extended indefinitely to east and west below the African continent.[33] This idea, which probably originated in the fancy of theorists bent on providing the countries of the northern hemisphere with a more or less symmetrical pendant,[34] received apparent confirmation from travellers who had found land to the south of Cape Guardafui – the point at which the African mainland was commonly supposed to reach its southern extremity – and from traders who had visited Ceylon but greatly exaggerated its dimensions from west to east.[35] It was rejected by Ptolemy, who correctly believed that the African coast extended continuously below Cape Guardafui, and mistakenly prolonged it in an easterly direction until it merged into China. Yet Ptolemy's authority did not suffice to discredit the southern Land. It reappears on maps of the seventh and eighth, and, even of the sixteenth and seventeenth centuries.[36] Moreover, Terra Australis imposed itself upon explorers in southern waters and upon geographers who utilized their reports. Mercator recognized it in Magellan's Tierra del Fuego, and it required Drake's voyage round this island to dispose of Mercator's theory. In mid seventeenth century Tasman imagined that he had found in the North Island of New Zealand the western edge of the phantom continent. It remained for Captain Cook to relegate Terra

Australis beyond the zone of habitable lands to the circumpolar regions.

But the most momentous consequence of ancient misconceptions about the globe is that they facilitated the discovery of America. The existence of an unknown continent to the west of Europe had been assumed by arm-chair explorers not later than the second century B.C. Crates and others imagined an equatorial and also a meridional ocean dividing the earth into four habitable islands, thus forestalling in theory the discovery of north America (land of the 'Perioeci'), of south America (the 'Antipodes'), and of Australia ('Antoeci', to the south and south-east of Africa), as pendants to the 'Oecumene', i.e., the known world on the fourth island.[37] In a geographical work belonging probably to the first century B.C. the opinion is expressed that the outer Atlantic might be studded with islands.[38] Strabo was inclined to believe that there were several habitable worlds in the temperate zone, especially in the latitude of Athens and the Straits of Gibraltar.[39]

These guesses were reinforced by false reasonings, and by correct inferences from imperfect data. In Aristotle's time an opinion was current that west Africa could not be far from India, because there were elephants in both these regions.[40] Eratosthenes, who had correctly measured the earth's circumference, but wrongly extended the Eurasian continent across more than one-half but less than two-thirds of the northern hemisphere, maintained that a voyage from Gibraltar to India along the residual arc would be feasible, if the distance were not so great.[41] Poseidonius, who reduced the width of Eurasia to 180°, but concomitantly scaled down the earth's perimeter from 250,000 to 180,000 stades, reckoned that on the latitude of the Mediterranean the voyage from Gades to India would amount to not more than 70,000 stades (c. 7,700 miles).[42] On the strength of such conjectures and calculations Seneca (in the reign of Nero) described the distance between Spain and India as a matter of 'very few days',[43] and prophesied the discovery of a new Atlantic land which would displace Thule as the end of all things.[44] Lastly, the great authority of Ptolemy was enlisted in support of this hope. Ptolemy not only extended the known part of Asia much too far eastward, but imagined its unexplored regions as stretching on indefinitely.[45] It was the false optimism engendered by these misprises of the ancients that determined Columbus to set out for the Spice Lands and India by way of the Atlantic. Thus by their failures as well as by their successes ancient explorers and geographers affected the course of early modern discovery.

CHAPTER 10

IMAGINARY DISCOVERIES

1. FICTITIOUS EXPLORATIONS

In antiquity, as in later ages, human imagination sometimes outran the records of actual exploration and invented tales of lands and seas that still awaited discovery. We shall first notice fictitious explorations. These come under three heads, and in all may be seen the working of imaginative minds possessed of a half-knowledge of distant lands and seas, though there is a lurking under-current of national pride. The three kinds are (*a*) explorations by mythical persons; (*b*) by half-mythical persons; (*c*) by historical persons.

(*a*) Of the first kind we have examples from three great cultured peoples. The ancient Egyptians had, besides their true history of merchants who went to Punt for trade, mere tales of men who went thither. Typical of these is the story told on a papyrus of the Middle Kingdom about one who was wrecked on his voyage and described his adventures. The story resembles somewhat an episode in the *Odyssey* and the voyages of Sindbad in the *Thousand and One Nights*. It begins with rejoicing that the ship has at last reached home from the ends of Wawat and Senmout. It tells how at the Pharaoh's command a ship set sail, was caught by a sudden wind-storm, and was wrecked with loss of all on board except the teller of the tale, whom a wave cast upon an island where he found plenty of good fruits, fishes, birds, and the like. He made himself king in this well-stocked Robinson Crusoe land. While he was making offering of thanks he heard a sudden noise which shook the earth; it was a huge snake which threatened him in human speech, took him unharmed to its lair, and threatened again. The castaway told his dismal story: the snake comforted him, for nothing was lacking on that enchanted island which would soon be changed into waves; in due course a ship would come and take him home again. The castaway said he would reward the snake handsomely in time to come. The snake said he was the Prince of Punt. The ship came as foretold; the snake sent him off loaded with good wares, assuring him that he would reach Egypt in two months, which indeed took place, and the returned mariner handed over the wares to the king of Egypt.

In this story we have the prototype (or something near it) of the Utopia which the Greeks later pictured to themselves, as we shall see. Everything points to Socotra as the scene of this happy land[1] (see pp. 72, 240–2).

The civilizations of Mesopotamia likewise had their stories of people who made strange explorations in lands far off. We are safe in seeing nothing historical in the personality of the deified hero Gilgamesh (who appears to be a Cassite or some other Elamite) of irresistible power, conqueror of Uruk or Erech in southern Babylonia. By desert wastes he apparently travelled, according to the story, to south Arabia and beyond it crossed a great sea (the Arabian Sea near Aden?) to Socotra and perhaps Bab el-Mandeb.[2]

Again, the great legendary queen Semiramis (Sammuramat), child of the fish-goddess of Ascalon, and in reality a form of Astarte, was made a heroine of Asiatic conquest. She captured Bactra, and as queen (with Ninus) travelled (*c.* 800) all over her Assyrian Empire and conquered Egypt and much of Ethiopia, but failed against India. It became a custom to ascribe every wonderful monument in Iran to her.[3]

But it is to the keen minds of the Greeks that we look to find a more plentiful fund of fiction in the matter of wanderings over the earth. The epic of Gilgamesh has many topics in common with the *Odyssey*, and may have exerted an indirect influence upon the structure of this Greek poem.[4] Gilgamesh, like Odysseus, visits the underworld, suffers shipwreck, builds himself a raft, and is eventually borne home on a magic ship. Be this as it may, the world's prototype of travel romance is the *Odyssey*, which is too familiar to need description. One of the earliest Greek legends of exploration was the story of the Argonauts, which was known perhaps in 'Homer's' time. It may have had its origin in the worship of Zeus Laphystius; it may be a myth about the sun; it may be a reflection of the earliest voyages taken from the Aegean through the Black Sea. In the good ship Argo of fifty oars Jason and his crew went to Colchis along the southern shore of the Black Sea until they reached the Caucasus and the Phasis.

IMAGINARY DISCOVERIES

Of the return various legends were told. Pindar tells how they sailed from Phasis through the northern Ocean to the Mediterranean; others brought them along the northern side of the Black Sea; or across Europe to the northern waters and then round to Gibraltar Straits; and so on.[5]

When the story had spread through all the Greek world, other heroes were made a part of the travellers, especially the mythical but national Greek Heracles (Latin Hercules), who met with adventures on the Mediterranean coasts, set up his pillars at the western entrance of the sea, and went to the Hesperides to bring back the golden apples (see p. 242). As the knowledge of the Greeks was widened, Heracles was not unnaturally made to travel in Ethiopia, Arabia, and other parts of Asia.[6] So too was Dionysus in his journey through the world; he was naturally stated to have visited the wondrous land of India, which was one of the regions where the Greeks put Nysa, the place where he was brought up by the nymphs, when Alexander found a place of that name on his advance into India, and so made the man march in the god's footsteps. After Alexander's campaigns Dionysus was made a conqueror of the East as far as the Ganges, and in Lydia a yearly festival celebrated his triumphant return from India.[7]

(b) The second class of false explorations was inspired by national pride which caused an already false story to be so twisted in tradition that the personality of its genuine hero took on a half-mythical shape, or became fused in the personalities of other equally historical persons. There is one well-known example. Thus, to satisfy vanity and pride, the ancient Egyptians fashioned a myth about one whom the Greeks called Sesostris or Sesōōsis, who was a mixture of several kings such as Seti I and Ramses II (nineteenth dynasty) and especially Senwosri III (twelfth dynasty), the prototype being probably the Hyksos king Sweserenre Khian. No Egyptian king went far beyond the Euphrates or conquered Asia Minor; or set an invading foot in Europe. Yet the Greeks told how Sesostris subdued much of the known world, conquering first the peoples along the Erythraean Sea, both Arabians and also Africans right round to west Africa, then those in Asia, and then the Thracians and Scythians of Europe.[8]

(c) Of the third class of false explorations, that is, those attributed to a single real person, we shall give three examples. Thus Tearchus

(Tirhaqa) of the twenty-fifth dynasty, an Ethiopian of the seventh century B.C., who hardly went out of the valley of the Nile and was vanquished twice by the Assyrians, was, by the fourth century, reported to have conquered eastwards to India and westwards to the Straits of Gibraltar – a tale which rose out of the not extraordinary political and commercial relations of that ruler.[9] Again, Nebuchadrezzar was given the credit of a westward push as far as the Straits of Gibraltar.[10]

But most remarkable (though not widely believed) was the false exploration and conquest deliberately attributed to Alexander the Great, firstly as a piece of flattery, then as a further glorification of a great man's memory. When Alexander reached the Hindu Kush a number of his men believed that he had reached the real Caucasus, and similarly when he reached the Iaxartes, that he had reached the Tanais or Don. This gave a great chance later to would-be storytellers; these shifted the name of the Caucasus to India in order to connect the great conqueror with the range famous for the legends of Prometheus and Jason. Similarly with the name of the Tanais. This transference still has its traces if 'Hindu Kush' means 'Indian Caucasus'. Thus Alexander was credited with exploration to and conquest of the Caucasus and the Don. He was also credited, falsely, with having reached the Ganges. This falsification was quite different from the romance of Alexander which has grown out of an anonymous Greek book written in Egypt in the second century A.D.[11]

2. WONDERLANDS AND UTOPIAS

The peoples of ancient civilizations tended to picture the lands just outside their own personal knowledge or of their nearest intermediaries as filled with wonderful animals and plants and so on, and peopled with ideally virtuous folk free from the troubles and turmoils of the wicked world. We have just seen a little of it in the story of the Egyptian Sindbad; in the 'Phaeacians' of the *Odyssey* we have an ideal society of peaceful traders who provide a contrast and perhaps a challenge to the ordinary Homeric type of city-sacking Achaeans;[12] the Indians too had their Blessed Lands, for instance, in the snowy heights of the Himalayas (see p. 239). But, again, it is to the developed

IMAGINARY DISCOVERIES

Greeks we look to find Utopias worked out systematically in half-known or hardly known regions of the earth, regions which were hazy in the minds of ordinary men, and in regions which were utterly unexplored. The Greek writers distributed their wonderful things with fair impartiality over the North, East, and West, but their fancy failed them somewhat when they turned to the South.

(a) In the North

Beyond the northern limits of detailed knowledge of the Greeks were reported peoples who had some substance of distorted reality. Thus a general term for nomads of the Steppes of eastern Europe and west-central Asia was 'Hippemolgoi' or 'Mare-milkers', who ate milk made into cheese.[13] More remarkable were the Arimaspi, of whom the unbelieving Herodotus reported that they were one-eyed folk; living above the Issedones, they stole masses of gold away from the griffins who guarded it in the north of Europe.[14] Men believed in these, just as they did in the Hyperboreans (which we shall deal with below), relying on the work of Aristeas (see pp. 160, 166), the source of information being the Issedones and the kernel of truth the gold on the sides of the Urals and of the Altais which shows itself in the streams.[15] The gold-guarding griffins and mysterious 'Big-headed' men go back to the Hesiodic poems.[16]

Beyond these, in completely unexplored regions of northern Europe and Asia, were the 'Hyperboreans' (men beyond the north wind), a mythical, long-living, virtuous, vegetarian people who dwelt happily in a paradise blessed with a perfect climate beyond the north wind, where all was sunshine and fertility, without disease or war. Hesiod and a later hymn to Dionysus are the first to mention these.[17] They were probably put at one time in Thrace, but were pushed farther north in legend as exploration failed to show them. By the fifth century B.C. they were put beyond the ever-snowy 'Rhipaean ("Gusty") Mountains' and stretched to the northern Ocean.[18] Says Pindar: 'Not going by ship nor on foot could you find the wonderful way to the gathering of the Hyperboreans'.[19] Many like Herodotus knew that the whole story about them was a fable. Granted the existence, he said, of men beyond the north wind, there must also be men beyond the south wind (Hypernotians). But belief in them lasted all through the ages of Greece and Rome because no one explored so far as to prove the tales untrue.[20] The absence of any way of reaching the Hyperboreans did not prevent the Greeks from telling stories about the northern Ocean beyond the northernmost limits of Europe and Asia. Thus in the Oeonae Islands dwelt people who lived on birds' eggs and oats; and in other islands men with horses' feet, and in others again men whose naked bodies were modestly covered by their colossal ears. Even men like the hardheaded Julius Caesar could listen with a believing ear to tales about the unexplored North. He reported that in northern Europe were animals called *alces*, which were elks. Since Providence had seen fit to create these poor beasts without certain important joints in their limbs, they were forced to roost, as it were, among the trees, leaning upright against the trunks. Unkind stag-hunters took advantage of this weakness, and, knowing the places where the elks were wont to take their sleep, sawed the trunks nearly through, or uprooted the trees altogether; so that the elks, when they came to rest themselves in the dark, leant against ghost-trees, thus falling easy victims of the booby-traps, and were found helpless on the ground in the morning.[21]

(b) In India and Central Asia

The Persian Empire brought the mysterious Indians to the knowledge of the Greeks, and stories were soon being told about them. These Indians got much gold from a sandy desert inhabited by huge ants bigger than foxes which burrowed in the sand and pursued all who came to get the gold. This was an Indian tale appearing in Herodotus, repeated and supported through the Greek and Roman ages by alleged specimens of the living insects, and of their skins and feelers, and believed until the middle of the sixteenth century A.D. Megasthenes put the source of gold on a table-land among the Derdae in the lofty tracts on the borders of Kafiristan and Tibet.[22] But India as a land of wonders was a very ordinary affair in Herodotus's time, though the animals and birds there were larger than anywhere else.[23] Ctesias later did much to 'enlighten' the Greeks and filled India with marvels as well as with real things. Three examples must suffice. Thus the Indus, not less than four miles at its narrowest, and twenty miles at its broadest, produced only one animal – a huge worm seven cubits long, which fed on oxen, and even camels: the Indian king went to battle with 100,000 elephants and an extra 3,000 specially chosen to destroy fortifications. No wonder the Greeks were not all ready to accept these tales, nor those about 'Shady-feet' (Sciapodes), men with feet so large that their owners could use them as sunshades when they lay flat on their backs. Hecataeus had placed these in Ethiopia: Ctesias put them in India.[24]

During the last centuries before Christ the Greeks, and through them the Romans, became acquainted with peoples of Central Asia, among whom they included the Seres or Chinese, without knowing how far they stretched away in the most distant East. They were reputed to be very gentle and righteous long-living men dwelling in a land filled with cattle, trees, and all kinds of fruits which ripened in a serene and lovely climate. They were tall of stature and knew no disease; they had blue eyes, red hair, and rough voice. They shunned the intercourse of strangers. The Greeks adopted also a mythical paradise belonging to Indian story – 'Uttarakuru' in the *Mahabharata* – which they peopled with blessed Attacori and put in central Asia. They have their place in Ptolemy's Geography as Ottorocorrae on the snow-capped mountains near Kashgar.[25]

(c) In Africa

The tendency to make the unknown parts of Africa uninhabitable because of great heat caused a scarcity of strange men and beasts in the far South. But nearer the known parts both were found. Thus to the south of Egypt (said the Egyptians of Herodotus's time) were tall, handsome, long-living Ethiopians who had an altogether special exhibit – the table of the sun. They washed in a wonderful rejuvenating fountain, took only meat and milk as food, enclosed their dead in crystal pillars, and possessed vast quantities of gold. And in other parts vague glimpses and distorted reports from the natives gave shape to monstrous men, some being of the Shady-feet variety; others blessed with dog's faces; or with one eye in the breast in place of the more usual head; others having no tongue, or no nostrils; others again being furnished with legs bent too much to be used for walking.[26] For the rest, such wonder-tales as were derived from Africa mostly related to horned donkeys and dangerous beasts, especially serpents, with which the steppe-country between the seaboard and the desert was in actual fact infested. A repertory of these reptiles will be found in Lucan's gruesome phantasy of Roman soldiers rapidly decomposing at the touch of the venom.[27] The oddest of the imaginary fauna of Africa was the basilisk, an ambulating and half-upright monster whose nearest living relative may be found among the spitting snakes of that continent. Eye-witnesses declared that the basilisk breathed a poison-gas which was fatal to all living creatures but that he dissolved like a ghost at the sound of a cock-crow.[28] Of the python species, which belong to central rather than to northern Africa, fewer yarns were told. But one Roman historian related that the army of Regulus in the First Punic War was held up in Tunisia by a huge serpent which had to be battered to death by an artillery bombardment; it was subsequently found to measure 120 feet.[29]

(d) In the Erythraean Sea

From Alexander's time onwards a story grew up that there was in the 'Erythraean Sea' an island of the blest which seems to have arisen out of Arabian reports about Dvipa Sukhadhara (island abode of bliss, the modern Socotra) and native Egyptian tradition about Pa-anch. Here was placed a real Utopia, descriptions of which were hung around the peg of a fictitious exploration. Euhemerus (c. 300) told how he visited Panchaea, but there can be no doubt that the journey was a tale. The same is to be assumed in the story of Iambulus, told by Diodorus[30] in connexion with an island found in the Ocean towards the south. Iambulus tells the story of himself. He was a merchant, like his father, and went up through Arabia to the spice-country, but was captured by pirates. He was taken to the sea-coast of Ethiopia; placed after a while in a small boat with one passenger and food for six months, he was ordered under threats of punishment to sail southward, where he would come to a happy island peopled by good men. Iambulus and his companion, after a storm-tossed voyage of four months, reached the island, 5,000 stadia round, where they were well received by the inhabitants, whose physical make was very strange but attractive. Pleasant too was their climate, not too hot or too cold, but always autumn, with day equal to night and the sun overhead. They were learned, especially in astrology, and had their own alphabet out of which they wrote words in upright columns. In their happy Sun State they lived long in good health, to 150 years, drugging the blemished or aged to death. The oldest man in each of their 'systems' was chief, and in all the people lived and worked equally, and all knew equally; the state was a communistic state; even women and children they had in common. They had curiously formed small animals and birds, which they used in unbelievable ways. The island produced all the food they required, in particular a reed bearing eatable fruit, and warm and cold health-giving waters. The surrounding sea with great tides tasted sweet and contained seven other islands with similar inhabitants, who worshipped the heavenly bodies, caught fish, and hunted birds. Olives, wines, and snakes also provided food. A wonderful mixture of down and oysters provided their purple clothes. Different foods were allotted to different days; in turn they hunted and served and did hand-work, and so on. The dead they buried in the sand at low-tide. There is no doubt that Iambulus, sick of class-war, was dreaming of a classless state without wealth, ambition, learning, or any other thing which makes men unequal. Seven years was he among these and then he was expelled as a blackguard. Four months' drifting brought him to India, where his companion was drowned; but Iambulus was led by natives to Palibothra, where the king was a cultured Greek-lover. He later returned to Greece through Persis.

Elsewhere[31] Diodorus speaks of Panchaea which traded in incense and myrrh with Arabia. It had several cities, the greatest being Panara, prosperous with a great temple. This was in lovely surroundings of trees and sweet water and pretty song-birds and gardens and meadows. There was beauty and plenty everywhere to the glory of the god. The whole sacred spot is described in detail. There were elephants, lions, leopards, and other animals of curious look. The men were warlike and used chariots; there were three classes – priests, and artisans; husbandmen; soldiers and shepherds; of whom the luxurious priests held the power. At the head of all was a government of three, annually chosen. The only private possessions were generally the house and garden. They dressed in sheep-wool, and both sexes wore gold decorations. The land was rich in all precious metals, none of which left the island. All this is another less marvellously impossible conception of the same Utopia dating before the exploration of the island of Socotra and neighbouring waters.

(e) In the West

Ever since ancient peoples learnt to dissociate the dead from their graves, the region of the setting sun occupied their fancy as being the natural goal of man's last journey. It was in the Far West that Gilgamesh found the entrance to the underworld, and here too, 'on the borders of the Ocean', that Odysseus passed out into Hades.[32] In one of the latest works of Latin literature, the poems of Claudian, the home of the dead was located at the western extremity of Gaul, whence the sound of their wailings was borne across to Britain.[33] According to the Byzantine historian Procopius the Scandinavian tribe of the Heruli imagined that the approach to the nether world lay through 'Brittia', which for this reason rather than from actual experience they deemed to be a land of fog and gloom.[34]

In the earlier stages of ancient eschatological speculation the abodes set apart for the Chosen Few among the dead were not far distant from the common limbo. The 'Elysian field' of Homer stretched to Ocean and was swept by its balmy zephyrs.[35] Similarly the Isles of the Blest in Hesiod and Pindar lay on the Ocean border.[36] No doubt it was by association with this Land of the Departed that the Greeks also transported the 'Garden of the Hesperides', where a dragon guarded the golden apples, to the Far West.[37] The discovery of Madeira would naturally confirm the belief that the Earthly Paradise lay out in the Atlantic; but in view of the rapid oblivion which overtook the group of isles we need not be surprised that by the time of Plutarch (c. A.D. 100) the 'isle of Cronus', as it came to be called, was transferred to a point five days west from Britain.[38] It may be something more than a coincidence that in the Middle Ages the Isles of St Brandan were usually located either to the west of Ireland, or near the Canary-Madeira archipelago.[39]

The Far West was also selected by Plato as the seat of the most famous of ancient Utopias, the island of Atlantis. According to Plato the western Ocean had at a remote period, more than 9,000 years before his own time, contained an archipelago of islands, of which one, situated 'at the mouth of the Atlantic', was called Atlantis. Besides being larger than Africa and Asia combined, it was a land of Heart's Desire, with a profusion of useful plants and animals and metals. Not content with this natural wealth, its inhabitants were accomplished traders and navigators, and furnished their city, which lay at a distance of five miles from the sea, with an elaborate artificial harbour. Furthermore, they were excellent architects and artists, who forestalled Giotto's campanile by building in variegated patterns of white, black, and red, and erected a grand and costly temple to Poseidon. Unfortunately the Atlantians, having conquered 'Libya as far as Egypt and Europe as far as Tyrrenia', attempted to capture Athens in their stride, but suffered a disastrous defeat. Worse still, Poseidon repaid all the art and wealth which had been lavished upon his temple by sending earthquakes and floods which engulfed Atlantis and all the other islands.[40]

In penning this *jeu d'esprit*, Plato lighted a false flare that has wrecked whole armadas. And still the quest for Atlantis goes on.[41] Most of the identifications which have been put forward are plainly contrary to Plato's description; indeed, the only one worth consideration supposes Atlantis to refer to the north Atlantic continent which once joined Ireland to America. But, in view of the fact that this continent disappeared many times 9,000 years before Plato's time, it is inconceivable that Plato could have laid hold of any record of this infinitely remote event, or could have reconstructed it by the methods of modern geology. The utmost that can be done to salvage Atlantis is to recognize in the landscapes known to fourth-century Greeks various elements which Plato might have transferred to the western Ocean and there pieced together into a composite continent.[42] The fragment of Atlantis which modern scholars have most commonly set out to recover is the artificial harbour.[43] A promising clue seems to be provided by Plato when he remarks that the temple of Poseidon in the chief city of Atlantis contained two fountains, one with hot water, the other with cold. Two such sources have been traced in the temple of Poseidon at Gades, which incidentally lies on an island at the mouth of the Atlantic, albeit a tiny one.[44] But we cannot go farther than say that Plato *may* have based a portion of his myth upon real landscapes like that of Gades. Atlantis as a whole is irrecoverable, and Jules Verne wrote better than he knew when he described its exploration by one Captain *Nemo*.

3. HUMOROUS REACTION

The exploration of the old world by the Greeks and Romans helped to produce a suitable background or scene for the early writers of prose fiction of the romantic type. The scene of these was normally on known ground: thus Iamblichus of Syria (second century A.D.) in his *Babyloniaca* laid the scene of his loves of Sinonis and Rhodanes in known Asia with Babylon as a centre: Philostratus in his account of Apollonius of Tyana takes his hero over Alexander's land-route to India and then back by Nearchus's coasting route. Heliodorus a little later, in his attractive romance called *Aethiopica*, concerns himself at start and finish with known Ethiopia and Meroe. Diogenes touches another type of story; he chose as a title for a tale *On the Unbelievable Things beyond Thule*,[45] wherein we can perhaps discern a tendency to laugh at the travellers' tales and gross exaggerations of truth in them during the past, and in fact witty Greek writers did succeed in making people laugh over these. For the fictitious explorations and Utopias which we have been describing led to a playful reaction, and we may conclude by looking at these efforts of Greek humorists.

It is probable that Aristophanes meant to picture a new Athens cleansed of wickedness when he produced in 414 B.C. the 'Birds', a play in which two Athenian citizens persuade the birds to build 'Cloud-Cuckoo-Town' in mid-air, so as to cut off the gods from men. Again, in another direction (but whether in humour is not certain) Dionysodorus of Melos was said to have left in his tomb a letter saying that after death he had penetrated to the centre of the earth and that the distance thither was 42,000 stadia. Pliny did not believe[46] in this masterly achievement of 'bematism'; nor do we. It is the result ('cooked' a little perhaps, for appearance's sake) of calculation of the length of the earth's radius from roughly correct estimates of the circumference.

But for a complete example of literary play we must turn to Lucian of the second century A.D. The Greeks did not create Utopias for purposes of satire, as Swift and Montesquieu did, but Lucian's *True*

IMAGINARY DISCOVERIES

History did something to inspire not only Swift but also Rabelais and Cyrano de Bergerac. The *True History* aims at poking fun at the story-tellers of the past, particularly Iambulus, Ctesias, and Herodotus (!), whom Lucian saw in his mind's eye more terribly punished in Hades than any other sinners. He set out with fifty companions in a ship which, leaving the Pillars of Heracles, sailed the Atlantic for eighty days and found an enchanted island, where an inscription in Greek marked the limits reached by Heracles and Dionysus. Soon a whirlwind carried the ship to the skies, where the voyagers were seized by huge 'horse-vultures' and taken to the man in the moon, who turned out to be Endymion at war with the people of the sun about the colonization of the planet Venus. They joined the moonites, of whom Lucian gives amusing accounts. After further voyaging through the Zodiac the ship came down safely to the sea again, and after long and laughable adventures in the belly of a whale or sea-serpent one hundred miles long, the voyagers killed it by lighting a forest-fire inside, and came to the Fortunate Islands, where they found great men of old times. The rest of their adventures included a visit to the regions below. Like Herodotus, Lucian will tell the rest 'in the next books'.

NOTES

CHAPTER ONE: INTRODUCTORY

1. Caesar, *Bell. Gall.*, v, 13; Cassius Dio, LXXVI, 13.

2. The eventual triumph of mercenary motives to the exclusion of all others is specially evident in the exploration of Asia and the adjacent seas (Chaps. 4 and 7).

3. Among the ancient Eldorados was numbered the island of Britain, which raised false hopes among Roman fortune-hunters in the days of Julius Caesar, and even of Agricola.

4. On ancient marching records see esp. W. Riepl, *Das Nachrichtenwesen der Alten*, p. 129 ff.

5. Witness the achievement of Shackleton's and of Scott's ponies on their expeditions to the South Pole.

6. It is not certain whether the foresail was known before the fourth century B.C. For other details see C. Torr, *Ancient Ships*; A. Köster, *Das Antike Seewesen*.

7. Columbus's flagship measured 100 tons, Vasco da Gama's 200, Magellan's 110. Drake's *Golden Hind* was of 150 tons, Capt. Cook's *Endeavour* of 360.

8. Lateen sails, though now universal in the Mediterranean, are not known to have been in use there before *c.* A.D. 900 (Laird Clowes, *Geogr. Journal*, March 1927, pp. 216–34).

9. Acts, Chap. 27. The angle to which ancient ships could sail into the wind is a matter of dispute.

10. Measuring ropes, water-levels, and sighting instruments were used by Greek and Roman engineers and surveyors, and a Greek man of science (Ctesibius or Hero) invented a 'hodometer' or taximeter (A. Laussedat, *Recherches sur les instruments topographiques*, Vol. 1). But none of these formed part of explorers' equipment.

11. Confusion has been caused among geographers, both ancient and modern, by the concurrent use of different ancient units of measurement by land and sea. Among the Greeks the norms were the stadium of 600 feet on land, and a distance of 1,000 or 10,000 fathoms on sea. But each of these measures varied according to the diverse lengths of the unit foot or fathom. Eventually a stadium equivalent to one-eighth of a Roman mile became standard for land measurements; and there is reason to believe that the present-day nautical mile (= 10 of the standardized stadia) came into common use in the late Greek or the Roman era. But it is doubtful whether geographers like Pliny or Ptolemy were able to reduce the various stadia of their sources to a uniform scale. – K. Miller, *Die Erdmessung im Altertum*; O. Cuntz, *Die*

NOTES

Geographie des Ptolemaeos, p. 110 ff. See Jüthner, *Real-Incyclopäde der klassischen Altertumswissenschaft*, III A 2, cols. 1931–63.

12. Known as γνώμονες or πόλοι. The πόλος probably differed from the γνώμων in having a movable pointer like the astrolabe. (G. Callegari, *Rivista di Storia Antica*, Vol. IX, pp. 636–8.)

13. Strabo, I, p. 7.

14. See the report of Mr Rodd's travels in the western Sahara. (*The Times*, 20 March 1928.)

15. The floating compass was known in China by the second century A.D., and in Italy (probably as the result of independent discovery) by A.D. 1100. The fixed naval compass was used by Italian mariners from c. A.D. 1250. (Bertelli, *Rivista Geografica Italiana*, Vol. IX, p. 281 ff.; Vol. X, p. 1 ff., 105 ff. C. Errera, *L'Epoca delle grandi scoperte geografiche*, p. 187 n.)

16. Plato, *Ion*, Chap. VII, pp. 535–6; Lucretius, VI, 906–1064.

17. Until the time of Pytheas (p. 49) the Greeks steered by the Great Bear. (*Od.*, V, 274; D'Arcy Thompson, *Class. Assoc. Proc.*, April 1929, 28–9.)

18. For the explanation of Aristotle's wind rose, whose points lay at uniform distances of 30 degrees, see D'Arcy Thompson, *Classical Review*, 1918, pp. 49–56.

19. VII, 176, 201.

20. IV, 177.

21. Priscus, p. 172, ed. Bekker-Niebuhr.

22. W. Tarn, *Cambridge Ancient History*, Vol. VI, p. 13. For recent instances of travel in a circle, see Mrs Forbes *The Secret of the Sahara*, pp. 48–9; G. Haardt and L. Audouin-Dubreuil, *Across the Desert*, pp. 105–6.

23. For such ὑδροθῆκαι, cf. Athenaeus, V, 42, 208a. Cf. also Plutarch, *Quaestiones Graecae*, No. 54 (a merchant makes a huge profit selling water to other becalmed ships).

In the Civil War between Caesar and Pompey, the Pompeian admiral Bibulus died of the hardships suffered by him during a prolonged blockade of the Albanian coast (Caesar, *De Bello Civili*, III, 18, 1).

24. The book of sailing-directions known as the *Stadiasmus Maris Magni* (*Geographici Graeci Minores*, Vol. I, p. 427 ff.) anticipates our Admiralty Pilots in noting the water-points along the coasts described.

25. A. Schmidt, *Drogen und Drogenhandel im Altertum*.

26. No statistics on this point are available. But it is noteworthy that the African explorer Eudoxus, despite the great care with which he had equipped his expedition, was unable to keep the high seas for long because of his crews' discomfort (p. 125). If any ancient journey of exploration had extended over three years, like the cruises of Magellan and Anson, we may be sure that its roll of casualties would likewise have amounted to full eighty per cent of the ships' complements.

NOTES

CHAPTER TWO: THE MEDITERRANEAN AND THE BLACK SEA

1. *The Mediterranean Pilot*, esp. Vol. I, pp. 12–15; Vol. IV, pp. 11–17.
2. Among Mediterranean seamen the use of the compass and other scientific instruments did not become general until the nineteenth century: empiric knowledge of well defined trade-routes was an adequate substitute. – V. Bérard, *Les Phéniciens et l'Odyssée* (2nd ed.), Vol. I, pp. 279–80.

It is no mere accident that another great school of early navigation was formed in the Malay Archipelago.

3. Elliot Smith, *Ships as Evidence of Ancient Culture*; Keble Chatterton, *Sailing Ships* (2nd ed.), pp. 21–2.
4. J. R. Breasted, *Ancient Records*, Vol. I, Nos. 146–7. For evidence of proto-dynastic intercourse with Phoenicia see H. Frankfort, *Journal of Egyptian Archaeology*, 1926, p. 83.
5. Sir Arthur Evans, *The Palace of Minos*, Vol. I, pp. 55, 65–6.
6. Breasted, *History of Egypt* (2nd ed.), p. 261.
7. So Schröder, *Keilschrifttexte aus Assur*, No. 92, l. 41 (quoted by Schulten and Bosch, *Avieni Ora Maritima*, p. 126).
8. E. Assmann, in *Festschrift Lehmann-Haupt*, pp. 1–7.
9. S. H. Langdon, *Camb. Anc. Hist.*, Vol. I, p. 405; ed. Meyer, *Geschichte des Altertums* (3rd ed.), Vol. I, pt 2, opp. 519–20.
10. H. R. Hall, *Camb. Anc. Hist.*, Vol. I, p. 587.
11. Evans, op. cit., Vol. II, pp. 229–52; A. Köster, *Schiffahrt und Handelsfahrt des östlichen Mittelmeeres*, p. 28. Evans, *Scripta Minoa*, Vol. I, pp. 203–4, shows a two-master on an early Cretan gem.
12. Evans, *Palace of Minos*, Vol. I, pp. 286–300, Vol. II, pp. 176–80; G. Glotz, *Aegean Civilization*, pp. 185–226.
13. See the illustrations in *Essays in Aegean Archaeology*, presented to Sir Arthur Evans, pp. 31–41.
14. G. Jondet, in *Mémoires présentés à l'Institut Égyptien*, Vol. IX, 1916; R. Weill, *Bulletin de l'Institut français d'archéologie orientale* (Cairo), 1919, pp. 1–37; Evans, Vol. I, pp. 292–7.
15. The failure of Strabo to mention these harbour-works suggests that they soon went out of use.
16. Breasted, *Ancient Records*, Vol. II, No. 492.
17. So Hall, *Essays in Aegean Archaeology*, pp. 31–41; Breasted, *History of Egypt*, p. 261; Glotz, op. cit., pp. 62–3; ed. Meyer, Vol. II, pt 1, pp. 107–8. But see J. L. Myres, *Camb. Anc. Hist.*, Vol. III, p. 635.
18. Evans, Vol. II, p. 60 ff.
19. Myres, loc. cit.; Glotz, pp. 39, 213.
20. Minoan influence is affirmed by Evans, Vol. II, pp. 180–90; Glotz, pp. 220–6; A. Schulten, *Tartessos*, p. 9 ff.; V. G. Childe, *The Dawn of European Civilization*, chaps. VI–VIII; R. Hennig, *Historische Zeitschrift*, Vol. CXXXIX,

NOTES

pp. 1-9. Doubts are raised by E. T. Leeds, *Camb. Anc. Hist.*, Vol. II, p. 590.

21. Evans, Vol. I, p. 87.

22. T. E. Peet, *Camb. Anc. Hist.*, Vol. II, p. 567; J. Déchelette, *Manuel d'Archéologie préhistorique*, Vol. II, p. 398.

23. Myres, *Camb. Anc. Hist.*, Vol. I, pp. 105-6; Childe, op. cit., pp. 33, 84-5, 125.

24. Glotz, p. 190.

25. Peet, *The Stone and Bronze Ages in Sicily and Italy*, pp. 490-1, 511-15.

26. P. Paris, *Essai sur l'art et l'industrie de l'Espagne primitive*, Vol. II, pp. 46-100.

27. There is some doubt as to the provenance of the Minoan pottery at Torcello near Venice (published by R. M. Dawkins, *Journal of Hellenic Studies*, 1904, pp. 125-8). On the amber trade, see Childe, p. 83.

28. It is tempting to suppose with Schulten (loc. cit.) that the Cretans fetched Atlantic tin from Tartessus. But an alternative prehistoric source of tin may be found in Bohemia. (Childe, pp. 33, 57.)

For the possible location of Tartessus on the site of Seville, see A. Berthelot, *Festus Rufus Avienus, Ora Maritima*, p. 83.

29. Myres, *Camb. Anc. Hist.*, Vol. III, pp. 635, 670; ed. Meyer, Vol. II, pt 1, pp. 546-59.

30. So Glotz, pp. 211-12.

31. Meyer, loc. cit. For a good characterization of the Achaeans, see W. Leaf, *Homer and History*, chap. II.

32. For a good summary of the points at issue, cf. How and Leigh, *Commentary of Herodotus*, Vol. I, pp. 347-50.

33. 1 Kings 10: 22 (Hiram of Tyre); Isa. 2: 16; 60: 9; 66: 19; Jer. 10: 9; Ezek. 27: 12; 38: 13.

34. III, 150.

35. First ed., in two vols., 1902; second ed., in six vols., 1927 ff.

36. See esp. K. J. Beloch, *Rheinisches Museum*, 1894, pp. 111-32.

37. A. W. Gomme, *Annual of the British School at Athens*, Vol. XVIII, pp. 189-210.

38. Gomme, *Journ. Hell. Stud.*, 1913, pp. 53-72; ed. Meyer, Vol. II, pt 1, p. 254 n. 3.

39. Breasted, *History of Egypt*, p. 307.

40. ibid., p. 514 (the story of Wen-Amon); H. R. Hall, *Ancient History of the Near East* (7th ed.), p. 321 n. 1.

41. So Leaf, op. cit., pp. 61-2.

42. See A. Blakeway, *Annual of the British School at Athens*, 1932-3, pp. 170-208.

43. Velleius Paterculus I, 2, 4; Pliny XVI, 216.

44. 1 Kings 10: 22.

45. Blakeway, op. cit., p. 180; T. J. Dunbabin, *The Western Greeks*, pp. 20-2.

46. The estimates of Greek and Roman authors for the foundation date of

NOTES

Carthage vary between *c*. 850 and 750 B.C., but converge on 814 B.C. The remains of the city, so far as unearthed, date back at least as far as 650 B.C. (S. Gsell, *Histoire ancienne de l'Afrique du Nord*, I, p. 401).

47. Kahrstedt, *Klio*, 1912, pp. 461-73 (from *c*. 800 to 650 B.C.).

48. On the Phoenicians in Spain, see P. Bosch-Gimpera, *Klio*, 1928, pp. 345 ff. (This scholar dates the Phoenician colonies after 800 B.C.)

49. Bérard (*Les Phéniciens*, 2nd ed., I, p. 53) aptly quotes the case of the modern Albanian immigrants into Greece, whom sheer necessity has turned into good seamen.

On early Greek acquaintance with the western lands, and with Italy in particular, see E. Wiken, *Die Kunde der Hellenen von dem Lande und den Völkern der Apenninhalbinsel bis 300 v. Chr.*, esp. pp. 20-41.

50. See esp. Bérard, 1st ed., I, p. 491 ff., and Vol. II.

51. Similarly the author of 'Sindbad' has some correct information about the giant birds of Madagascar and the precious stones of India or Ceylon, but gives no idea of the relative position of these lands.

52. This point is well emphasized by Sir Edward Bunbury (*History of Ancient Geography*, Vol. I, pp. 64-6), and by H. Berger (*Mythische Kosmographie der Griechen*, pp. 25-33).

53. Polybius, XII, 2.

54. This is the most felicitous of Bérard's identifications. (1st ed., Vol. II, pp. 187-8.)

55. The location of the Phaeacian land in north-west Sicily may be considered the most serious contribution to Homeric studies in Samuel Butler's adventurous *Authoress of the Odyssey*.

56. Bérard, pp. 567-8.

57. Hogarth, *Ionia and the East*, p. 101.

58. On Greek colonization in general, see A. Gwynn, *Journ. Hell. Stud.*, 1918, pp. 88-123; Myres, *Camb. Anc. Hist.*, Vol. III, chap. XXV.

59. A. S. Pease, *Classical Philology*, 1917, pp. 1-20.

60. H. R. Hall, *Camb. Anc. Hist.*, Vol. III, pp. 291-2.

61. P. N. Ure, *The Origin of Tyranny*, pp. 103-24.

62. H. Weld-Blundell, *Annual Brit. School, Athens*, Vol. II, p. 115.

63. IV, 150-9. For a similar story of the manner in which the Phoenicians, under the guidance of their god Melkarth, groped their way to Gades, see Strabo, I, 169-70.

64. VI, 2.

65. See Blakeway, op. cit., pp. 192 ff.; *Journal of Roman Studies*, 1935, pp. 129-42.

66. See Dunbabin, op. cit., Appendix I.

67. Herodotus IV, 152. Summer 'Levanters' are not common in the eastern Mediterranean, but often blow in the western half. Later instances of involuntary exploration under the stress of devious winds include the discovery of Labrador by Leif Ericsson (*c*. A.D. 1000), of Brazil by Pedro Cabral (1500), and of West Australia by Dirk Hartogszoon (1616). Similarly after passing

NOTES

through the Straits of Magellan Francis Drake was blown to south of the Tierra del Fuego and within sight of the archipelago of C. Horn (1578).

68. With Colaeus we may compare the Portuguese merchant who brought back gold to the value of 100,000 gulden from a first journey to Japan. (E. Kämpffer, *Beschreibung der japonischen Inseln*, chap. VII.)

69. *The Scholar Gypsy*, last two stanzas.

70. R. Carpenter, *The Greeks in Spain*, pp. 19 ff., 117 ff.

71. Pseudo-Aristotle, *De Mirabilibus Auscultationibus*, chap. 100.

72. Schulten, *Tartessos*, p. 28.

73. E. Pais, *Ancient Italy* (trans. C. D. Curtis), pp. 374–8. The use of the letter-form sigma (in lieu of san) in this inscription points to an Ionian writer.

74. Clerc, pp. 90–1.

75. Strabo correctly points out that Maenace was distinct from Malaga (III, 156). Cf. Schulten and Bosch, *Avieni Ora Maritima*, pp. 426–7.

On the question of a pre-Phoenician Greek colony on the site of Abdera, see Th. Reinach, *Revue des Études grecques*, 1898, p. 54.

76. Clerc, pp. 115–31.

77. Justin, XLIII, chap. III, sects. 5–12 (probably from Timaeus). On this story see Clerc, *Revue des Études anciennes*, 1905, p. 329 ff.

78. Schulten, p. 26; Carpenter, op. cit., p. 47 ff.; Clerc, p. 249 ff.

79. Hecataeus (F. Jacoby, *Die Fragmente der griechischen Historiker*, Vol. I); fr. 339: Εὐδείπνη; fr. 342: Φοινικοῦσσαι; fr. 343: Κύβος πόλις Ἰώνων; fr. 344: Μεταγώνιον; fr. 349: Κρομμύων; fr. 335: Φασηλοῦσσαι. 'Scylax', *Periplus* (C. Müller, *Geographici Graeci Minores*, Vol. I), 111: Ψέγας, νῆσοι Ναξικαὶ πολλαί, Πιθηκοῦσσαι, Εὔβοια, Ψάμαθος.

80. Herodotus, I, 165–6. The red-figure vases found in Corsica (Clerc, *Revue des Études anciennes*, 1905, p. 335) may have been imported by the Etruscans, who took over the island.

81. See R. S. Conway and S. Casson, *Camb. Anc. Hist.*, Vol. IV, chap. XII; M. Pallottino, *The Etruscans*.

82. Kahrstedt, *Klio*, 1912, p. 461 ff.

83. On the early history of the Adriatic, see R. L. Beaumont, *Journal of Hellenic Studies*, 1936, pp. 159 ff.

84. Herodotus I, 163.

85. It is not certain which author first suggested this identification. The claims of Pherecydes are disputed by Jacoby (op. cit.), commentary on fr. 74.

86. Frs. 90, 92–105, (ed. Jacoby).

87. Writing *c.* 350 B.C., Scylax showed a good knowledge of the Adriatic (chaps. XV–XXVI).

88. *The Black Sea Pilot*, pp. 11–22.

89. Miss J. R. Bacon, *The Voyage of the Argonauts*; J. Friedländer, *Rheinisches Museum*, 1914, pp. 299–317; Jessen, in Pauly-Wissowa, *Real-Encyclopädie der klassischen Altertumswissenschaft*, s.v. Argonautai.

90. *Iliad*, VII, 467; *Odyssey*, XII, 69–72. In this version Jason perhaps discovered the sea of Marmora. (Bury, *Camb. Anc. Hist.*, II, p. 475.)

NOTES

91. W. Leaf, *Troy*, chap. VI; F. Sartiaux, *La Guerre de Troie*, pp. 192–6.

92. T. W. Allen, *The Homeric Catalogue of Ships*, pp. 175–7; J. B. Bury, *Camb. Anc. Hist.*, II, p. 493.

93. Wilamowitz-Möllendorff, *Homerische Untersuchungen*, pt I, chap. VIII; J. B. Bury, *History of Greece*, pp. 89–90; and many others.

94. W. Tomaschek, *Sitzungsberichte der Wiener Akademie*, Philosophisch-historische Klasse, Vol. CXVI, p. 729.

95. Leaf, op. cit., p. 295.

96. Bacon, op. cit., p. 166. The 'anchor of Jason', which was shown to Arrian at Phasis, but did not deceive him (*Periplus Ponti Euxini*, chap. XI), is on a par with the 'anchor of Columbus' displayed at the Wembley exhibition in 1924.

97. W. Leonhard, *Hettiter und Amazonen*, pp. 203, 230.

98. For Egypt's exports in gold *c.* 1400 B.C., cf. the Tell-el-Amarna letters, Nos. 2–35.

99. ibid., No. 21.

100. See Pl. 10 in Th. Makridy Bey, *Mitteilungen der vorderasiatischen Gesellschaft*, Vol. XII, pt 4.

101. M. Rostovtzeff, *Iranians and Greeks in S. Russia*, chaps. II and III; A. J. Wace, *Camb. Anc. Hist.*, Vol. II, pp. 471–2.

102. e.g. at Callatis and Tanais. (K. Neumann, *Die Hellenen im Skythenlande*, Vol. I, p. 341.)

103. e.g. Salmydessus, Odessus, Ordessus. Rostovtzeff (op. cit., p. 61) suggests joint expeditions of Achaeans and Carians *c.* 1000 B.C. in quest of iron and gold. But most of the Carian settlements were on the west coast, which was not metalliferous.

104. Herodotus, II, 152 (*c.* 660 B.C.). On the alleged thalassocracy of the Carians in the seventh century, see A. R. Burn (*Journ. Hell. Stud.*, 1927, p. 165 ff.), who argues forcibly that the original compiler of the list of thalassocrats wrote, not Κᾶρες, but Μεγαρε(ῖ)ς.

105. E. v. Stern, *Klio*, 1909, pp. 142–4; V. Parvan, *Dacia*, pp. 74–5. (Excavations at Berezan, Apollonia, and Istria).

106. Neumann, op. cit., pp. 346–7. On Sinope, see Leaf, *Journ. Hell. Stud.*, 1916, pp. 1–15.

107. *Theogony*, 339.

108. e.g. at the sites mentioned in note 105. Also at Byzantium (Tacitus, *Annals*, XII, 63, 2), and at Cyzicus (tunny on coin types). Cf. Rostovtzeff, p. 43.

109. Strabo, XI, 499. The Black Sea episodes were probably introduced into the legend by Arctinus of Miletus, after the Milesian explorations. (Friedländer, op. cit.)

110. Rostovtzeff, p. 63.

111. The belief that it was the pirates of the Black Sea that caused the Greeks to dub it Πόντος Ἄξεινος ('The Inhospitable Sea') is probably mistaken. It seems more likely that Ἄξεινος is a corruption of an Iranian word 'achshaenas', which means 'dark', or 'northern'. Some Greeks translated this

NOTES

word instead of transliterating it, whence the name Μέλας Πόντος. (W. Vosmer, quoted by H. Hirt, in *Geographische Zeitschrift*, 1926, pp. 430–1.) The alternative name Πόντος Εὔξεινος ('Hospitable Sea'), which became the most usual, was perhaps invented *ad captandum*, like the 'Greenland' of the Medieval Norsemen.

112. Myres, *Camb. Anc. Hist.*, Vol. III, p. 659.
113. Burn, loc. cit.; Hogarth, *Camb. Anc. Hist.*, Vol. II, p. 561.

CHAPTER THREE: THE ATLANTIC

1. *The Admiralty Pilot for the West Coasts of France, Spain, and Portugal*, pp. 15–17.
2. Avienus, l. 98–107. The coracle discovered at Brigg in Lincolnshire was 50 feet in length.
3. Witness the rock-carvings at Bohuslan in southern Sweden. (Nansen, *In Northern Mists*, pp. 236–8.) For other details about north European ships, see Keble Chatterton, *Sailing Ships* (2nd ed.), chap. IV; W. Vogel, *Geschichte der deutschen Seeschiffahrt*, Vol. I, bk 1.
4. *Bellum Gallicum*, III, 8 and 13.
5. R. A. S. Macalister, *The Archaeology of Ireland*, p. 92. The supposed exports of tin from Ireland are a myth (ibid., p. 55).
6. V. G. Childe, *The Dawn of European Civilization*, pp. 118–19, 136–7.
7. Childe, loc. cit.; A. Schulten, *Tartessos*, pp. 9–15; H. Schuchhardt, *Alteuropa*, pp. 57–8; H. Peake, *The Bronze Age and the Celtic World*, pp. 41–5. A. Lucas (*Journal of Egyptian Archaeology*, 1928, pp. 97–105) contends that Atlantic tin was not used before 1000 B.C. But this conclusion is based upon a somewhat precarious *a priori* argument.
8. *Odyssey* X, 81–6. The context in which the Laestrygones here appear shows that they are to be sought in the western Mediterranean.
9. On the Cimmerians, see *Od.* XI, 13–19; J. B. Bury, *Klio*, 1906, pp. 79–88. We have seen on p. 38 that Homer's Cimmerians cannot be located in the Black Sea. Neither can they be placed near the Straits of Gibraltar, for this is by no means a region of fog and gloom.
10. The Eridanus is first mentioned in Hesiod (*Theogony*, l. 338) without being located. Herodotus (III, 115), while disbelieving in its existence, referred to it as a stream in north-west Europe, from which came the amber. R. Hennig (*Historische Zeitschrift*, Vol. CXXXIX, pp. 17–19) points out that the singing swans mentioned by Hesiod (Shield of Heracles, ll. 315–17) are only found in northerly latitudes.
11. See the edition by A. Schulten and P. Bosch-Gimpera.
12. Avienus, 95–115. Oestrymnis is generally admitted to be Brittany. The above passage tells against those who assign the prehistoric tin-trade to the Phoenicians. (So L. Siret, *L'Anthropologie*, 1908–10.)

NOTES

13. Sacra Insula = Ἱερὴ Νῆσος (in Ionian Greek) = Ἰέρνη.

14. 'Albion' was a pre-Celtic name which was driven out of use by the Celtic 'Britannia'. The present passage shows that c. 500 B.C. the older name was still prevalent.

15. Herodotus, I, 163.

16. Pliny, VII, 197. This passage is probably derived from Hellanicus, a good authority of the fifth century B.C.

17. Schulten, *Tartessos*, pp. 25-6.

18. M. Cary, *Journal of Hellenic Studies*, 1924, pp. 169-70.

19. Schulten and Bosch, on Avienus, ll. 160 ff. cf. the name Κοτινοῦσσα for the island of Gades (Pliny, IV, 120).

20. Greek writers of the fifth century spoke of the Pillars of Heracles as the natural and divinely appointed terminus for sea-travel. Pindar, *Olympia*, 3, ll. 43-4; *Nemea*, 3, ll. 20-1; 4, l. 69. Euripides, *Hippolytus*, 743-7. Herodotus (III, 115) professes complete agnosticism on the subject of the Tin Islands.

21. Schulten (*Tartessos*, p. 17) suggests that Tartessus was under Phoenician rule from 1100 to 700 B.C. But the allusions to Tartessus in the O.T. are far from proving such a domination; and the passage from Strabo which he quotes (III, 149) may be nothing more than an anticipation of the later Carthaginian supremacy.

22. It is last mentioned in Hecataeus, fr. 38, as a town in being (c. 500 B.C.). Later writers often confused it with Gades. Schulten (op. cit.) has clearly distinguished between the two cities.

23. II, 169.

24. ll. 114-34, 406-15.

25. The tin of Brittany was probably exhausted in historical times. (Siret, *L'Anthropologie*, 1908, pp. 139-44.)

26. C. Jullian (*Histoire de la Gaule*, Vol. I, p. 388) suggests that Himilco reached the amber coast of Jutland.

27. So Martin Beheim in Columbus's day, followed by Jullian, op cit., Vol. I, pp. 385-8; H. F. Tozer, *History of Ancient Geography*, pp. 109-12; Clerc, *Massalia*, p. 406.

28. In the 'forties' heavy weather nearly always comes from the west. But occasional summer gales blow from the east (*Bay of Biscay Pilot*, pp. 2-3).

29. The Riff pirates knew this to their advantage.

30. *Pilot*, pp. 183-93 (shoals between Gibraltar and C. St Vincent). A. Blazquez (*El Periplo de Himilco*, p. 62) attests the presence of sandbanks and weedbeds on Barreta Isle (off the south Portuguese coast).

31. *De miris auscultationibus*, 136.

32. IX, 12.

33. Pliny, II, 169.

34. Strabo, III, 175.

35. The Scillies now number about thirty, whereas the Cassiterides, according to Strabo (loc. cit.) were only ten. But some of the islands may have been split by erosion since Strabo's day (O. G. S. Crawford, *Antiquity*, March 1927,

NOTES

pp. 5–14). On the Tin Islands in general, see T. R. Holmes, *Ancient Britain and the Invasions of Julius Caesar*, pp. 483–98; F. J. Haverfield, in Pauly-Wissowa, s.v. Kassiterides.

36. Witness Marco Polo's 'Island of Zanzibar' (Central and South Africa); Torres' 'Australian Archipelago' (the Australian continent). Similarly North America was not certified a mainland until Vancouver's cruise up the west coast in 1791. It remained an open question whether Lake Victoria Nyanza was a single great sheet of water or a cluster of lesser lakes, until Stanley completely circumnavigated it.

Similarly Graham Land, on the Antarctic Circle, has recently been found to be a cluster of islands (Sir George Wilkins, as reported in the *Daily Mail*, 18 February 1929).

37. On Pytheas's life and circumstances, see Strabo, II, 104.

38. *Commentariorum in Aratum Reliquiae* (ed. E. Maass), pp. 70–1.

39. 43° 3', instead of 43° 17'. Pytheas omitted to add the angle subtending the semi-diameter of the sun. – See Sir Clements Markham, *Geographical Journal*, Vol. I, p. 512.

40. Strabo, II, 75. For an explanation of this confused passage, see Tozer, op. cit., p. 154 ff.; H. Berger, *Die Geographischen Fragmente des Hipparch*, p. 30.

41. Of the vast literature on Pytheas it may suffice here to mention: Holmes, op. cit., pp. 217–27; H. F. Tozer, *A History of Ancient Geography*, pp. 154–64; J. Thomson, *History of Ancient Geography*, pp. 143–51; G. E. Broche, *Pythéas le Massaliote*. (Broche's detailed estimates of Pytheas's route and daily sailing distances differ somewhat from those assumed in the present text.)

42. Strabo, I, 64. The Ὠστίμιοι here mentioned are clearly the Osismi of Caesar, *Bell. Gall.*, I, 9, 9, and Οὐξισάμη = Ushant.

43. Strabo, loc. cit.; 'Pytheas asserts that Uxisame lies three days off.' – Presumably off Spain, which is mentioned just above.

44. Strabo, III, 148, explained by Müllenhoff, Vol. I, pp. 370–1.

45. Fr. 36, in C. Müller's *Fragmenta Historicorum Graecorum*, Vol. I.

46. Diodorus, v, 21–3. The observations of Pytheas referred to in Notes 41–5 make it practically certain that he coasted along Spain and western France, and did not embark at Le Havre after an overland journey along the Rhône and the Seine, as is suggested by R. Hennig (*Geographische Zeitschrift*, 1928, pp. 95–6).

47. Diodorus, v, 22, 2, 4. This passage almost certainly comes from Timaeus. A similar account is given in Pliny, IV, 104. Ictis has been identified with Thanet, the Isle of Wight, and St Michael's Mount, by Penzance. There can be little doubt that it was the last-named (Holmes, pp. 499–514). Pliny's statement that Ictis lay 'six days inwards', i.e. toward the Mediterranean, is probably due to confusion with Thule.

48. Diodorus, v, 21, 3. The three corners are here named Belerium, Cantium, and Orca.

49. Hergt, p. 41.

50. Holmes, p. 223.

NOTES

51. Ap. Strabo, II, p. 104.
52. Diodorus, V, 21. Except for the remark that Britain was very cold (which was probably inspired by some Greek *géographe de cabinet*), this account seems based on a record of actual travel, most probably on Pytheas by way of Timaeus.
53. Pliny, II, 217.
54. *North Sea Pilot*, Vol. I, p. 19. A. Blazquez (*Pyteas de Marsella*, pp. 28–30) explains the waves as due to high winds.
55. Polybius ap. Strabo, II, 104.
56. Strabo, II, 114; IV, 201; Pliny, II, 187; IV, 104; Geminus, *Elementa Astronomiae*, 6, 9, l. 22. Pliny inadvertently speaks of a six-month day, followed by a six-month night (IV, 104).
57. Ptolemy III, 3, 14; Tozer, p. 159.
58. The case for Iceland is now best presented by Broche.
59. The advocates of Norway include Holmes, Nansen (*In Northern Mists*, pp. 56–62), and G. Hergt, *Pytheas* (pp. 51–69).
60. IV, 96.
61. Strabo, III, 175.
62. Tozer, p. 163.
63. Markham, p. 519.
64. Ap. Strabo, II, 104.
65. Hergt (p. 74). Summer fogs are frequent off the Norwegian coast (*Norway Pilot*, p. 13).
66. This fact, generally unnoticed, has been duly recognized by Broche. It has also been confirmed for me in a letter which Professor W. Brown, formerly of Hong Kong University, has kindly sent to me.
67. Strabo, I, p. 63; II, p. 104.
68. Diodorus, V, 23; Pliny, IV, 94–5. All these passages are probably derived from Timaeus.
69. Rendel Harris, *Journ. Hell. Stud.*, 1925, p. 239.
70. Tacitus, *Germania*, chap. 44, sect. 1.
71. D. Detlefsen, *Die Entdeckung des germanischen Nordens im Altertum*, pp. 6–9.
72. Ap. Strabo, II, 104. This passage proves (against Hergt, p. 29 ff.) that Pytheas's voyage along the continent was subsequent to his cruise round Britain.
73. *Bay of Biscay Pilot*, pp. 4–5.
74. *Placita Philosophorum*, 3, 17, explained by Müllenhoff, Vol. I, p. 365.
75. Strabo, II, 114–15, etc.
76. Ap. Strabo, II, 104.
77. Müllenhoff (Vol. I, p. 377) suggests that the distance from Land's End was measured to the Gironde; Hergt (p. 29) assumes that the crossing from Kent was to the Rhine estuary.
78. Diodorus, V, 21, 3. Similar figures in Strabo, I, 63 (Britain over 20,000 stades in length), and Pliny, IV, 104 (a circuit of 4,875 miles – so the best MSS.).

NOTES

79. Hergt (p. 26); V. Chapot, *Revue des Études grecques*, 1921, p. 67 ff.

80. Herodotus, IV, 66 (11,000 stades vice 6,000); Arrian, *Indica*, chap. 25 ff. 22,700 stades vice 11,000). Pliny, writing at a time when Boulogne was a frequented port, reckons its distance from Britain at 50 miles (IV, 102). Similarly (IV, 24) he gives the Sea of Azov a circuit of 1,400 Roman miles (vice *c.* 500). On the whole subject, see Müllenhoff, Vol. I, p. 381.

81. The third-century coins of Tarentum discovered near Amiens probably travelled overland, like the other Greek coins in Gaul (pp. 124–5), and not by sea, as is supposed by R. Forrer (*Keltische Numismatik der Rhein- und Donauländer*, pp. 95–100).

82. Strabo, III, 175–6. This passage is plainly derived from Poseidonius, who wrote *c.* 80 B.C. We may assume that previous to the Pax Romana the Carthaginians did not invite interlopers to commit suicide, but unceremoniously hanged or drowned them, as did the Spaniards and Portuguese.

83. J. Hatzfeld, *Les Trafiquants italiens dans l'Orient hellénique*, pp. 238–56; T. Frank, *Economic History of Rome* (2nd ed.), pp. 305–7.

84. *Bell. Gall.*, IV, 20–21. When Scipio Aemilianus (*c.* 135 B.C.) inquired about Britain from merchants of Massilia, Narbo, and Corbilo (on the Loire estuary), he could obtain no more information than Caesar (Strabo, IV, 190). Plainly the trade-routes were secret.

85. *Bell. Gall.*, V, 12–15.

86. Strabo's probable authority for his statement about Crassus, the Greek historian Posidonius, wrote long before Caesar's invasion of Gaul. He had travelled in Spain and was well informed on mining affairs.

87. Cary, *Journ. Hell. Stud.*, 1924, pp. 167–8; F. J. Haverfield and M. V. Taylor, *Victoria County History*, Cornwall, Vol. V, pp. 19–21.

88. On this whole subject, see Holmes, chaps 6 and 7, and pp. 595–735.

89. *Bell. Gall.*, IV, 29.

90. ibid., V, 12–14. Caesar's figures for the British coast-lines are 500, 700, and 800 Roman miles.

91. Strabo, IV, pp. 199–200.

92. Ptolemy I, 2, 7, quoting Philemon, a writer of the first century B.C. Pliny (IV, 102) estimates Ireland to be 800 × 100 miles. These figures are probably also from Philemon.

Ptolemy described this estimate as a good example of the scientific incuriosity of travellers bent on trade, or of their aptitude for drawing the long bow.

93. Strabo, IV, 201.

94. III, 6, 53.

95. Juvenal, II, 160–1.

96. *Agricola*, 10, 5–6; 38, 5, with notes by J. G. C. Anderson in his edition of this work (pp. 141–2).

97. Pliny, XVI, 103; Mela, III, 6, 54 (30 to 40 Orcades, 7 Acmodae or Aemodae, 30 Hebudes).

98. The chief reason for this much-discussed error no doubt lay in Ptolemy's

NOTES

lack of reference-points for the west and north coasts of Scotland. For this, see T. G. Rylands, *The Geography of Ptolemy Elucidated*, pp. 78–80.

99. R. McElderry, *Classical Review*, 1909, pp. 460–1, on the strength of Juvenal, II, 159–60.

100. Bk. II, chap 2. cf. Haverfield, *English Historical Review*, 1913, pp. 1–12.

101. Solinus, XXII, 2. It is noteworthy that Ptolemy's data on Ireland accord well with those preserved in early Ulster tradition (J. MacNeill, *New Ireland Review*, 1906).

102. Strabo, VII, 291.

103. *Monumentum Ancyranum*, chap. 26; Velleius Paterculus, II, 106–7; Pliny, II, 167. The Roman imports in north-west Germany were probably brought by land (pp. 128–9).

104. Tacitus, *Annals*, I, 70; II, 23–4. (A.D. 15–16.) The poem of Albinovanus in Seneca, *Suasoriae* I, 15, probably alludes to these mishaps.

105. IV, 96. cf. Mela, III, 3, 31.

106. II, 3, 14.

107. Th. Fischer, *Die Seehäfen von Marokko*, p. 40.

108. This promontory was not rounded till 1434 (by Gil Eannes).

109. On Euthymenes, see esp. F. Jacoby in Pauly-Wissowa, s.v. It is not certain whether Herodotus's criticism of the Ocean origin of the Nile (II, 20–2) was directed against Euthymenes, for the same theory was known to the Egyptian priests (Diodorus, I, 19).

Columbus was similarly confirmed by the discovery of alligators in Cuba in the belief that he had reached the Indies; and Cardinal d'Ailly, following perhaps Aristotle, *De Caelo*, II, 14, 14, argued from the existence of elephants in West Africa that it could not be far distant from India.

110. *Les Phéniciens et l'Odyssée* (1st ed.), Vol. II, p. 87.

111. Odyssey, I, 23.

112. Pliny, II, 169.

113. Hecataeus (frs. 355, 357) mentions a river Lizas (= Lixus) and a town named Melissa which recur in Hanno's narrative. But Melissa is here called 'a city of the Libyans', not of the Phoenicians. Hecataeus's information therefore was obtained from some predecessor of Hanno, i.e. either from Euthymenes or from some earlier Phoenician mariner.

114. The text is given in C. Müller, *Geographici Graeci Minores*, Vol. I, pp. 1–14. The references to Hanno's voyage in later writers are sadly confused. The attempt by H. Tauxier (*Revue Africaine*, 1882, pp. 15–37) to prove Hanno's narrative a forgery has been generally discredited. The incidents which he finds incredible have been satisfactorily explained.

115. W. Aly, *Hermes*, 1927, pp. 317–30.

116. P. F. Gosselin, *Recherches sur la géographie des anciens*, Vol. I, pp. 70–106.

117. A. Mer, *Mémoire sur le Périple d'Hannon*, p. 46 ff. S. Gsell, *Histoire ancienne de l'Afrique du nord*, Vol. I, p. 472 ff. R. Hennig, *Geographische Zeitschrift*, 1927, pp. 378–92.

118. See especially the introduction and commentary to the text by C. Müller;

NOTES

also the works by Mer, Gsell, and Hennig, quoted above; C. T. Fischer, *Der Periplus des Hanno*; and R. Harris, *The Voyage of Hanno*.

119. 'In a line' probably means 'due south'. If Hanno really said this, he must have shared the general ancient view, that the African coast trended more or less uniformly eastward. But this is hardly credible. It is more likely that the Greek translator misunderstood the Phoenician text at this point.

120 If this is the Senegal estuary, the river has since obliterated the lake and islands.

121. *Africa Pilot*, Vol. I, p. 284.

122. A careful measurement of distances recently made by J. Carcopino (*Le Maroc antique*, pp. 119–29) confirms the location of Cerne in the Rio de Ouro bay.

123. Confusion has here been caused by writers ancient and modern who have taken the 'horn' for a cape, despite the plain words of the text.

124. Hennig, op. cit.

125. e.g. by Pedro de Cintra (*c*. A.D. 1450) and Mungo Park (*c*. A.D. 1800).

126. Mer (op. cit., p. 54 n. 1) points out that in the Yolof tongue 'gorhl' stands for any tall monkey. Marco Polo, who suspected that the mummies of 'Pygmies' traded in medieval Europe were really shaved monkeys, accepted the story of tailed men (orang-outangs?) in Sumatra (*Travels*, bk. III, chaps. 12, 15).

Gorillas, though generally peaceable, will fiercely resist any intruders upon their harems.

127. Arrian (*Indica*, 43, sect. 11) states that Hanno's journey lasted thirty-five days in all, and ended through lack of water. The former remark ignores Hanno's halts, the latter the 180 yearly inches of rain on the Sierra Leone coast. Pliny (II, 169) gaily asserts that Hanno reached 'the extremity of Arabia'.

128. Shipwreck of the Vivaldi brothers, 1291; Fernandez rediscovers Sierra Leone, 1446; Cadamosto doubles C. Palmas, 1455.

129. Herodotus, IV, 196. The 'silent haggling' there described was also observed by Idrisi (twelfth century) and Capt. Lyon (1820).

130. Pliny, II, 169. The West African trade is also mentioned in the Periplus of Scylax, written *c*. 350 B.C. (sect. 112).

131. Pliny, V, 9; VI, 199. Polybius was presumably inspired by Hanno.

132. We need not recognize an explorer in Ophelas, the author of a Periplus which gave the number of Phoenician colonies in West Africa as 300. We may concur with Strabo (XIII, 826) that he was a mere compiler of *mirabilia*.

133. Pliny, VI, 199–201; Ptolemy, IV, 6, 14. Strabo (I, 47) dismisses Cerne as a myth.

134. C. Müller, *Frag. Hist. Graecorum*, Vol. III, pp. 472–3; Pliny, VI, 202. Pliny's second source, Statius Sebosus, was probably a mere compiler.

135. These were in possession in 1312, when the Canaries were rediscovered by Malocello.

136. Diodorus, V, 19–20. A similar account in *De mir. auscult.*, chap. 84. Both stories were probably drawn from Timaeus.

NOTES

137. *Sertorius*, chap. 8. The double discovery of Madeira in antiquity is matched in its medieval history. In 1370 an English adventurer, Robert Machin, was blown out thither in a storm. After a period of oblivion it was re-explored in 1418 by Zarcho and Vaez. For a possible visit to it by the Greek pioneer Eudoxus, see p. 125.

138. So Martin Beheim, followed by P. Gaffarel, *Histoire de la découverte de l'Amérique*, pp. 50–6.

139. Strabo, III, 175.

140. B. V. Head, *Historia Numorum* (2nd ed.), pp. 877–81.

141. In 1341, by di Recco and Del Tegghia. R. Hennig (*Terrae Incognitae* I, pp. 109–19) defends the authenticity of the Cuervo find. Unfortunately not enough is known of the circumstances of its discovery and transmission to Europe.

142. See the careful discussion in Gaffarel, op. cit., pp. 48–172. Also R. Hennig, *Von rätselhaften Ländern*, pp. 162–75.

143. On an alleged coin of 'Caesar Augustus' in the Spanish Indies, see J. de Barros and D. do Couto, trans. by D. Ferguson, in *Journal of the Royal Asiatic Society*, Ceylon Branch, Vol. XX, p. 841, on the authority of a Spanish writer of mid-sixteenth century.

144. Mela, III, 5, 45; Pliny, II, 170 (on the authority of Cornelius Nepos).

145. Instances in Gaffarel, p. 167. In 1903 the Atlantic was crossed from Boston to Gibraltar in a rowing-boat 19 feet long; the time taken was 100 days (*The Times*, 21 Nov. 1903). Cf. also the journeys of the Malayans and Polynesians to New Zealand and Madagascar in diminutive catamarans. For further instances of ocean travel in open boats, see Hennig, op. cit., p. 237, and W. W. Hyde, *Ancient Greek Mariners*, p. 162, n. 90.

Recent examples of successful drifting across oceans are afforded by the Kon-Tiki cruise in the Pacific and the voyage of the *Égaré* II craft from Maine to Devon.

146. I, 23, 7. Similar tales seem to have been used for the plots of satyric plays at Athens (see E. Buschor, in *Athenische Mitteilungen*, 1927, p. 230 ff., on the strength of an Attic black-figure vase).

CHAPTER FOUR: INDIAN WATERS

1. See *Red Sea and Gulf of Aden Pilot*, 7th ed., pp. 1–50. Monsoons: *West Coast of India Pilot*, 7th ed., 46 ff., 37–8. Currents: id., 53–4.

2. Agatharch., *De Mari Erythraeo* (Müller, *Geog. Gr. Min.*, I), sect. 5. Phoenician settlements on Arabian coasts: Miles in *Geog. Journ.*, VII, 335–6. Hirt, in *Geogr. Zeitschr.*, 1926, pp. 430–1, says 'Red' Sea is a Persian name in which 'Red' = 'South', as among the Chinese.

3. G. Bénédite, *Fondation Eugène Piot, Mon. et Mém.*, XXII, 1916, p. 1 ff. (*Acad. des Inscr. et B.-L.*); R. Hall in *Camb. Anc. Hist.*, I, pp. 580–1; S. Smith,

NOTES

Early History of Assyria, pp. 52-3. See J. O. Thomson, *History of Ancient Geography*, p. 81.

4. *Proceedings of the Society of Antiquaries*, Dec. 1919, pp. 22 ff. Egypt and Sumer: L. Adametz, *Herkunft und Wanderungen der Hamiten*; Langdon in *Journ. of Egypt. Archaeol.*, VII, 133-153.

5. D. Nielsen, *Handbuch der altarabischen Altertumskunde*, Vol. I, p. 53 ff.; A. H. Keane, *The Gold of Ophir*, 72.

6. Pa-anch: W. Golenischef, *Sur un ancien conte égyptien*, 1881. See pp. 194-5 of this book. Amon Re (quoted): Breasted, *Ancient Records of Egypt*, II, 287, reading 'incense' instead of 'myrrh'. For the voyages to Punt, see *Records*, I, 161, 1, 8 (Sahure); 351 (under Isesi); 360, 361 (under Pepi II, Dyn. VI); see also 618, and for Hatshepsut see II, 428-95. For 'God's Land' see id., II, 288 and index, s.v.; ct. E. Glaser, *Punt und die südarabischen Reiche*. Egyptian Exploration Fund, *The Temple of Deir-el-Bahri*, III, 12, limits Hatshepsut's voyage to Somali coasts; cf. Schoff, *The Periplus*, pp. 218-19, 270, 272. For various wares got from Punt, etc.: see Breasted, op. cit., II, 109; cf. the Sesostris legend – Herod, II, 102; Strabo, XVI, 769, etc., p. 235 of this book. See also C. Conti Rossini, *Storia d'Etiopia*, 39 ff.

7. F. Hommel, *Ancient Hebrew Tradition* (English translation), 35-6 (Gilgamesh); id., 39 ff. E. Glaser, *Skizze*, II, 309 ff. Langdon in *Camb. Anc. Hist.*, I, 415-16, 427-8. Sir A. T. Wilson, *The Persian Gulf*, 25 ff. Dilmun in the Mesopotamian records is Daira or else Banna rather than Bahrein Island. For Magan as Sinai: H. Hall, id., 262, 583. See also S. Smith, *Early Hist. of Assyria*, 49, 86, 89, 99, 163.

8. 1 Kings 9: 28: 10, 11; 2 Chronicles 8: 18, 9: 10; Genesis 10: 29; 1 Chronicles 1, 23, etc. Ophir: Hastings, *Dict. of the Bible*, s.v. Ophir. N. H. Baynes, *Israel amongst the Nations*, 232. E. Glaser, *Skizze*, 357-88, gold in Arabia – id., 347-9. J. T. Bent's theories (*Ruined Cities of Mashonaland*) are obsolete; see also C. Peters, *King Solomon's Golden Ophir*, 1899; cf. his *Ophir nach den neuesten Forschungen*, 1908, and *Im Goldland des Altertums*, 1902. Keane, *The Gold of Ophir*, may be right so far as concerns Ophir as a town with Moscha (Khor Reiri) as port. But every argument that Havilah was Rhodesia is overthrown by the revelation that the Zimbabwe remains are only a few centuries old. R. Randall McIver, *Medieval Rhodesia*, 1906, *Journ. Anthr, Inst.*, XXXV; *Geog. Journ.*, 1906; opposed, however, by R. N. Hall, *Prehistoric Rhodesia*, 1909, cf. *Geog. Journ.*, Nov. 1909. See the sensible remarks of Schoff, *The Periplus*, pp. 96-9. R. Hennig, *Von rätselhaften Ländern*, 65-81, decides for Sofala = Ophir. Gertrude Caton-Thompson studied the site in 1928-9; see her *The Zimbabwe Culture*, and *Antiquity* III, 1939, pp. 424 ff.

9. H. Salt, *A Voyage to Abyssinia*, 103. Solomon's efforts were renewed by Jehoshaphat (1 Kings 22, 48-50) and Ahaziah (2 Chron. 20, 36-7), but were spoilt by shipwrecks.

10. Schoff, *The Periplus*, 147, 162; cf. Purali = Arabis, p. 63 of this chapter. J. R. Wellsted, *Travels in Arabia*, I, chap. 5. Sir T. Holdich, *The Gates of India*, 35, 370, 372. C. Lassen, *Indische Altertumskunde*, II, 187-91.

NOTES

11. J. Harrison, *Journ. Anthr. Inst.*, IV, 1874–5, 386 ff. T. de Lacouperie, *The Western Origin of Chinese Civilization*, 385–9.

12. *The Influence of Ancient Egyptian Civilization in the East and in America*, *Bull. John Rylands Library*, Jan.–March 1916. T. de Lacouperie, op. cit., is very speculative; for his views on elements of culture received by China from Babylonia and Elam, see p. 9 ff., 27, 378, cf. 70 ff., 29 ff.; connexion with early Egypt: 36–7, etc. The tendency has been, e.g., Sir H. Yule, *Cathay*, 8, to discredit him, but Elliott Smith, p. 24, is inclined to accept him. Chinese inscriptions found in Egypt were late forgeries. – H. Cordier, *Hist. gén. de la Chine*, I, 20–1.

13. See S. Smith, *Early Hist. of Assyria*, 371, cf. A. H. Sayce, *Hibbert Lectures*, 126–8; de Lacouperie, op. cit., 101 ff. Kennedy, *J.R.A.S.*, 1917, pp. 220–2; but see especially the evidence collected by him in *J.R.A.S.*, 1898, p. 241 ff.

14. Mookerji, *Indian Shipping*, has collected much information, but is uncritical as regards chronology. Cf. Rhys Davids, *J.R.A.S.*, 1899, 432. Schoff, *The Peripl.*, p. 229. Ezek. 27: 15, 19. Theophrast., *H.P.*, v, 4, 7, etc. *Peripl.*, 35–6.

15. Herod., II, 152, 154, 163; III, 4, 11; II, 178, 180–1.

16. Herod., II, 158–9 (a fleet). Strabo, XVII, 804.

17. Scylax: Herod., IV, 114, cf. III, 102. Canal route: Gray in *Camb. Anc. Hist.*, IV, p. 200. Herod., IV, 39; II, 158. Diodor., I, 33. Doubts: Vincent, I, 301–10. H. Tozer, *Hist. of Anc. Geog.*, 101–2. See also Myres, *Geog. Journ.*, VIII (1896), 623. Berger, 61, 75.

18. Hecatae. Frs. 174–9, Müller; 294–9, Jacoby. Herod., III, 107 ff.

19. Coral (?): Aesch., in Strabo, I, 33. What Aeschylus's Arabian city near the Caucasus was (*Prom. Vinct.* 420–3) one dare not guess. Tales: Herod., III, 107–13, cf. II, 73. Southern Sea: II, 11; III, 17. Happy city: Aristoph., *Birds*, 144–5. Arabia the happy and Asia: Eurip., *Bacchae*, 16–18. Somali: id., *Phaeton* in Strabo, I, 33. Islands in 'Red' Sea reported by Ephorus and others (referring to Persian Gulf?) – Pliny, VI, 198–9.

20. Aesch., *Suppl.*, 284 ff. Damastes in Strabo, I, 47. Herod., VII, 80, may allude (by hearsay) to the Bahreins. Canal route still in use: id., II, 158; IV, 39.

21. Nearchus's narrative: Arrian, *Indica*, 21–42. The distances are much overestimated, but the whole account is authentic. See Jacoby, *Die Fragmente der Griech. Historiker*, II, B1, 684–707, where passages from Strabo are also given. Onesicritus wrote a separate account – e.g. Pliny, VI, 96–100 (from Juba) – Jacoby, id., 734–5. For commentaries: see Bunbury, I, 525 ff.; Müller, *Geog. Gr. Min.*, I, p. 329 ff.; V. A. Smith, *Early History of India*, 110 ff.; Tomaschek, *Sitz. Wien. Ak., Phil.-Hist. Cl.*, 121 (1890), art. viii; Jacoby, II, BD2, 448; W. Vincent, *The Voyage of Nearchus*; Sir A. T. Wilson, *The Persian Gulf*, p. 36 ff.; Kempthorne, *Geog. Journ.*, V, 1835, 624 ff. The passages quoted are Arrian, *Indica*, 24, sects. 2–3 and 9 (savages); 31, sects. 2–5 (cf. Strabo, XV, 726 Astola; Holdich in *Geog. Journ.*, VII, 388); 30, sects. 2–6 (cf. Strabo, XV, 725, whales); 28, sects. 5, 8 (extorting supplies); Onesicritus and the turtle-eaters: Pliny, VI, 109. A 'prickly crab' is what is meant by κάραβος, which cannot be (as Liddell and Scott give) the Palinurus or 'Rock Lobster', for this is not found in Indian Seas.

NOTES

22. Alexander's schemes: Arr., *Anab.*, VII, I, 1–2; 19, 6; v, 26, 1–2. Archias: id., VII, 20, 7. Androsthenes: id., l.c., Strabo, XVI, 7 ff.; Berger in Pauly-Wissowa, I, 2172–3; Pliny, XII, 38–9 (from Juba). Athenae., III, 93, b–c, cf. d. Hieron: Arr., *Anab.*, VII, 20, 7–8, cf. *Indica*, 43, sect. 9. Arabian tribes: Theophrast., *H.P.*, IX, 4. Aromatics-trade: e.g. id., and Arr., VII, 20, 2. A ship reaches Yemen: Theophr., *H.P.*, IX, 4, 3–4, and Arr., *Indica*, 43, sect. 7, and Socotra? (as Panchaia) Virg., *G.*, I, 213. Pliny, VII, 19; X, 4. Strabo, II, 104; VII, 299. Diodor., V, 42–6. For the Utopia see pp. 240.

23. Ships: Pliny, VII, 208. Philo and chrysolites: Pliny, XXXVII, 108.

24. Strabo, XVII, 789; Pliny, VI, 167. For the new ports along the Red Sea, see: Strabo, VI, 768–70. Pliny, VI, 165–74. Ptol., IV, 58. E. Warmington, *The Commerce between the Roman Empire and India*, pp. 6–8 with notes. Satyrus: Strabo, XVI, 769, cf. Pliny, VI, 167. Elephant-hunts and stations: Agatharch., I (p. III), etc.; Diodor., III, 36, 3. Elephant-eaters: Agatharch., 53. Aristo: id., 85 (p. 175). Pythagoras: Athenaeus, IV, 183 f.; XIV, 634a; Ael., *H.A.*, XVII, 8–9. Pliny, XXXVII, 24. Eumedes: Hall, *Class. Rev.*, XII, 279. Hedjaz: Tarn, *Eg. Arch.* 1929, 9 ff.

25. Strabo, XVII, 789.

26. Agatharch., 41 (Müller, I, p. 135, and LXII); Diodor., III, 18, 4. Marcianus, 112.

27. Philo: Strabo, II, 77. Antigon., *Mirab.*, 145. Peitholaus, etc.: Strabo, XVI, 774. Other names: Müller, *G.G.M.*, I, LX. Inscriptions: Dittenberger, *O.G.I.*, I, 82 (Lichas), 86 (Alexandros). A private letter of cheer to men on hunt-duty: U. Wilcken, *Chrestomathie*, No. 452. Record of pay for Peitholaus's men: *Elephantine Papyri*, 28. Loan: Wilcken, *Puntfahrten*, in *Zeitschr. f. äg. Sprache*, LX, 1925, pp. 86–102.

28. See Pliny, IX, 6, where the Ptolemy must be Ptol. III; cf. XII, 76. We cannot connect Simmias with any of this, though cf. Müller, LXII, LXIII.

29. For Arabia as known to Eratosthenes at this time see Strabo, XVI, 767 ff. Of India he knew no more than Megasthenes did – id., XV, 685–720.

30. See Agatharch., 31 ff. (chiefly in Diodor., III, 14 ff.). Outside Red Sea: id., 41. Socotra – id., 110, 103. 'Fortunate Islands' must have included Socotra = Dvipa Sukhadhara, 'Island abode of bliss'. Nile: Strabo, XVII, 789, cf. XVI, 774; Agatharch., 50, p. 141; 84, p. 174; 112, pp. 194–5.

31. Sabaeans and exaggerated ideas of their well-being: Agatharch., 97 ff., 186 ff., cf. *Peripl.*, 26. Kuria Muria and Masira included in the 'Fortunate Islands' of Agatharch., 103 (cf. 110).

32. Strabo, II, 103.

33. id., 98–9. J. H. Thiel, *Eudoxus van Cyzicus* (as cited in chap. v, note 45), pp. 39 ff.

34. id., XVI, 774.

35. Egyptian trade not much spoilt by bad government: cf. Milne, *J. R. Stud.*, 1927, p. I ff. Juba writing in I B.C. knew very little about Africa outside Bab el-Mandeb. Pliny, VI, 174–6. Pliny and Arabia: Pliny, VI, 147–60. A. Klotz, *Quaestiones Plinianae Geographicae*, 198 ff.

NOTES

36. Ditt., *O.G.I.*, I, 186. *C.I.G.*, 4751.

37. Ditt., *O.G.I.*, I, 69, 70, 71, 74; *C.I.G.*, 4838, 4838c (62 B.C.). Strabo, XVII, 798, cf. II, 118.

38. Pliny, VI, 153; Cosmas, III, 169 B; *Peripl.*, 30.

39. Acila: Pliny, VI, 151 (not the Ocelis by Bab el-Mandeb); small vessels: *Peripl.*, 57. India in Strabo, XV, 685–720, and Pliny, VI, 56–80. Barygaza: Pliny, VI, 174.

40. Ditt., *O.G.I.*, I, 186, 190.

41. Pliny, XVI, 135.

42. Antiochus, III, Polyb., XIII, 9. Epiphanes: details in Pliny, VI, 147–9. Numenius: id., 152.

43. Pliny, VI, 149.

44. Strabo, II, 118; XV, 686; XVII, 798, 815.

45. id., XV, 686, 719–20. Cass. Dio., LIV, 9. Suet., *Aug.*, 41. Hor., *C.S.*, 55–6. *Odes*, I, 12, 56; IV, 14, 41–3, etc. Warmington, *Commerce*, etc., 35–8 with notes.

46. Tac., *Ann.*, II, 33; III, 53. Cass. Dio., LVII, 15.

47. *Peripl.*, 26 ff.

48. See chap. X, pp. 191–2.

49. Isidore: Müller, *G.G.M.*, I, Proleg., LXXX ff. Aden: *Peripl.*, 26. The dispute over this place is whether the person who destroyed it was really a Roman 'Caesar' or whether the reading is wrong – Warmington, 15–16; see Schoff, Fabricius, and Müller on the passage. Schwanbeck, *Rhein. Mus. Phil.*, VII, 1850, 352 ff., 328. Kornemann, *Jan.*, I, 61 ff. Rostovtzeff, *Archiv. f. Papyrusforschung*, IV, 308–9, and other references in Warmington, p. 334, n. 32.

50. Pliny, VI, 100–6. *Peripl.*, 57. His name survived in geographical features of the Erythraean Sea: Ptol., IV, 7, 12. *Itin. Alex.*, 110. His date is still disputed, e.g. Schoff, *Peripl.*, pp. 8, 227–8, C. Lassen, *Ind. Alt.*, III, 3–4; Kornemann, *Jan.*, I, 57–8, following Chwostow, *Historiia*, etc., 346–9, 360, 386–7, 433. Schur, in *Klio*, XX, 1926, 220. Pauly-Wissowa, VIII, 2, s.v. Hippalos, 1600–1. Warmington, pp. 44 ff.; M. P. Charlesworth, *Class. Quart.*, April 1928, pp. 94 ff. The general tendency now is to assign Hippalus to the later Ptolemaic period. So R. Hennig, *Terrae Incognitae*, pp. 226–8; H. Kortenbeutel, *Der ägyptische Süd- und Osthandel in der Politik der Ptolemäer und römischen Kaiser*, p. 48; W. W. Tarn, *The Greeks in Bactria and India*, pp. 368–9; Sir M. Wheeler, *Rome beyond the Imperial Frontiers* (Bell, 1954, pp. 126 ff.; Pelican Books, 1955, pp. 153 ff.); J. H. Thiel, op. cit., pp. 63 ff.

51. So Pliny, VI, 84–9 implies. But it may be earlier. A *graffito*, duplicated in Latin and Greek on a rock beside the old road from Coptos to Berenice, records how one Lyses, a slave of P. Annius Plocamus, passed that way in early July A.D. 6 – thirty-five years before Claudius's accession – D. Meredith, *Journ. Rom. Stud.*, XLIII (1953), p. 38; Sir M. Wheeler, op. cit., pp. 128–9 = 155–6.

52. *Peripl.*, 57, 'trachelizontes' – throwing the ship's head off the wind, Schoff translates, *The Peripl.*, pp. 45, 230, 232.

53. The passage quoted is from *Peripl.*, 40. Dangers less now – G. of Cutch:

NOTES

West Coast of India Pilot, 7th ed., 1926, p. 214; Nerbudda: id., 188; avoid the Indian coast: id., 63. See also Lucian, *Hermotim.*, 4.

54. Lucian, *Pseudomant.*, 44.

55. See Philipps, *Indian Antiquary*, 1903, p. 1 ff., 145 ff. Warmington, *Commerce, etc.*, 83.

56. *Peripl.*, 56, 10.

57. For Sophon (Subhanu, if this guess at the correct name is right) see Hultzsch in *J.R.A.S.*, 1904, p. 402. Some think he came to Egypt under the Ptolemies.

58. Sewell in *J.R.A.S.*, 1904, p. 623–37; Wheeler, op. cit., pp. 137 ff. = 164 ff. One hoard from Pudukottai is in the British Museum. Each one there has the emperor's head gashed across. This was done, as in other hoards, to put the coins out of circulation. For full details of Roman coins found in India and Ceylon, see Wheeler, *Ancient India*, No. 2 (as cited below), pp. 116–21.

59. This seems established. It is difficult to fix narrower dates. J. A. B. Palmer, in *Class. Quart.*, 1947, pp. 137–40, thinks that political conditions in India, as indicated in the Periplus, point to A.D. 110–15. For various opinions see Schoff, *The Periplus*, pp. 8–16, 290–4; and *J.R.A.S.*, 1917, 827–30; Kennedy, id., 1916, 836; 1918, 111 ff. Kornemann, *Jan.*, 1, 58 ff. Schur, *Klio*. Beih., XV, 1923, pp. 43–4; XX, 1926, 222. Warmington, 52 and 343, n. 51. Charlesworth, *Class. Quart.*, Apr. 1928, p. 92 ff.

60. *Peripl.*, 38–9, 47.

61. A. Maiuri, in *Le Arti*, 1938–9, pp. 111–15.

62. id., 41 ff., 46, quoted.

63. id., 53–4. Ptol., VII, 1, 7; 84. Malabar Pirates: Marco Polo, III, chap. 28.

64. *Peripl.*, 54 ff. *Tab. Peutinger* (Miller), 790.

65. *Peripl.*, 58–9; 61.

66. id., 59, 60, 62 ff.

67. Pliny, VI, 55, 57. Mela, III, 7, 70. *Peripl.*, 56, 63. Josephus, *Ant. Jud.*, VIII, 1, 64.

68. *Peripl.*, 4–14. Ptol., IV, 7.

69. *Peripl.*, 15–17. Menuthias: 15. Of the scanty remains of this traffic the best are the Roman coins (nearly all of the fourth century A.D. found at Port Durnford, 250 miles north-east of Mombasa. H. Mattingly, *Numismatic Chronicle*, 5th Series, XXI (1932).

70. *Peripl.*, 20–9, 32 ff.; Socotra: 30–1.

71. id., 32–6. Pliny, VI, 140, 145, 149.

72. Ptol., I, 9, 3–4.

73. Plut., *Pomp.*, 70. Statius, *S.*, IV, 1, 40–2; IV, 3, 155; III, 4, 57–63. Silius Ital., *Bell. Pun.*, III, 612–15; Mart., XII, 8, 8–10.

74. Cass. Dio, LXVIII, 28–9. Eutrop., VIII, 3. Zonar., XI, 22.

75. Cass. Dio, id., 15. *Hist. Aug.*, 'Hadrian', 21; 'Vict.', *Ep.* 15, 4, cf. App. *Praef.*, 7.

76. Ptol., VII, 1.

NOTES

77. *Oxyrhynchus Pap.* III, 413, pp. 41–57; Hultzsch, *Hermes*, 1904, p. 207; Barnet, *Journ. of Eg. Arch.*, 1926, pp. 13–15.

Discoveries of pottery, imitations of coins, etc.: Wheeler, op. cit., pp. 150–3 = 179–82.

78. V. A. Smith, *History of Fine Art in India and Ceylon*, esp. 178–9. Sir M. Wheeler, op. cit., pp. 165 ff. = 195 ff.

79. V. Kanakasabhai Pillai, *The Tamils Eighteen Hundred Years Ago*, pp. 12, 16, 19, 37–8, 24–5. See also P. T. Srinivas Iyengar, *History of the Tamils from the Earliest Times to* 600 A.D.; also the *Silappaditaram*, translated by V. Ramachandra Dikshitar, 110.

80. Ptol., VII, 4.

81. H.W. Codrington, *Ceylon Coins and Currency*, 31 ff.; J. Still, in *J.R.A.S.*, Ceylon Branch, 1907, 161–90.

82. For Arikamedu, see Wheeler, in *Ancient India*, No. 2, pp. 17–124 – Arikamedu: An Indo-Roman trading station on the east coast of India, with contributions by A. Ghosh and Krishna Deva; J. M. Casal, *Fouilles de Virampatnam-Arikamedu*, especially pp. 16 ff.

83. Ptol., I, 14, 1, cf. 13, 5–9; VII, 1, 12, 16; VII, 2, 3; I, 15; 3, 6; I, 14, 1. Marco Polo was told there were 12,700 Maldives – Marco Polo, bk. III, chap. 39.

84. Hirth, *China and the Roman Orient*, 47.

85. For all regions beyond India, see Ptol., VII, 2, and Gerini in *J.R.A.S.*, 1897, 551–7, with map and tables, and also his *Researches on Ptol.'s Geog. of E. Asia*; A. Herrmann, *Zeits. d. Gesellschaft für Erdkunde*, 1913, 771–87. Warmington, 125 ff.

86. Oc-ceo: *Bulletin de l'École Française d'Extrême-Orient*, XLV, fasc. 1 (1951), pp. 75 ff. P'ong-Tuk: G. Coedès, in *Journ. of the Siam Soc.*, XXI, part 3 (1928), pp. 204 ff. Sir M. Wheeler, in *Aspects of Archaeology, Essays presented to O.G.S. Crawford*, ed. W. Grimes, p. 361.

87. A. Herrmann, *Verkehrswege*, 8, and id. *Zeits d. G. f. Erdk.*, 1913, 553–61 Kan Ying: Hirth, 39.

88. For the name Seres (from the word for silk) as applied to the Chinese by overland travellers – see chap. 7.

89. Hirth, op. cit., 16, 39 ff., 68, 147, 167 ff., 272–5, 306 ff. F. Teggart, *Rome and China*, p. 145. He argues that the reason why every barbarian disturbance in Europe between 58 B.C. and A.D. 107 followed a war on Rome's eastern frontiers or China's western regions was interruption of an international trade, the important factor being the silk route. Thinae: Schoff, *Peripl. of the Outer Sea* (Marcian), p. 50. Habit of Malays to bring 'royal presents' to China: Vidal de Lablache, *C.-R. de L'Acad. d. Inscr. et B.-L.* 1897, p. 525.

90. China: C. G. Seligman, in *Antiquity* XI, 1937, p. 5 ff.

91. M. P. Charlesworth, *Studies in Roman Economic and Social History*, ed. P. Coleman-Norton, p. 140.

92. Warmington, pp. 136–40.

NOTES

CHAPTER FIVE: THE CIRCUMNAVIGATION OF AFRICA

1. Homer, *Il.*, XIV, 246; XVIII, 607; XXI, 194, etc. Gem., *Elem. Astron.*, 13. Strabo, I, 3–5.
2. Hecatae., fr. 302 (Jacoby), 187 (Müller); Hesiod, fr. 64 (Rzach); Schol. ad Apollon, Rhod., IV, 259.
3. Herod., II, 154, 178.
4. For the view that before this attempt or success the circumnavigation had been made, see Bougainville, 'Reflexions sur le commerce de Carthage', in *Mém. de l'Acad. des Inscr. et B.-L.*, XXVIII, 308, 9.
5. cf. Herod,. II, 158.
6. Herod., IV, 42.
7. W. Müller, *Die Umsegelung Afrikas*; English readers should consult W. Vincent, *The Periplus of the Erythraean Sea*, I, 173 ff.; Rennell's *Geograph. System of Herod.*, 672–714; J. Wheeler, *The Geog. of Herod.*, pp. 335–46; Bunbury, I, 289–96, 317; H. Tozer, 99–101, 106; cf. H. Berger, *Geschichte der wissenschaftlichen Erdkunde*, 62 ff. For the very adverse view, see E. J. Webb, in *English Historical Review*, Jan. 1907.
8. The same objection applies to Herodotus's notice about Scylax. See p. 79.
9. cf. Vivien de St Martin, *Hist. de la Géogr.*, 30.
10. Müller, 61. But see Wheeler, op. cit.
11. Müller, 52, 58, 60.
12. cf. Müller, 62. Herodotus is sometimes reticent; e.g. must he not really have known more about the Carthaginians than he tells us?
13. In Bredow, *Untersuchungen*, II, 344. Vincent, I, 177–8. Gaffarel in *Mém. d. l. Soc. d'Émulation du Doubs*, IV Sér., Vol. VII, 1872, p. 48. Müller, 77.
14. Bunbury, I, 294; Gosselin, *Recherches s.l. géogr.*, I, 214; against which cf. Gaffarel, 53, Müller, 69–70.
15. Müller, 97, 99 ff. Wheeler, 344.
16. Vincent, I, 176–7, denies the possibility.
17. Müller, 72–3. Gaffarel, 50; and see instances given in n. 142 of chap. 3, p. 261.
18. See Müller, 14 ff., 20–1, 23 ff., esp. 25–6.
19. Gosselin, *Rech.*, I, 213. But see Gaffarel, 50–1. Müller, 92.
20. See pp. 82–3, Ptol., I, 14, 1 ff.; cf. 13, 1–9.
21. Bunbury, I, 294. Gosselin, *Rech.*, 215, 350. Müller, 96.
22. Müller, 88, 89 ff. Keane, 28, 181.
23. Müller, 36, cf. 45–8, develops this view.
24. id. 41–3 (choice); 40–1 (Kosseir); 38 (Suez); 72 (penteconters against the merchant-ships of Wheeler, 377); 78 ff. (what grain? wheat, 84); 65–6 (quoted in translation).
25. Müller, 66–96. For the sun: see Müller, 97, 99 ff, 107; Wheeler, 340;

NOTES

Bunbury, I, 293-4; Tozer, 100-1. Currents: J. Rennel, *An Investigation of the Currents of the Atlantic Ocean*, chaps. III, IV, cf. pp. 7 ff. For winds and currents of West Africa, see also *The Africa Pilot*, pt I, 8th ed., 1920, pp. 1 ff.; winds and weather of S. African coasts: id., pt III, 8th ed., 1915, pp. 25-42. There is no evidence (Junker 'Die Umschiffung Afrikas durch die Phöniker', in *Neue Jahrbücher f. Philol. u. Pädagog.*, Suppl.-B. VII, p. 365) that the voyagers called at the port of Aden.

26. J. Rennel, in Bredow, *Untersuch*, II, 693 ff., and id., *Geogr. of Herod.*, 672-714. Wheeler, 339-44. Other writers will be found in Bunbury, I, 317. Wheeler's scheme is: Suez, August 613 B.C. – Indian Ocean, Oct. – tropic of Capricorn, end of Jan. 612 B.C. – round Cape, April – N. to equator, July or later – Senegal, March 611 B.C.; until harvest Sept. (Senegal would certainly be a better region for growing corn than Morocco) – Egypt, Feb. 610 B.C. Suggests the first rest was in Angola since the voyagers may, through previous trade relations, have kept themselves well supplied as far as Sofala.

27. Pliny stupidly makes Hanno sail right round. – See II, 169.

28. Strabo, II, 98, Meineke's text has μάγον. Gaffarel, 55, suggests that Herodotus had told about this in the lost work on Assyria.

29. cf. Herod., II, 34.

30. cf. Herod., II, 32.

31. Herod., IV, 43. C. Fischer, *Der Periplus des Hanno*, 84 ff.

32. Gaffarel, 57, cf. Berger, 61, 75.

33. Junker, 384. cf. Klotz, in *Klio*, 1937, pp. 343 ff.

34. Wheeler, 344-5. Ships going south have to face unfavourable winds all the summer off Guinea – cf. Tozer, 103.

35. Aesch., *Suppl.*, 384-6. Even Aristotle had doubts about this.

36. Plato, *Timae.*, 6, 3, 25 – but this is connected with Atlantis, for which see pp. 242-4.

37. Herod., I, 203; II, 28; IV, 42. Arist. *Meteor.*, II, 5.

38. So the present writer (E.H.W.) connects 'Scylax', 122, with Ephorus in Pliny, VI, 199.

39. Aristotle, *Meteor.*, II, 1, 12. 'Scylax', 112.

40. Arist., *Meteor.*, II, 5.

41. Arr., *Anab.*, VII, 1, 2; V, 26, 2. Plut., *Alex.*, 68.

42. Strabo, II, 114 (heat), 118.

43. Letronne, 'Discussion de l'opinion d'Hipparque sur le prolongement de l'Afrique', *Journ. des Savants*, 1831, pp. 476-80. Polyb., III, 38, 1.

44. Pliny, II, 169.

45. Strabo, II, 98-102. Gaffarel, as in note 13, pp. 13-100. Berger, 569 ff., cf. 71 ff. J. H. Thiel, *Eudoxus van Cyzicus* (Mededeelingen der Koninglijke Nederlandsche Akademie van Wetenschapen, afdeeling Letterkunde, Nieuwe Reeks, Decr. 2, no. 8); R. Hennig, *Terrae Incognitae*, I, pp. 218 ff.

46. Deduced from Mela, III, 9. Pliny, VI, 188.

47. This last statement caused some later writers, e.g. Nepos (Mela, III, 9, 90; Pliny, II, 67), to think that Eudoxus had sailed right round Africa; Nepos

NOTES

even reversed the successful voyage to one from east to west, thus confusing two actual voyages not taken all round.

48. Gaffarel, pp. 91 ff., would have it (wrongly) that Eudoxus had sailed along the coasts of Sahara, Senegambia, and a part of Guinea. Awkward wind – *Africa Pilot*, as in note 25, 1, pp. 2–3. Thiel, pp. 33 ff., 54 ff.

49. Naturally Gaffarel, 88–9, suggests islands too far south.

50. Mela, III, 9. Gaffarel, 85–88.

51. Strabo, II, 102. Settlements destroyed: id., XVII, 3, 7, 8; I, 3, 2,; Ezek, 17, 10.

52. Gaffarel, 38–40. Nor even did Eudoxus settle at Gades itself the identity of his prow.

53. See Warmington, pp. 4, 74.

54. Cf. Bougainville, *Mém. de l'Acad. d. Inscr. et. B.–L.*, XXVIII, 314. For the brothers Vivaldi see C. R. Beazley, *The Dawn of Modern Geography*, III, 413–19, 551. Maros Jimenez's ed. of 'Conocimiento de todos los Reinos', *Boletin* of the Geog. Soc. of Madrid, II, 2 (1877), 111, 113, 117–18.

55. Strabo, II, 132, 133, 120; I, 22; XVII, 825, etc. Pliny, V, 5.

56. Strabo, I, 32.

57. Pliny, VI, 175.

58. Pliny, II, 168.

59. *Peripl.*, 15, 18. Keane, 127–9.

60. Ptol., I, 9, 4. Zimbabwe: Charlesworth, *Trade-Routes and Commerce of the R.E.*, 255 (top); Madagascar: E. Gautier, *Quatenus Indici Oceanis – pars patuerit*, 60. Congo: *Rivista Italiana di Numismatica*, VI, 1893, 505. cf. also M. P. Charlesworth, in *Num. Chron.*, 6th series, vol. IX, (1949), p. 107.

61. For the whole see Warmington, in *The Cambridge History of the British Empire*, Vol. VIII, chap. 3, 'Africa in Ancient and Medieval Times', pp. 51 ff.

CHAPTER SIX: EUROPE

1. On the contribution of the European peoples to early civilization, see esp. V. G. Childe, *The Dawn of European Civilization*; J. Déchelette, *Manuel d'archéologie préhistorique*; C. Schuchhardt, *Alteuropa*.

2. ll. 178–82. See Schulten and Bosch, op. cit., ad. loc.

3. ll. 267–70.

4. R. Carpenter, *The Greeks in Spain*, pp. 37–46.

5. *Meteorologica*, I, 13, 19. Aristotle may have been misled by the alternative name 'Hiberus' for the Tartessus river (Avienus, 248), which invited confusion with the Hiberus of northern Spain (the Ebro).

6. That this was Hannibal's route may be inferred from the description of his foray across the Sierras in Livy, XXI, 5.

7. The Don still marked the boundary between Europe and Asia in the days of William of Rubrouck (c. A.D. 1250).

NOTES

8. K. Neumann, *Die Hellenen im Skythenlande*, Vol. I, p. 74 ff. Hence the summer rains of Russia in antiquity (Herod., IV, 28).

9. *Iliad*, XIII, 5-6.

10. E. v. Stern, *Klio*, 1909, p. 141; M. Rostovtzeff, *Iranians and Greeks in South Russia*, chap. 4 and map; V. Parvan, *Dacia*, p. 76.

11. Herod., IV, 52-3. No mention is made here of the cataracts of the Dnieper, or of its eastward bend. This no doubt suffices to show that Herodotus had not explored the Dnieper in person, but it proves nothing against his informants.

The volume of water carried by the Russian rivers also excited the admiration of the William of Rubrouck.

12. Herod., IV, 104-6. For the location of the tribes in question, see J. T. Wheeler, *The Geography of Herodotus*, pp. 179-80; W. Tomaschek, *Sitzungsberichte der Wiener Akademie*, Philosophisch-historische Klasse, Vol. CXVII, pp. 3-13.

13. See esp. R. W. Macan, *Herodotus, Books 4-6*, Vol. II, pp. 25-8. Similarly the rubber traders in Brazil became but half acquainted with the ramifications of the Amazon, and left it to an ex-president of U.S.A. to discover the Rio Roosevelt for them.

14. Herod., IV, 51-2, 54-5.

15. III, 115; IV, 17, 31, 45.

16. The Rhipaean Mountains are mentioned by Herodotus's contemporaries Damastes (fr. 1, Jacoby) and Hellanicus (fr. 187, Jacoby).

On the alleged finds of early Greek coins at Schubin in Posnania, see C. T. Seltman, *Athens*, pp. 133-5; C. Fredrich, *Zeitschrift der historischen Gesellschaft für die Provinz Posen*, 1909, pp. 193-247 (very sceptical).

The Greek metal-work found at Vettersfelde in Brandenburg is explained by A. Furtwängler (*Kleine Schriften*, Vol. I, pp. 512-16) as the property of a Scythian chief who trekked westward during Darius's Scythian expedition.

The theory of an amber route from the Baltic to Olbia (Rendel Harris, *Journ. Hell. Stud.*, 1925, p. 230) is not supported by archaeological evidence.

17. ll. 799-859.

18. VII, 306.

19. Rostovtzeff, op. cit., p. 215.

20. I, 31, 2.

21. V. Sadowski, *Die Handelstrassen der Griechen und Römer durch das Flussgebiet der Oder und Weichsel*, pp. 193-4.

22. III, 5, 5, 14.

23. G. Jacob, *Der nordisch-baltische Handel der Araber im Mittelalter*; Sir T. W. Arnold, in A. P. Newton, *Travel and Travellers in the Middle Ages*, pp. 94-5; A. Meyendorff, ib., chap. 6.

24. Childe, op. cit., chaps. 10-12.

25. Parvan, op. cit., chap. 2.

26. C. T. Seltman, *Classical Quarterly*, 1928, pp. 155-9.

NOTES

27. Parvan, chap. 3. The longitude of Bucharest may be taken as the rough limit of the more intensive Greek penetration.

28. Parvan, p. 91, ff.; R. Forrer, *Keltische Numismatik der Rhein und Donauländer*, p. 210 ff.

29. IV, 48–50. Herodotus's transpositions are probably due to confusion between the northern and southern Haemus, i.e. the Carpathians and the main Balkan range. (Brandis, Pauly-Wissowa, s.v. Danuvius.)

30. How and Wells, *Commentary on Herodotus*, on IV, 49. Herodotus, not knowing of the great north-south bend of the middle Danube, would naturally infer that tributaries from the right must flow northward. Strabo mentions Ἄλβια Ὄρη in upper Dalmatia (VII, 314): hence perhaps the name Alpis for the Save.

31. Herodotus, II, 33; Ephorus, ap. ps.-Scymnus, ll. 191–5; Timagetus (an otherwise little-known writer), ap. Schol. Apoll. Rhod., IV, 259; Timaeus ap. *De mirab. auscult.*, chap. 105.

32. Scylax, chap. 20; Theopompus ap. Strabo. VII, 317 (a subterranean connexion). The belief lingered on in Cornelius Nepos (ap. Pliny, III, 127), and Mela (II, 4, 57). It probably originated in the discovery of a tribe named Histri at the head of the Adriatic, which suggested propinquity to the Ister (the Greek name for Danube). Greek geographers, who were acquainted with a good many river deltas, would have no difficulty in assuming a bifurcation of the Cassiquiare type. See J. Partsch, *Sitzungsberichte der sächsischen Akademie*, philosophisch-historische Klasse, 1919).

33. e.g. of the Mississippi, the Amazon, the Murray, the Niger, the Congo, and (to a large extent) the Nile.

34. VII, 289. Strabo errs in making the Colapis (Culpa) flow into the Danube, instead of the Save (314). Diodorus, V, 25 (who here probably followed Timaeus), stated that the Danube (as distinct from the Ister) flowed into the Atlantic. The problem of the Danuvius-Ister was akin to several river-puzzles which have vexed modern geographers, e.g. those of the Lualuba-Zaire (the Congo), the Joliba-Quorra (the Niger), and the Tsanpo-Brahmaputra.

35. S. Casson, *Macedonia, Thrace and Illyria*, p. 3.

36. Frs. 166–83. In fr. 166 Hecataeus mentions a Persian fort, in frs. 169–72 he apparently marks off the stages in Darius's march to the Danube.

37. V, 3–8.

38. *Anabasis*, VII, 1 ff.

39. The fine bronze-ware found in tombs at Trebenishte in Upper Macedonia perhaps belonged to enriched Greek mercenary officers of c. 500 B.C. (B. Filow, *Die archaische Nekropole von Trebenischte*, pp. 97–108). But S. Casson (*Journ. Hell. Stud.*, 1928, p. 270) prefers the view that the interments were of native chiefs.

40. Arrian, *Anabasis*, I, 4. Alexander's point of departure, Amphipolis, suggests an advance up the Struma.

41. Forrer, op. cit., p. 270.

NOTES

42. Strabo, VII, 317. According to (Aristotle), *De mirab. auscult.*, chap. 104, the Balkanic 'peak in Darien' lay in the hinterland of Istria.

43. For the name of the mountain, see F. W. Walbank, *Philip V of Macedonia*, p. 249 and n. 4.

44. Livy, XL, 21, 2. On ancient mountaineering in general, see Tozer, *History of Ancient Geography*, pp. 313-26.

45. Polybius retained the current belief, but was corrected by Strabo (VII, 313). The error lingered on in Mela (2, 2, 17).

46. Cassius Dio, LI, 23-7.

47. *De mirab. auscult.*, chap. 104; Theopompus ap. Strabo, VII, 317.

48. T. R. Holmes, *The Architect of the Roman Empire*, pp. 130-5.

49. H. Stuart Jones, *Papers of the British School at Rome*, No. 5, pp. 452-5.

50. cf. Ptolemy's enumeration of Balkan towns (III, 9-11). The Roman Itineraries do not mention a road up the Vardar valley.

51. Sadowski, op. cit.; J. Partsch, *Schlesien*, Vol. I, p. 329 ff. Rendel Harris (*Journ. Hell. Stud.*, 1925, p. 236-8) traces alternative routes across the Carpathians.

52. Parvan, chap. 1.

53. J. M. De Navarro, *Geographical Journal*, 1925, p. 484 ff. Pytheas, as we have seen, found the Heligolanders using their amber as fuel.

54. O. Montelius, *Praehistorische Zeitschrift*, 1910, p. 249 ff.

55. Partsch, op. cit., p. 337. These finds are at any rate well authenticated.

56. Parvan, pp. 138-9, 155.

57. Parvan, chap. 2.

58. De Navarro, p. 502; Sadowski, p. 167 ff.

59. Parvan, p. 151 ff.

60. Polybius ap. Strabo, IV, 208. By the time of Horace the reputation of the Styrian iron was well established (*Odes*, I, 16, 11).

61. Sadowski, p. 179 ff.; Partsch, pp. 335-8; M. Jahn, *Praehistorische Zeitschrift*, 1918, p. 80 ff.

62. Philemon ap. Pliny, XXXVII, 33. Philemon was probably a contemporary of Augustus (E. Norden and H. Philipp, in *Festschrift Lehmann-Haupt*, p. 182 ff.)

63. Pliny, XXXVII, 45.

64. Sadowski, loc. cit.; Fredrich (op. cit., p. 202 ff.) records that the Roman coins found in Posnania mostly date from A.D. 70 to 160. He suggests that they were looted rather than traded. But this overlooks the defensive power of the early Roman empire.

65. H. Willers, *Die römischen Bronzeeimer von Hemmoor*, p. 191 ff.; *Neue Untersuchungen*, p. 104; P. Hauberg, *Aarbøger for nordisk Oldkyndighed og Historie*, 1894, pp. 325-76. Hauberg reports 3,748 coins on Gothland, Willers 4,200.

The Hanseatic station at Wisby owed its importance to eastern rather than to trans-European commerce.

66. Pliny, XXXVII, 43.

NOTES

67. II, 11, 16. K. Malone (*American Journal of Philology*, 1924, p. 362) convincingly emends Σόυλωνες into Σουίωνες.

68. II, 11, 12, 16; III, 5, 1.

69. Tacitus, *Germania*, XLI, sect. 2; '*Albis notum olim flumen, nunc tantum auditur.*'

70. In A.D. 101 Trajan attempted a pincer operation from the west (probably by the Iron Gate Pass) and from the south (by the Vulcan or the Red Tower Pass). In subsequent campaigns he made direct drives through one or other of the western passes. See G. A. T. Davies, *Journ. Rom. Stud.*, 1920, pp. 1-28; E. T. Salmon, *Transactions of the American Philological Association*, 1936, p. 81 ff.

71. V. Pârvan, *Dacia*, pp. 192-3.

72. De Navarro, p. 484 ff.

73. Montelius, loc. cit.

74. H. Genthe, *Über den etruskischen Tauschhandel nach dem Norden*, p. 120 ff., and map. F. Stähelin, *Die Schweiz in römischer Zeit.*, pp. 10-12.

75. Déchelette, *Manuel d'Archéologie préhistorique*, Vol. II, pt 3, pp. 1596-8.

76. F. von Duhn, *Neue Heidelberger Jahrbücher*, 1892, p. 55 ff. On the subject of Alpine commerce, see also U. Kahrstedt, *Göttingen Nachrichten*, 1927, pp. 11-22.

77. Stähelin, op. cit., pp. 331-8. It is noteworthy that the St Gothard Pass was not brought into use by the Romans.

78. So Apollonius Rhodius, IV, 640 (probably from Timaeus). Caesar (*Bell. Gall.*, VI, 25) placed the forest from hearsay on the north bank of the Danube. Strabo (VII, 292) located it at some distance to north of the river.

79. Déchelette, Vol. II, pt 2, p. 582. But see J. M. De Navarro, *Antiquity*, 1928, pp. 437-41.

80. IV, 49.

81. ll. 188-91.

82. IV, 625-9. According to C. Jullian (*Revue des Études anciennes*, 1906, pp. 117-18) Apollonius intended to convey that the Argonauts shouldered Argo across the Alps. But the poet says that 'the waters roared' in the connecting passage.

Euripides imagined the Rhône and the Po as meeting somewhere near the Adriatic. (Ap. Pliny, XXXVII, 32.)

83. Jullian, *Histoire de la Gaule*, Vol. I, pp. 46-8. It is here suggested that the legend of Heracles' journey across the Alps (Diodorus, IV, 19; V, 22; *De mirab. auscult.*, chap. 85) indicates an early use of the Mt Genèvre Pass. But granted that the legend has an historic basis, it fits all the western passes equally well.

84. Polybius, III, 50-6; Livy, XXI, 32-7. Of the vast but indecisive modern literature it will suffice to mention here: Spenser Wilkinson, *Hannibal's March* (Col du Clapier); G. De Sanctis, in *Storia dei Romani*, III, pt 2, pp. 65-84 (Mt Genèvre); D. Freshfield, *Hannibal Once More* (Col d'Argentière); F. W. Walbank, *Journ. Rom. Stud.*, 1956, pp. 37-45 (a carefully considered article in favour of a pass of the Mt Cenis group).

NOTES

85. III, 47–8. Polybius imagined the Alps as extending uniformly from west to east.

86. Ap. Strabo, IV, 208–9.

87. Strabo, loc. cit.; Servius ad Vergil, *Aeneid*, X, 13. On Rome's Alpine roads in general, see W. W. Hyde, *Memorials of the American Philological Association*, Vol. II, 1935.

88. Sallust, *Epistula ad Senatum*, 4.

89. Strabo, IV, 204.

90. Tacitus, *Hist.*, I, 70, 4. The Simplon was adapted for local traffic only (Stähelin, p. 329).

91. Déchelette, Vol. II, pt I, pp. 398–400.

92. Cary, *Journ. Hell. Stud.*, 1924, pp. 172–7. The Ligurian amber mentioned by Pliny (XXXVII, 33) was excavated *in situ*.

93. Déchelette, p. 395; De Navarro, *Antiquity*, 1928, pp. 430–2, and map on p. 428. P. Jacobsthal, *Germania*, 1939, pp. 14 ff.

94. ll. 641 ff., with notes by Schulten and Bosch. The 'gaping cavern' is clearly a crevasse in the Rhône Glacier. The 'Sun Mountain' may be the Finsteraarhorn (Clerc, p. 154, n. 3). The 'narrow gorge' is the Perte du Rhône below Geneva (also known to Aristotle, *Meteorol.*, I, 13, 30). Despite Avienus's use of the term 'palus' which ill describes a deep moraine-groove, the preceding lake can only be Lake Leman. (So Clerc, as against Schulten and Bosch, who identify it with the inundation area above Arles.)

95. Ap. Pliny, XXXVII, 32.

96. Von Duhn, loc. cit.; Forrer, pp. 83–93; Clerc, p. 352 ff.; A. Blanchet, *Revue belge de Numismatique*, 1913, pp. 315–17. The merchants of the Sigynni (presumably a Ligurian people), who plied from Venetia to the Rhône Valley, no doubt used the same track (Herod., V, 9).

97. Stähelin, pp. 40–1; Blanchet, p. 314; E. Norden, *Die germanische Urgeschichte in Tacitus Germania*, pp. 225–32. The abundance of gold in Helvetia is noticed by Strabo (VII, 293).

98. Blanchet, pp. 317–22; Forrer, pp. 66–79. For the finds of Greek pottery, see De Navarro, p. 428.

99. Strabo, IV, p. 190.

100. Diodorus, V, 22. This passage is derived from Poseidonius, a writer of the early first century (Cary, p. 175).

101. Cary, p. 178.

102. IV, ll. 634–44 (no doubt an embroidery upon Timaeus).

103. Strabo, IV, 190.

104. Cicero, *Pro Fonteio*, sect. 11 (delivered in 63 B.C.). It is not clear from Cicero's words how far the Italian money-lenders went beyond the borders of the province of Gallia Narbonensis.

105. Diodorus, V, 26 (wine bartered against slaves). Dry areas among the Nervii (by the Sambre) and in Germany: Caesar, *Bell. Gall.*, II, 15, 4; IV, 2, 6.

106. Forrer, p. 103.

NOTES

107. The Rhine is mentioned in *De Mirab. Auscult.*, chap, 168. The date of this passage is uncertain.

108. In 56 B.C. Cicero declared that not a day passed, but Caesar's letters and dispatches filled his ears with novel names of peoples and places. (*De Provinciis Consularibus*, 22.)

109. This delusion is implicit in all that Strabo says about the Atlantic seaboard.

110. On the details of penetration in Germany, see Sir Mortimer Wheeler, *Rome beyond the Imperial Frontiers* (Pelican Books), pp. 21–117.

111. Forrer, op. cit., *passim*; Stähelin, p. 39; Jullian, *Revue des Études anciennes*, 1906, pp. 113–15.

112. Forrer, pp. 316–43.

113. Caesar's bridge across the Rhine was probably built in the swift-flowing reach between Koblenz and Andernach. See T. R. Holmes, *Caesar's Conquest of Gaul* (2nd ed.), pp. 706–10.

114. On the campaigns of Drusus and Tiberius, see Cassius Dio, LV, 1; Strabo, VII, 292; Velleius Paterculus, II, 105–10.

115. On Roman trade in Germany, see Mrs O. Brogan, *Journ. Rom. Stud.*, 1936, pp. 195–222; S. Bolin, *Eynden ev romerska mynt i det fria Germania*.

On the upper Danube German traders were licensed to visit the Roman station at Augsburg (*Germania*, chap. 41). But Tacitus emphasizes that this arrangement was exceptional.

116. Thus Ptolemy, instead of locating the Hercynian Forest (by which he means the forest belt of northern Bohemia) in reference to the Elbe, states that it lay above the Quadi, who lived above the forest of Luna, which lay above the Baemi, who lived above the Danube (II, 11, 11).

117. Malone, loc. cit., pp. 362–70.

CHAPTER SEVEN: ASIA

1. S. Smith, *Early History of Assyria*, 277 ff., 298 ff., and his Chapter x on Cappadocian trade, etc. R. W. Rogers, *A History of Babylonia and Assyria*, II, 322 ff. Tarn, in *Journ. Eg. Arch.* 1929, 11–12.

2. Egyptians in Asia: Breasted, *Ancient Records*, I, 236, 250, 267, 315, 360, 681, 707; II, 30, 101; III, 20, 453, 471, 490; IV, 119, 122; Euphrates: id., II, 68, 73, 478, 479 (p. 203), 588 (quoted); cf. 574 ff.

3. See pp. 37–42 and 234.

4. Bunbury, I, 45. The Maeonians and Phrygians are the furthest peoples mentioned – Homer, *Il.*, III, 401; XVIII, 291. The name Ἀσία occurs in a local sense only: *Il.*, II, 461.

5. See p. 42.

6. Hesiod, *Theog.*, 340.

7. Alcman the poet, in Steph. Byz., s.v. Ἰσσηδόνες.

8. Herod., IV, 13, 32. Suidas, s.v. Ἀριστέας. Did he hear remotely of the Arctic Ocean? – See IV, 13. W. Thomaschek in *Sitzungsberichte der Philosoph.-*

NOTES

Hist. Klasse der Kais. Akad. der Wissensch., Wien, 1888, p. 757 ff., is favourable on Aristeas. The Phoenicians are mentioned even in Homer, *Il.*, VI, 288-91.

9. Ctesias, reproduced by Diodorus, II, 23-7.

10. Strabo, XIII, 617. The alleged visit of Pythagoras need not be believed.

11. Strabo, XI, 517. Herod., I, 202, 204.

12. Herod., IV, 44; III, 97.

13. Williams Jackson, in *Camb. Anc. Hist.*, I, 341; Herod., III, 97 ff.; IV, 44; Athenae., II, 70c. Arabians: Herod., III, 7-8; IV, 88; II, 75, 155; III, 4-9, 91, 97; *Camb. Anc. Hist.*, VI, 181-2. Hecataeus, frs. 174-8 Müller = frs. 294-9 Jacoby (*Die Fragmente*), may have relied on Scylax's report. See also Arr., *Anab.*, I, 43, 4-5. Scythian exped.: Herod., IV, 83 ff., 116; the Oarus is not the Volga? - See Bunbury, I, 203-4.

14. Herod., V, 100, cf. 31, 54.

15. Herod., V, 52-4. For the confusion, see Calder in *Class. Rev.*, XXXIX, 1925, pp. 7-11; and for the older views see also Macan, *Herod.*, Vol. II, Append. XIII; Ramsay, *Journ. Hellen. Stud.*, XL, 1920, 89 ff.; Kiepert, *Monatsber. Berl. Akad.*, 1857, 123 ff.; Bunbury, I, 249 ff., 259-60; Tozer, 90-1; cf. Herod., I, 73-9, VII, 26 – the looping route again.

16. Herod., V, 49, 50.

17. Hecataeus in Jacoby, frs. 196-299, and cf. Bunbury, I, 134 ff., for details. Wells in *Journ. Hellen. Stud.*, XXIX, 1909, 41 ff., disputes the authenticity. Races in Xerxes' army: Herod., VII, 61 ff.

18. Aesch., *Prom. Vinct.*, 411 ff., 709 ff., 730 ff., 2; *Prom. Sol.*, fr. 1; *Persae*, passim, cf. Gow, in *Journ. Hellen. Stud.*, 1928, 133 ff.; *Seven against Thebes*, 817-23; *Suppl.*, 547 ff., 284-6; cf. Euripides, *Bacchae*, 15, and in Strabo, XI, 519-20; I, 27; Aristoph., *Birds*, 552.

19. Compare ideas of Sophocles, in Pliny, XXXVII, 40, and (*Triptol.*) in Strabo, I, 27.

20. Herod., I, 4.

21. Pliny, XXXIV, 68.

22. Herod., III, 129-38.

23. Strabo, I, 47.

24. Strabo, I, 49; XII, 570, etc. Democritus: Cic., *De Fin.*, V, 19; Strabo, XVI, 703.

25. Herod., II, 104, 44; I, 181-3 (evidence of travels in Asia). Modern scholars, e.g. E. Meyer, Lehmann-Haupt, H. Stein, Busolt, differ as to the dates. The satrapies: Herod., III, 90 ff.; the Euphrates, etc.: I, 180 ff.-189, 194; Asia Minor: I, 6, 22, 28, 72, 93; IV, 39, etc.; II, 34. Waist: I, 72; II, 34; cf. *Hellen. Oxyrh.*, XVII, 4; Pseudo-Scylax, 102; Bunbury, I, 233. Araxes: Herod., I, 202, 204. Details about Iran: e.g. I, 95-123, 125, 178 ff.; IV, 37; VIII, 117. India: III, 98, 106; IV, 40, 44; its cotton and gold: III, 106, cf. VII, 65.

26. For Ctesias, see J. W. McCrindle, *Ancient India as described by Ktesias*.

27. Herod., I, 203 (Caucasus); 202-3, Caspian, fifteen days long, eight days across at its broadest. Was it then joined to the Aral? Scythians and Greeks: IV, 24.

NOTES

28. id., IV, 21, 116–17. The Tanais had been sailed on a little; one tributary was heard of, and the source was said to be a large lake (see Herod., IV, 57); it is, in fact, a small lake.

29. id., IV, 21, 22, 108–9.

30. id., IV, 108–9. The details given are curious. But can we accept the Greek settlement as a fact?

31. Steph. Byz., s.v. Ὑπερβόρεοι.

32. Herod., IV, 24; cf. I, 204.

33. id., IV, 16, 13.

34. id., IV, 22.

35. id., IV, 23, 24.

36. id., IV, 25–7; cf. Hippocrates, *De Aere*, 17 ff. So far the identifications given are those fixed on by Mr S. Casson in *Brit. Sch. at Athens*, XXIII, 183 ff., 190. For other views, see Bunbury, I, 193–202; Tozer, 86–8; J. T. Wheeler, *The Geog. of Herodotus*, 181 ff.; E. H. Minns in *Encycl. Brit.*, 11th ed., s.v. Scythia, and in his *Scythians and Greeks*, pp. 105–14; Berger, 226 ff. For the Issedones see Kiessling, in Pauly-Wissowa, s.v.

37. Herod., I, 204.

38. id., IV, 25; Bunbury, I, 198.

39. id., IV, 31.

40. E. H. Minns, *Scythians and Greeks*; M. Rostovtzeff, *Iranians and Greeks in South Russia*.

41. These, like the tribes in the Taurus range, and others in Zagros, inner Elam, and North Media, were a constant trouble to the Persian Government.

42. Xenophon, *Anabasis*, I, 1–2 (Bunbury, I, 346).

43. id., I, 3–4 (Bunbury, I, 364–5).

44. id., I, 4, sects. 9–11. At Issus there came over to Cyrus 400 Greeks who had already seen the Euphrates while serving under Abrocomas.

45. id., I, 4–8, 10 ff.

46. id., II, 4, sects. 14 ff.–III, 1.

47. id., III, 3, sects. 6 ff.

48. id., IV, 1–7.

49. id., IV, 5, sects. 12–14 (quoted); IV, 7, sects. 19–28, sect. 22; F. Segl, *Vom Kentrites bis Trapezus*. He gives his own ideas and summarizes former solutions. See also Bunbury, I, 344–5, n. 8; Tozer, 112–18. Tarn, in *Camb. Anc. Hist.*, VI, 4 ff.

50. Xen., id., IV, 5, sects. 25–7.

51. id., V, 2–VII, 1.

52. See Isocrates, *Panegyr.* (280 B.C.); *Philippus* (346); *Letter I* (368?); *Letter IX* (356?).

53. Ephorus in Jacoby, frs. 43 ff.; 180–5. Bunbury, I, 382.

54. Strabo, II, 76.

55. Aristotle, *Meteorol.*, I, 13, 15 ff.; Ephorus in Jacoby, frs. 30, 158, etc.

56. Aristotle, *Meteorol.*, II, 1, 10.

57. Strabo, I, 14. Those interested in Alexander should read W. W. Tarn,

NOTES

Alexander the Great, Vol. I (narrative) and II (sources and studies); U. Wilcken, *Alexander der Grosse* (translated as *Alexander the Great* by G. C. Richards); A. Weigall, *Alexander the Great*.

58. Arrain, *Anabasis*, I, 13, 1 ff.; 17, 3 ff.; I, 29; II, 3, 1 ff.

59. id., II, 4, 1 ff., 6–11, 15.

60. id., II, 15 ff.; III, 1 ff.

61. id., III, 8–14.

62. id., III, 15 ff. So well was Alexander served that he was able to send Parmenio along the main road while he himself by a mountain-climb forced the Persian Gates, Jan. 330 – id., III, 17, 18.

63. id., III, 19–21.

64. id., III, 23; 19, 7; 24, 1–3 (Curt., VI, 5, 13; Plut., *Alex.*, 44); III, 25, 1. A. F. v. Stahl, in *Geog. Journ.*, LXIV, 1924, 312.

65. Arr., op. cit., III, 25, 3 (Strabo, XV, 724–5).

66. id., III, 25, 6 (Pliny, VI, 61, 93; Strabo, XI, 514, 516; XV, 723).

67. id., III, 25, 8; 28, 1.

68. Ptol., IV, 10, 4. Isidore, *M.P.*, 19.

69. Arr., op. cit., III, 28, 4; IV, 22, 5. (Diodor., XVII, 87; Curt., VII, 3, 23.)

70. id., III, 29, 1.

71. id., III, 29, 2 ff.

72. id., III, 30, 1–3.

73. id., III, 30, 6 ff. (cf. Strabo, XI, 505–6, 509). For Chodjend, see id., IV, 1, 3 ff. (cf. App, *Syr.*, 57; Pliny, VI, 49, etc. Whether Nearchus saw *Chinese* silk – Strabo, XV, 1, 20 – we do not know.)

74. For Alexander in Sogdiana, see F. v. Schwarz, *Alexanders des Grossen Feldzüge in Turkestan*. Strabo, XI, 513, and Pliny, VI, 18, and some modern followers make Zariaspa and Bactra the same. But Arrian, op. cit., IV, 7, 22, makes them different, and his sources were good – Aristobulus and Ptolemy son of Lagus. We therefore follow him as do Bury and others. Read if you wish F. Reuss, *Baktra und Zariaspa*, in *Rhein. Mus.*, LXII, 1907, p. 591.

75. Arr., op. cit., IV, 15, 7.

76. Curt., VII, 40; Pliny, VI, 47; Steph. Byz., s.v. Ἀλεξάνδρεια. Alexander seems to have founded up to twelve in these regions.

77. Strabo, XI, 509–10, cf. II, 73. Arr., op. cit., IV, 15 ff. Aral: echoes of it in Pliny, VI, 51 (authority: Alexander the Great), cf. Curt., VI, 12; Mela, III, 5. Also the Maeotis of Strabo, XI, 509, 510, and Polyb., X, 48. See Tarn., *Journ. Hellen. Stud.*, XXI, 1901, pp. 21–2.

78. Strabo, XI, 518.

79. Arr., op. cit., IV, 15, 4–6; VII, 1, 3. (Berger, 55 ff., 72).

80. id., IV, 5, 6. Schwarz, op. cit., 44–5. (Lack of water in Bactria – cf. Marco Polo, I, 22.)

81. For the Indian campaigns, see Bevan, in *Camb. Hist. of Ind.*, Vol. I, chaps. 15–16; J. W. McCrindle, *The Invasion of India by Alexander the Great*; V. A. Smith, *The Early History of India*, p. 52 ff.; Sir T. Holdich, *The Gates of India*, 66 ff., 94 ff.

NOTES

82. Arr., *Anab.*, v, 26. 1–2 (Strabo, xi, 511; xv, 689; Berger, 75 ff.). Geographical ideas: P. Bolchert, *Aristoteles Erdkunde von Asien u. Libyen*; H. Endres, *Geographischer Horizont und Politik bei Alexander den Grossen*.

83. Arr., *Anab.*, iv, 22. 3 ff. (Strabo, xv, 697–8). Indus: id., iv, 22, 7; 28, 5; v, 3, 5; v, 7. A. Foucher, *Notes sur la géog. anc. du Gandhara*.

84. Arr., *Anab.*, iv, 23; 25, 5 ff.

85. id., iv, 28. Sir A. Stein, *Alexander's Campaigns on the Indian N.W. Frontier*, Geog. Journ., Nov.–Dec. 1927, pp. 417 ff., 515 ff.; he decides for Pirsar – 515 ff. – against the identification with Mahaban – 517. Cf. his *Report of Archaeological Survey Work ... Jan. 2, 1904–Mar. 31, 1905*; and his *Alexander the Great*, in *The Times*, 25, 26 Oct, 1926; and *On Alexander's Track*, 143.

86. Arr., *Anab.*, v, 3, 6; 8, 2; 4 ff., 15 ff. (Strabo, xv, 270; Curt., ix, 1, 6. Diodor., xviii, 89.)

87. Chenab.: id., v, 20, 8–9. Ravi: v, 21, 4–6; Cathaeans and suttee: Aristobul. in Strabo, xv, 714. Diodor., xvii, 91; cf. Arr., *Anab.*, v, 22–4. Beas: id., v, 24, 8 ff.

88. Refusal: id., v, 25–8. Sandrocottus: Plut., *Alex.*, 62. How much Alexander had heard about the Ganges is really uncertain; cf. Tarn, *Journ. Hellen. Stud.*, xlii, 1922, 93 ff., and *Camb. Anc. Hist.*, vi, p. 410. Course of Ganges: Strabo, xv, 719, 689, 690, 698. Craterus falsely said that Alexander had reached the Ganges and claimed to have seen it – id., xv, 702, cf. *Peripl.*, 47.

89. Arr., *Anab.*, vi, 1, 1–6.

90. id., vi, 1–5.

91. id., vi, 6–11; 13 ff.

92. id., vi, 15, 2; 17, 5.

93. id., vi, 18, 2–20; 21, 3 (cf. id., *Indica*, 18 ff.).

94. Strabo, xv, 721 ff.

95. Arr., *Anab.*, vi, 21.

96. id., vi, 23, cf. 25; 24, 1; 27, 1.

97. id., vi, 27, 3. (Strabo, xv, l. c.; cf. Holdich, op. cit., 147–8.)

98. id., vi, 28, 5. (Pliny, vi, 107; Ptol., vi, 8, 14.)

99. id., vii, 15.

100. Pliny, vi, 183 ff.

101. Arr., *Anab.*, vii, 16, 1–3; Curt., vi, 4, 18; Diodor., xvii, 75, 3; Pliny, vi, 36–8; Strabo, xi, 509–10.

102. Of course he cannot have intended to make Arabia the seat of his empire: Strabo, xvi, 785.

103. See W. Tarn in *Camb. Anc. Hist.*, Vol. vi, chap. 13, pp. 423 ff., esp. pp. 431 ff, and in his *Alexander the Great* (see above). For his new cities, see V. Tscherikower, *Die Hellenistischen Städtegründungen*, index, s.v. Alexandreia, and esp. pp. 138–54.

104. Not official, thinks Bunbury, i, 481–3. But Eratosthenes was able to draw on a *register* of these surveys: Strabo, xv, 689. See Jacoby, *Die Fragmente*, etc., ii, bd2, 406–7.

280

NOTES

105. Pliny, VI, 61–2; Strabo, XV, 697–8, 723–4. For Philonides, see Pliny, I, 4; II, 181 (in Europe).

106. Pliny, VI, 44–5; Strabo, XI, 514; cf. II, 80.

107. Pliny, VII, 11, makes this likely; cf. id., 69.

108. Eratosthenes in Strabo, II, 79, 80, 71; Phasis surveyed: II, 92. Roads between Thapsacus and Armenian hills not surveyed: II, 79, 80.

109. Strabo, XV, 723, cf. XI, 514; Athenae., XI, 102, 500d; XII, 39, 529e-530a; II, 74, 67a; X, 59, 442b; XII, 9, 514 f.; Ael., *N.A.*, XVII, 17; V, 14.

110. Strabo, XV, 689–90; Pliny, II, 184–5; VII, 28; Arrian, *Indica*, 3, 6. Ceylon: Pliny, VI, 81; Strabo, XV, 691. C. Lassen, *Ind. Alt.*, III, 21 ff.; Taurus was looked on as a great backbone of Asia running from Asia Minor to the E. Ocean at 'Tabis' – Strabo, II, 68; XI, 519; cf. Mela, III, 7; Pliny, VI, 53.

111. Strabo, II, 96. India: id., XV, 685 ff.; Pliny, VI, 56 ff.

112. cf. W. Tarn in *Camb. Anc. Hist.*, Vol. VI, pp. 432–6.

113. See W. Tarn (and G. T. Griffith), *Hellenic Civilization* (3rd edition, 1952), for a solid survey of this era of Greek History.

114. Tarn, op. cit., 122–3, 194–6. A glance at the map facing p. 155 of *Camb. Anc. Hist.*, Vol. VII, will show what parts of Asia Minor became best known by the Greeks of this age.

115. Strabo, XII, 538.

116. Polyb., X, 27; Strabo, XI, 525; e.g. Rhagae, new town of Seleucus – Strabo, XI, 524; Pliny, VI, 43.

117. Tarn, op. cit., 120. V. Tscherikower, as in note 103, 174–6.

118. Pliny, VI, 49.

119. Strabo, XV, 724; Appian, *Syr.*, 55; Justin, XV, 4. Plut., *Alex.*, 62; Athenae, I, 18d.

120. Pliny, VI, 47. Strabo, XIV, 516.

121. Pliny, VI, 58. Gibbon (Vol. V, p. 42, n. 9, ed. Bury) suggests that Patrocles and his fleet went down the Oxus to the Caspian.

122. Strabo, II, 69.

123. See note 77.

124. cf. Strabo, XI, 511 ff.

125. For the whole question, see Strabo, XI, 509, 518; II, 68, 74; Diodor., XIX, 100; Plut., *Demetr.*, 47; Phot., *Cod.*, 224. W. Tarn, on Patrocles and the Caspian, in *Journ. Hellen. Stud.*, XXI, 1901, pp. 10 ff., whose views have largely been adopted here. C. Müller, *Fragmenta Historicorum Graecorum*, II, 442. Did Patrocles mistake the Atrek for the Oxus? See Tarn, *The Greeks in Bactria and India* (2nd ed., 1951), pp. 112–13, 444, 488–90; J. O. Thomson, *History of Ancient Geography*, pp. 127–9.

126. Seleucus' scheme: Pliny, VI, 31. Neumann, in *Hermes*, XIX, 183. Polemon, ap. Athenae., IX, 387 f., cf. Pliny, XXVI, 43; Strabo, II, 91. Tanais: id., 492–3. Tanais mart: id., 493; Steph. Byz., s.v. Τάναϊς. Eratosthenes near the end of the century summed up the knowledge of his time.

127. Strabo, II, 70; XV, 702. Excavation has confirmed Megasthenes' remarks about Patna. See B. C. J. Timmer, *Megasthenes en de indische*

NOTES

Maatschappij (Amsterdam, 1930); Stein, in Pauly-Wissowa, *Real-Encyclopädie*, article 'Megasthenes'.

128. id., XV, 719; Pliny, VI, 62.

129. id., XV, 690, 719; Arrian, *Indica*, 4.

130. Pliny, VII, 25; Strabo, XV, 705–6.

131. See Strabo, XV, 712 ff., 703 ff. Arr., *Ind.*, 4, 2 ff., 5 ff., 7, 1; Clem. Alex., *Strom.*, I, 15, 17, perhaps from Megasthenes.

132. Arr., *Ind.*, 7; Strabo, pp. 68–70, 76–7, 686–7, 689, 690, 693, 702–7, 709–13; Arr., *Ind.*, I, 1, 3, 7–8; Pliny, VI, 81; Diodor., II, 39, 5; Arr., *Ind.*, I, 10, 8–9. India still pointing eastwards: Strabo, III, 1, 34.

133. Strabo, XV, 689, surely from Megasthenes ultimately; in Pliny, VI, 69, does not Diamasa = the later Damirice, the west coast of the Tamils?

134. Strabo, II, 70, cf. 76.

135. V. A. Smith, *Early History of India*, 193.

136. Polyb., X, 28 ff., 48 ff.; XI, 34; Justin, XLI, 5.

137. Polyb., XI, 34; X, 49. For A. and the Persian Gulf, see pp. 72–3.

138. See, e.g. Artemidorus in Strabo, XIV, 663–4, and cf. the interest taken by Polyhistor, c. 82–60, Müller, *F.H.G.*, III, 206 ff.

139. Trog., *Prol.*, 41; Justin, XLI, 4, 5; Strabo, XI, 515.

140. Polyb., XI, 34, 2; Strabo, XI, 516; Isid., *M.P.*, 19.

141. Justin, XLI, 6, 8. V. A. Smith, *Early Hist. of India*, 240–1, 263 ff. Strabo, XI, 511.

142. Apollodorus in Strabo, XI, 516; II, 118; cf. XV, 685–6 and 719, where read Ἰομάνης (the Jumna). In 685–6 the Ἰσαμός might be the river Son. V. A. Smith, op. cit., 210, 227.

143. *Peripl.*, 47.

144. Strabo, XI, 516; XV, 701–2. Tomaschek, *Über das Arim.-Gedicht*, p. 769, reads, in Strabo, 516, Φοῦνοι = Huns. Cf. Pliny, VI, 55, *Phuni*. The Seres as a name meant the Chinese and their middle-men as approached by the land-routes.

145. Sir A. Stein, *Sand-buried Ruins of Khotan*, 396.

146. V. A. Smith, op. cit., 238, n. 3, Dittenberger, *Syll.*, 588, l. 109. For the Graeco-Bactrians in general, see Tarn, *The Greeks in Bactria and India*; V. A. Smith, op. cit., 233 ff., 210, 227; H. H. Wilson, *Ariana Antiqua*, 215 ff.; Cunningham in *Numism. Chron.*, N.S., VIII–XII (1868–72); P. Gardner *Coins of the Greeks and Scythian Kings of India* (Catal. of Greek coins in B.M., X); W. Tarn, Notes on Hellenism in Bactria and India, in *Journ. Hellen. Stud.*, XXII, 1902, 268 ff.; Clarke in *Classical Philology*, 1919, 310 ff. D'Alviella, *Ce que l'Inde doit à la Grèce*; G. N. Banerjee, *Hellenism in Ancient India*; A. v. Sallet, *Die Nachfolger Alexanders des grossen in Baktrien u. Indien*, Zeitschr. f. Numism., VI, 165–209; A. v. Gutschmid, *Geschichte Irans*, 37 ff.; Bunbury, II, 46–7, 102–3; H. G. Rawlinson, *Bactria*, 50 ff.

An important new book, written by an Indian scholar from the Indian point of view, is A. J. Narain, *The Indo-Greeks*. (*A numismatic and historical study*.)

147. Strabo, I, 14; XI, 508.

NOTES

148. Strabo, I, 14.

149. Bunbury, II, 104–5. Plut., *Luc.*, 23–9; App., *Mithr.*, 84–6; Memnon, 46, 56, 57; Eutrop., VI, 9. Livy, *Ep.*, XCVIII, Cass. Dio, XXV, 2; Strabo, XI, 532.

150. Plut., *Luc.*, 30–2; App., *Mithr.*, 87; Cass. Dio, XXXV, 4–7. T. R. Holmes, *The Roman Republic*, I, 407 ff. Eckhart in *Klio*, X, (1910), 72 ff., 192 ff. The site of Tigranocerta is uncertain. See e.g. Holmes, op. cit., 409 ff.

151. App., *Mithr.*, 97–102, 107; Cass. Dio, XXXVI, 46 ff.; Plut., *Pomp.*, 32; Livy, *Ep.*, CI; Oros., VI, 4. Strabo, XI, 526 ff. (details of Armenia), cf. 496–7, XII, 555.

152. Plut., *Pomp.*, 36. A good deal of Theophanes' work is embodied in Strabo. Metrodorus and Hypsicrates: Strabo, XI, 504.

153. Strabo, XI, 496, 503–4; XIII, 617; XI, 491–2. Darial: XI, 500. Suram: 497; Kur (and Araxes), 500, 501 (before long the Araxes joined the Kur – see Tozer, 221). Iberi: 499–501. Albani: 501 ff. Dioscurias: 497–8; Pliny, VI, 15. Piracy: Strabo, XI, 495.

154. Pliny, VI, 52. Tobogganing, etc.: Strabo, XI, 506.

155. Plut., *Pomp.*, 38–9; Cass. Dio, XXXVII, 15; App., *Syr.*, 18. Pontus became a province in 65.

156. Plut., *Comp. Nic. et Crass.*, 4; *Luc.*, 30; *Crass.*, 16.

157. Plut., *Crass.*, 17 ff.; Cass. Dio, XL, 12 ff.; Pliny, VI, 47.

158. Plut., *Caes.*, 58; Cass. Dio, XLIII, 51; Suet., *Iul.*, 44.

159. Vell., II, 82; Livy, *Ep.*, CXXX–CXXXI; Plut., *Ant.*, 31, 50; Cass. Dio, XLIX, 28 ff., 33.

160. E. Warmington, *The Commerce between the Roman Empire and India*, 334, n. 31.

161. Strabo, XVI, 766, 767 ff. Bunbury, I, 648.

162. Pliny, VI, 141 ff., 147 ff.

163. Strabo, XVI, 780–2; II, 118; Pliny, VI, 160–2; Cass. Dio, LIII, 29; *Mon. Ancyr.* (Lat.), V, 18–23; Virg., *Aen.*, VIII, 705.

164. Bunbury, II, 179 ff., authorities on pp. 204–6; Warmington, 334, n. 29. Tkač in Pauly-Wissowa, s.v. Saba, 1343 ff. C. M. Doughty, *Travels in Arabia Deserta* (ed. Warner, 1921), Vol. II, pp. 175, 176 (quoted). Augustus himself tells us that the army reached the borders of the Sabaeans and Mariba – *Mon. Ancyr.* (Lat.), V, 18–23.

Some deny that the Mariba, Mariaba, etc., of this expedition was the Sabaean capital at Mar'ib – see Tkač, op. cit., 1354 ff., and id., 1368, 1370–1 (on Ilisaros); 1432 ff.

165. Charax: Pliny, VI, 140. Jordan, etc.: id., V, 77, 80, 71–3, 88. Trajan and Nabataeans: Cass. Dio, LXVIII, 14; LXXV, 1, 2. Eutrop., VIII, 18. Road-building: Warmington, 349, n. 31.

166. Ptol., VI, 7, 1–47.

167. Travellers: Strabo, III, 166; Pliny, V, 12. Better chances: Strabo, XI, 508 (cf. 507). That the traveller Pausanias in the second century A.D. had met no one who had seen Susa or Babylon is no sign of inactivity; under the Empire

NOTES

most people travelled for gain. Who would want to visit Parthian places like these, Babylon already being a city of the past?

168. Isidore, *M.P.*, 1-19; Strabo, XVI, 748-9; Cass. Dio, LXVIII, 30.

169. Hor., *Odes.*, I, 12, 56; III, 29, 27; IV, 15, 23; Virg., *G.*, II, 121; Propert., IV, 3, 8; and see Warmington, 366, n. 63.

170. cf. Warmington, 24-5.

171. Diog. Laert., II, 17; Pliny, I, 37, etc. Jacoby, II, B1, 629-31; Pliny, XXXVII, 46.

172. Veget., *De re militari*, III, 6-military painted maps. Large maps: Vitruv., VIII, 2, 6. Agrippa: Pliny, III, 17. Had Caesar begun this? – Bunbury, II, 693, 706.

173. Pliny, VI, 127-8.

174. Mela, I, 1, 11; III, 8, 60 – a notice from the land-trade, not the sea-trade as is that of the *Peripl.*, 65. See also Pliny, VI, 54; XXXIV, 145; Dionys, *Per.*, 881-93; cf. Sen., *De Ben.*, VII, 9, 5. Trajan: see Warmington, 93-4. Expectation of further conquests: Stat., *S.*, IV, 1, 40-2; cf. IV, 3, 155; Sil. Ital., *B.P.*, III, 612-15; Mart., XII, 8, 8-10; Plut., *Pomp.*, 70.

175. Philostrat., *Apollon.*, I, 19 ff., 40; II, 4 ff.; III, 53 ff. O. de B. Priaulx, *The Indian Travels of Apollonius*, 1-61. The present writer once thought he saw fact in the story – Warmington, 78, 88.

176. Stat., *S.*, III, 2, 136-8.

177. Dionys. Perieg., 713-14.

178. Warmington, pp. 110-11. Greek penetration into India from most of the marts on the coast was not very far-reaching; it therefore finds a place in the chapter on Indian Waters, pp. 100-2.

179. The routes to China as reported by Maes: Ptol., I, 11, 4-8; 12, 2-10; VI, 13, 1; 16, 1-8; cf. Ammianus Marcellinus XXIII, 6, 60, 67-8; Ptol., IV, 13, 3; 14, 1, 3; VI, 10, 4; 11, 9; 13, 23; VI, 14, 1. Gerini in *J.R.A.S.*, 1897, Tab. V and Tab. XI, in connexion with pp. 551 ff. For other modern authorities, see Warmington, p. 336, n. 58; 357, notes 130-3; and especially Sir H. Yule, *Cathay and the Way Thither*, I, pp. 1-66, 187 ff. cf. also J. O. Thomson, *Hist. of Anc. Geog.*, pp. 306-19. Bactrians, etc.: Warmington, p. 105; Bactrian embassies: *Historie Augusta Vita Hadriani*, chap. 21; 'Vict.', *Ep.*, 15, 4; cf. Appian, *Praef.*, chap. 7. Stations: Cosmos, II, 97b. Coins at Singanfu: *Academy*, 1886, no. 730, p.316. Silk a moth-product: Paus., VI, 26, 4, 6-8. The fear of Asia = Mela, III, 29; Pliny, VI, 53. Nothing in the ancient West can be traced to Japan, except possibly the Bantam fowls which Romans received along the land-routes – Pliny, X, 146, 156.

180. J. Hackin, *Recherches archéologiques à Bégram* (*Mém. de la Délégation Archéologique Française en Afghanistan*, IX, 1939); R. Ghirshman, *Bégram, Recherches archéologiques et historiques sur les Kouchans* (Cairo, 1946); J. Marshall, *Excavations at Taxila*, pp. 6 ff.; id., *Guide to Taxila*, pp. 10 ff., 26 ff.; the same author's *Taxila*, in 3 vols. (Cambridge, 1951). cf. Sir M. Wheeler, *Rome beyond the Imperial Frontiers*, pp. 157 ff. = 187 ff.

181. Sir M. A. Stein, *Ruins of Desert Cathay*, I, pp. 410-11, 457-8, 467,

NOTES

471–2, 480, 486–7, 492–3; cf. p. 381. Also his *Serindia, Detailed Report of Explorations in Central Asia and Westernmost China*.

182. Strabo, XI, 509, 518. Nepos: Pliny, II, 170; Mela, III, 5, 45; cf. Pliny, II, 167. Indus is a European name in some instances, with no connexion with the Indians of the East.

183. Warmington, 341, n. 23.

184. id., 348, n. 4. Schur, *Klio*, Beih., XV, 1923, 62–9.

185. Tac., *Ann.*, XIII, 7 ff.

186. Mela, III, 38; cf. anon., *De Mundo*, 3, sect. 12. Scymnus in Schol. ap. Apoll. Rhod., IV, 284.

187. Dionys. Perieg., 305, 308 (Hadrian's time, probably).

188. Arrain, *Peripl.*, esp. 9, 3 and 5; 10, 3; 11, 2–3; *Anth. Pal.*, IX, 210; Ptol., V, 10, 2; Joseph., *B.J.*, II, 16; *Cagn.-Laf.*, *I.G.R.*, III, 133.

189. Dionys. Perieg., 730.

190. Ptol., VII, 5, 4; III, 5, 14; V, 9, 12, 13, 17, 19, 21; VI, 14, 4 (Caspian and Volga); V, 9, 13 ff. (Russia in Asia); VI, 12, 1 ff., etc. (Iaxartes).

191. Gibbon, Vol. IV, pp. 354–5, ed. Bury.

CHAPTER EIGHT: AFRICA

1. Herod., II, 96, 41, 179. Juv., XV, 129, etc. Diodor., I, 96. C. Torr, *Ancient Ships*, 2, 9, 56, 78, 106–7, 137.

2. For Egypt and the lands to the south, see the original sources in Breasted, *Ancient Records of Egypt*, e.g. I, 333, 336 (Harkhuf); 366 ff. (Sebni); 370, 640 ff. (Nubia); II, 121, 271, etc. (Cush); cf. Sir E. Budge, *The Egyptian Sudan*, I. 505 ff. For Cush, see also C. Conti Rossini, *Storia d'Etiopia*, 39 ff. Jebel Moya – Brit. Assn. Rep., 1912.

3. Menelaus: Homer, *Od.*, IV, 83–5, 125–30; cf. 226–30; XIV, 246. Aegyptus river: *Od.*, IV, 581; XIV, 257.

4. *Il.*, XXIII, 205–6. *Od.*, I, 22–4; V, 282; cf. Strabo, I, 1, 3, 6, 7, 24.

5. *Il.*, III, 3–6. Bunbury, I, 50. Cranes migrant from Europe – Aristotle, *H.A.*, I, 8, 2, rightly as shown by P. du Chaillu, *The Country of the Dwarfs*, chap. 25, pp. 261 ff., wrongly denied by G. Schweinfurth, *Im Herzen von Afrika*, chap. 16, pp. 133 ff.

6. Hesiod., *Theog.*, 985. Nile: id., *Theog.*, 338. Strabo, I, 29.

7. Strabo, I, 36. Herod., II, 5 (Hecatae, Jacoby, fr. 301).

8. H. Röhl, *Inscr. Gr. Ant.*, 482. Herod., II, 154, 178.

9. Pythagoras: Isocrates, *Laud. Bus.*, 28. Hecataeus: Herod., II, 143. Solon: id., I. 30. Alcaeus: Strabo, I, 37. Democritus: Diodor., I, 98.

10. Herod., III, 17–26; cf. 114, 97.

11. The difficulty pointed out by Bunbury, I, 271 ff., is largely removed if we take μακρόβιοι as Africans extending from the Nile to the Gulf of Aden; in II, 8, Herodotus refers to a huge mountain stretching from the Nile to the 'Red' Sea and incense regions; here again Nubia, Abyssinia and Somaliland

NOTES

are vaguely connected. Strabo, XVII, 70, confuses the two campaigns, but rightly regards Meroe as an object of Cambyses's attacks, a fact not realized by Herodotus. For the desert, see Hoskins, *Travels in Ethiopia*, 19–32; K. Lepsius, *Briefe aus Ägypten*, 124–36. For further comments on Cambyses: see Sir E. Budge, *The Egyptian Sudan*, II, 93–4 (but Nastesen is now known to be of later date). For the Ethiopians as 'long-living', see pp. 206, 239.

12. Herod., III, 26.

13. Indians and Ethiopians: Aeschylus, *Suppl.*, 284–6, cf. statement about Nile sources in *Prom. Vinct.*, 807 ff. Canobus is ἐσχάτη χθονός, id., 846. The Athenian expedition to Egypt, 459–454, did not go beyond Memphis – Thuc., I, 104, 110.

14. See the account of Egypt, Herod., II. He reached Thebes (II, 3) and Elephantine (II, 29).

15. Herod., II, 9, 17.

16. id., 22, 23.

17. id., 22, 28 (quoted). That there was much later a spot called the 'fountain of the Nile' near Elephantine is shown by a mention of it in an inscription of Ptol., VIII: Ditt., *O.G.I.*, I, 168, p. 244. The deep holes were no fiction, as was found by Lobo and Bruce at different periods. One was shown to Germanicus in Tiberius' reign – Tac., *Ann.*, II, 61.

18. Herod., II, 29–31, 34.

19. W. D. Cooley, *Claudius Ptolemy and the Nile*, 20–3. V. de St Martin, *Le Nord de l'Afrique dans l'antiquité*, 24–9. Nile flooding: Herod., II, 19–25; *Aristotelis Fragmenta*, pp. 213–14, Didot; Tozer, 62–3.

20. Ephorus is wrong about the Nile floodings: fr. 108 Müller, = 65 Jacoby. Diodor., I, 37, 4; 39, 7–13.

21. Aristotle on rivers Aegon and Nyses, and Nile rising in the Silver Mountain: *Meteorol.*, I, 13, 21. Sudd: id., *H. A.*, VIII, 12, 2.

22. Arrian, *Anab.*, III, 4. Curt., IV, 33. Strabo, XVII, 813–14; cf. Herod., II 32; III, 25. Aristoph., *Birds*, 619, 716. Nile: see *Camb. Anc. Hist.*, Vol. VI, p. 596.

23. Agatharch., 50, 84, 112. Coloe = Kohaito: Bent. *Sacred City of the Ethiopians*, chap. XII. Route Meroe – Red Sea: Strabo, XVI, 771.

24. Dalion: Pliny, VI, 183, 194. Aristocreon: id., 183, 191. Ptolemy Phil.: Diodor., I, 37; III, 36; cf. Theocrit., XVII, 87. For Nubian Ethiopia, see G. Reisner in *Journal of Egyptian Archaeology*, IX (1923), 34 ff.

25. Pliny, VI, 183.

26. Diodor., III, 6 (cf. Strabo, XVII, 823), dated under Ptol. Phil., cf. F. Ll. Griffith, *Meroitic Inscript.*, II, p. 24. But cf. also Reisner in *Journ. Eg. Arch.*, IX, 1923, 75; A. Kammerer, *Essai sur l'histoire antique d'Abyssinie*, 72.

27. See E. Bevan, *A History of Egypt under the Ptolemaic Dynasty*, pp. 246–7.

28. Agatharch., p. 119, cf. A. Peyron, *Papyri Graeci*, I, 5, 1, 27. Weigall, *Antiquities of Lower Nubia*, 42.

29. We all know this geographical information from Eratosthenes in Strabo, XVII, 786–7, 821–2; XVI, 771; and it is clear that his information came from

NOTES

explorers and writers who had penetrated far up the Nile. Heat was the inevitable explanation of why exploration had limits towards the south.

30. Pliny, VI, 178, 180, 183, 191, 193 (lists of names chiefly from Bion), 183 (Basilis). For Basilis: see also Athenae., IX, 390b; Agatharch., 64 (p. 156); and Bion: Diog. Laert., IV, 58; Athenae, XIII, 566c. Damocritus and Charon also wrote on Ethiopia.

31. Agatharch., 23–9 (Müller, *G.G.M.*, I, pp. 123–8).

32. Bevan, op. cit., 293–5.

33. Strabo, XVII, 824–5, 839; II, 95; cf. II, 113–14, etc. Diodor., I, 30. The Cinnamon country was really reached only by sea, but under this name Abyssinia might be vaguely included.

34. Strabo, XVII, 813, 799, 791.

35. Pliny, VI, 179.

36. Strabo, XVII, 820–1; Pliny, VI, 181. Cass. Dio, LIV, 5. R. Cagnat, *I.G.R.*, 1359, Ditt., *O.G.I.*, I, 205. Sir E. Budge, *A History of Ethiopia*, I, 58–60. Kammerer, op. cit., 72 ff. Candace from Ktke is 'Queen' or 'King's wife' – Reisner, op. cit., 67, 73–4.

37. Pliny, VI, 181, 184–6; Seneca, *N.Q.*, VI, 8, 3–5; cf. Cass. Dio, LXIII, 8. Ptolemy's list of names hardly goes beyond Meroe: Ptol., IV, 7, 21.

38. Pliny, V, 81. Then follows Juba's curious theory echoed by Seneca.

39. Pliny, XII, 19. Shooting the rapids: Seneca, *N.Q.*, IV, 2, 6.

40. For the limit reached, see V. de St Martin, *Hist. de la géogr.*, 178 ff., and most later writers agree – Berger, 587, 601; Sir E. Budge, *A History of Ethiopia*, I, 99–100; cf. his *Egyptian Sudan*, II, 172–4.

41. For the exploration up the Blue Nile as shown in Ptolemy, see the good maps in C. Müller's edition of Ptolemy's Geography, Tabulae, IV, pp. 27–8, and also the map in Sir E. Budge, *Hist. of Ethiopia*, I, facing p. 120.

42. Ptolemy is our sole authority. Diogenes: Ptol., I, 9, 3–4. He was probably the ultimate source of all Ptolemy's knowledge about the sources of the Nile. See Ptol., IV, 8, 3; IV, 8, 6; IV, 7, 23–4. Consult also Bunbury, II, 612 ff.; Berger, 598; Tozer, 352; J. O. Thomson, *Hist. of Anc. Geog.*, pp. 275–7; R. Hennig, *Terrae Incognitae*, I, p. 346; J. N. L. Baker, *History of Geographic Discovery*, p. 29; W. Hyde, *Ancient Mariners*, pp. 285–7; Warmington, in *The Cambridge History of the British Empire*, Vol. VIII, pp. 65–8. Sir H. Johnston, *The Nile Quest*, p. 28, decides for Ruwenzori as the range. He would identify Rhapta (ruled by Arabians from Arabia, *Peripl.*, 16) with Pangani at the mouth of the Rufu – id., 22. See also the maps mentioned in the preceding note.

Most classical references to Nubia and Meroe will be found in Sir E. Budge, *Hist. of Ethiopia*, I, 62 ff.; cf. Paulitschke, *Die geographische Erforschung des Afrikanischen Continents* (Vienna, 1880).

42. All who are interested should read R. Brown, *The Story of Africa*, Vols. II and III, and Sir H. Johnston, *The Nile Quest*. For the Kagera as the ultimate source of the river, see R. Kandt, *Caput Nili*.

44. The present maximum distances between wells amount to 110 miles in the western, 180 miles in the eastern Sahara. – E. F. Gautier, *Le Sahara*, p. 92.

NOTES

On the North African desert and steppe lands in general, see A. Bertholot, *L'Afrique Saharienne et Soudanaise*.

45. Gautier, op. cit.; A. G. Martin, *Les Oasis sahariennes*; H. Schirmer, *Le Sahara*, chap. 7; S. Gsell, *Histoire ancienne de l'Afrique du nord*, Vol. I, pt 1, chap. 3. There is abundant evidence that the Sahara once possessed the flora and fauna of a steppe, but it had certainly lost these before the beginning of historical records. The ruins of citadels on the rocky heights of the western Sahara do not necessarily prove continuous habitation.

46. Camels in Tunisia in 46 B.C.: *Bellum Africanum*, chap. 68. Strabo (XVII, 828) mentions that horses travelled in the western Sahara with water-skins tied beneath them.

47. Elephants in Morocco and in the Algerian hinterland: Pliny, V, 5, 12.

48. Mrs Forbes, *The Secret of the Sahara*, pp. 126–40 (a waterless march from 25 Dec. to 2 Jan.).

49. IV, 168–95. The description of the Sahara as a 'ridge' tells against Miss E. G. R. Taylor's attractive theory that Herodotus's line of oases (4, 181–5) ran from Aujila to Jerma and Mt Hoggar, and that Mt Hoggar was Herodotus's 'Atlas' (*Scottish Geographical Magazine*, 1928, p. 129 ff.).

50. IV, 173.

51. II, 32–3.

52. IV, 172.

53. Murzuk was used as an advanced base for the exploration of both these routes in 1850–5 by H. Barth.

54. IV, 183.

55. Mr F. Rodd, in *The Times*, 20 March, 1928.

56. II, 22, p. 44e.

57. H. Barth, *Travels in Northern and Central Africa*, Vol. I, p. 274.

58. Pliny, V, 36.

59. Barth, Vol. I, pp. 156–7.

60. Ptol., I, 8, 4. This expedition was no doubt a sequel to the Garamantian raid recorded in Pliny, V, 38, and Tacitus, *Histories*, IV, 50, 6.

61. Sir Harry Johnston, *The Nile Quest*, p. 20.

62. Ptol., I, 8, 4.

63. The Chad route is suggested by Stein in Pauly-Wissowa, s.v. Iulius, No. 347; the Asben route by Ch. de la Roncière, *La Découverte de l'Afrique au Moyen Âge* (*Mémoires de la Société Royale de Géographie d'Égypte*, Vols. V and VI), Vol. V, p. 76. Sir Harry Johnston (loc. cit.) leaves the question open.

64. The route Ghadames-Insala-Timbuctu was explored by Major Laing in 1825–6, the track from Tangier to Timbuctu by Caillié in 1828.

65. No such stream can have traversed the Sahara from west to east except in a remote geological period. – See J. Ball, *Geographical Journal*, July 1927, p. 21 ff.

66. Pliny, V, 14–15. Cassius Dio (LI, 9) attributes a similar desert march to Suetonius's successor Hosidius Geta. Hosidius may have repeated Suetonius's performance, but more probably Dio mistook the names.

NOTES

67. G. Rohlfs, *Reise durch Marokko*, pp. 36 ff., 69–71; Ch. de Foucauld, *Reconnaissance au Maroc*, pp. 98–9.

68. Herod., II, 33. cf. the identifications of the Nile with the Senegal (p. 63) and the Indus (p. 179). Juba also found crocodiles in his Moroccan 'Nile'.

69. IV, 6, 4.

70. See C. Müller's note in his edition of Ptolemy, Vol. II, pp. 737–40. Spates on the Igharghar: Gautier, op. cit., p. 48 ff.

71. For a view of modern theories, see C. Müller, loc. cit.; Bunbury, *History of Ancient Geography*, chap. 29, pt. 2. The names 'Gir' and 'Nigir' were no doubt derived from the native forms 'Gher' and 'N'Gher', which survive in modern Tuareg. Both simply mean 'water'. (Ch. Tissot, *Géographie comparée de la Province romaine d'Afrique*, Vol. I, p. 91 ff.)

CHAPTER NINE: RESULTS OF ANCIENT EXPLORATIONS

1. The extent of ancient explorations as defined above is somewhat larger than that which is marked on map 1 of A. Oppel's *Terra Incognita*. This map assumes a general exploration of Germany and Russia to south-west of a line from the Niemen to the Caspian. On the other hand, it does not show any ancient exploration to the north of the Caspian, to the east of C. Comorin, to the south of C. Guardafui, across the Sahara, or beyond lat. 20°N. on the West African coast.

2. Marcian, in Müller's *Geog. Gr. Min.*, I, p. 565, 112; cf. W. H. Schoff, *The Peripl. of the Outer Sea* (Marcian), p. 11. Thuc., VI, 1; II, 97; Strabo, VIII, 332, etc.; Ael., *N.A.*, XII, 1. Nearchus perhaps rightly called his report a παράπλους, 'sailing along' – Jacoby II, BD2, p. 445.

3. For the text of this guide-book, and of the others mentioned below, see C. Müller, *Geographici Graeci Minores*, Vol. I.

4. A recent translation and full commentary by W. H. Schoff, *The Periplus of the Erythraean Sea*; text by H. Friske.

5. W. H. Schoff, *Parthian Stations*.

6. Texts of the Antonine and Jerusalem Itineraries in the edition of Parthey and Pinder, and in O. Cuntz, *Itineraria Romana*, Vol. I; of the Gades-Rome Itinerary in *Corp. Inscr. Lat.*, XI, 3281–4.

On the whole subject, see also K. Miller, *Itineraria Romana*, and W. Kubitschek in Pauly-Wissowa, s.v. Itinerarien.

7. F. Jacoby, *Die Fragmente der griechischen Historiker*, Vol. I, pp. 16–47; do., in Pauly-Wiissowa, s.v. Hekataios.

8. For the fragments of Poseidonius, see Jacoby, op. cit., Vol. II, pt A., pp. 252–85.

9. C. Müller, op. cit., Vol. I, pp. 111–95.

10. D. Detlefsen, *Ursprung und Bedeutung der Erdkarte Agrippas*.

11. H. Berger, *Die geographischen Fragmente des Eratosthenes*; do., *Die geographischen Fragmente des Hipparchos*.

NOTES

12. Text and Commentary by C. Müller, *Cl. Ptolemaei Geographia*. A revised text for the European section by O. Cuntz, *Die Geographie des Ptolemaeus*, and for the Indian by Renou, *Texte Établi*.

13. For map-making, see the various figures in the text of J. O. Thomson, *History of Ancient Geography*. For the development of theory, see the same work, pp. 94 ff., 152 ff., 202 ff., 320 ff. See also W. A. Heidel, *The Frame of the Greek Maps*.

14. On Dicaearchus, see H. Berger, *Geschichte der wissenschaftlichen Erdkunde der Griechen*, 2nd ed., p. 166.

15. S. Smith, *Early History of Assyria*, 86, shows a map on a Babylonian tablet.

Map-making is not beyond the means of primitive peoples. See the examples of Esquimaux maps in C. F. Hall, *Life with the Esquimaux*. These maps were good enough to be of use to the explorers Parry and Ross.

16. K. Lepsius, *Urkundenbuch*, Pl. XXII.

17. R. Lanciani, *Forma Urbis Romae*.

18. Strabo, I, 7. Herod., IV, 36; V, 49. On early Greek maps, see J. L. Myres, *Geographical Journal*, Vol. VIII, pp. 605–31.

19. For fifth-century maps, see Aristophanes, *Clouds*, ll. 201 ff.; Aelian, *Varia Historia*, III, 28. On ancient maps in general, see Kubitschek, in Pauly-Wissowa, s.v. Karten.

20. Strabo, II, 116. For the earth as a globe, see Thomson, Index s.v. Globe.

21. See K. Kretschmer, *Die italischen Portolani des Mittelmeeres*, pp. 51, 99. No accurate coastal maps could be made before the discovery of the compass.

22. Specimens of these maps in C. Müller's edition; H. Kiepert, *Formae Orbis Antiqui*, pls. 35 and 36; J. Fischer, *Denkmäler der Wiener Akademie der Wissenschaften*, Philosophisch-historische Klasse, 1916; Thomson, pp. 346–7, 412.

23. The most recent edition is K. Miller, *Die Weltkarte des Castorius, genannt die Peutingeresche Tafel*. See also J. R. Wartena, *Inleiding op een Uitgave der Tabula Peutingeriana*, Thomson, pp. 379–81.

24. For the possibility that Nearchus was acquainted with Scylax's report, see Jacoby, II, BD2, 445, 478.

25. Pliny, VI, 100, 172; Ptolemy, IV, 7, 12; *Itin. Alexandri*, 110.

26. For references to the terrors of the Atlantic, see J. Partsch, in Pauly-Wissowa, s.v. Atlantis.

27. For a drastic criticism of Ptolemy, see Bunbury, *History of Ancient Geography*, Vol. II, p. 546 ff.

28. Occasional medieval scholars, such as Bede, retained the belief that the earth was spherical, and had fairly accurate ideas of the earth's surface. See C. R. Beazley, *The Dawn of Modern Geography*, esp. Vol. II, chap. 7; A. P. Newton, *Travel and Travellers of the Middle Ages*, p. 5 ff.

29. On the imperfections of reports by traders, see the remarks of Ptolemy (I, 4, 2).

30. Similar mistakes occurred in early modern times. Credence was given

NOTES

to the Zeno brothers' claim to have touched North America; and to the story of the sailors from Dieppe who were said to have found the Guinea Coast in 1364 (on whom see Ch. de la Roncière, *Mémoires de la Société Royale de Géographie d'Égypte*, Vol. VI, pp. 6-17). Conversely Sebastian Cabot had some difficulty in dispelling doubts as to his voyage to North America.

31. On classical influences in medieval and early modern geography, see Beazley, op. cit.; Newton, op. cit., Introduction.

32. Writing c. A.D. 1050, Adam of Bremen repeated the ancient fables about the Atlantic in preference to the reports on actual travel by the Norsemen.

33. On this subject in general, see G. A. Wood, *The Discovery of Australia*, chap. 1; Thomson, pp. 277-8; W. Hyde, *Ancient Greek Mariners*, p. 306 ff.

34. A separate Terra Australis is first outlined in Theopompus, fr. 74 (ed. Grenfell and Hunt).

35. Hipparchus ap. Mela, III, 7, 70. Marco Polo (Bk III, chap. 37) mentions an 'island of Zenzibar', reputed to have a circuit of 2,000 miles.

36. Wood, op. cit., *passim*; *Encyclopaedia Britannica*, article 'Map', p. 639. C. R. Beazley, *The Dawn of Modern Geography*, II, 549 ff.

37. Pauly-Wissowa, s.v. Krates, 1637.

38. Ps. Aristotle, *De Mundo*, chap. 3.

39. I, 64.

40. Aristotle, *De Caelo*, 2, 14, 14.

41. Strabo, I, 64-5.

42. Strabo, II, 102.

43. *Quaestiones naturales*, prologue, sect. 13.

44. *Medea*, ll. 375-80.

45. Ptol., VII, 3, 6.

CHAPTER TEN: IMAGINARY DISCOVERIES

1. W. Golenischef, *Sur un ancien conte égyptien*, 1881. The story is translated there on pp. 4-8. Cf. Maspero, *Popular Stories in Ancient Egypt*, trans. Mrs C. W. H. Johns.

2. M. Jastrow, *The Religion of Babylonia and Assyria*, pp. 472-92. F. Hommel, *The Ancient Hebrew Tradition* (English trans.), 36.

3. On this point, see P. Jensen, *Das Gilgamesch Epos* (*Ex Oriente Lux*, Vol. III, pt 1).

4. Diodor., II, 1-20 (from Ctesias); Ael., *V.H.*, VII, 1, etc. A. Sayce, *The Legend of Semiramis*, *His. Rev.*, Jan. 1888; F. Lenormant, *La Légende de Sémiramis*.

5. Homer, *Od.*, XII, 69, and cf. *Il.*, VII, 467, etc.; IV, 211-12, 251. Mimnermus ap. Strabo, I, 46; Hecatae., schol. ap. Apollon. Rhod., IV, 259. The four books of this work and also pseudo-Orpheus, Apollod. (I, 9), and Valerius Flaccus are all extant accounts of these voyagers. cf. Rohde, *Der Griechische Roman*, 3rd ed., 184-5.

6. Apollod., II, 5-11. Diodor., IV, 26, etc.

NOTES

7. Diodor., IV, 4, etc. In Asia: Strabo, XI, 505; XV, 687; Eurip., *Bacchae*, 13; Diodor., III, 63; IV, 3; II, 38. Arr., *Ind.*, 5, etc. Nysa: Arr., *Anab.*, V, 1, 1 ff.

8. Herod., II, 102–11. Diodor., I, 53–9. Strabo, XV, 686–7; XVI, 769, 790. Tac., *Ann.*, II, 59. C. H. Sethe, *Sesostris*, in *Unters. z. Gesch. u. Altertumsk. Aegyptens*, II. Heft I, esp. p. 18.

9. Strabo, I, 61; XV, 86–7. To the ordinary man of the Graeco-Roman world India often meant simply Ethiopia + Abyssinia + Somali coasts.

10. Strabo, XV, 687.

11. Strabo, XI, 505, 506; Arrian, *Anab.*, III, 30, 7 ff.; V, 3, 3. Romance of Alexander: Rohde, 197 ff. Ganges: false claim by Craterus: Strabo, XV, 702. Innocent belief during Nero's reign: *Peripl.*, 47.

12. R. Hennig, *Von rätselhaften Ländern*, equates this with Tartessus, but fails to establish any striking analogies.

13. Homer, *Il.*, XIII, 4–6. Strabo, VII, 300–2, etc. cf. Hesiod (ed. Rzach), fr. 55 – included among the Scythians; cf. Ptol., VI, 14, 12.

14. Herod., III, 116; IV, 13, 14, 27.

15. See on this gold F. H. A. v. Humboldt, *Asie Centrale*, I, 330–408. A suggestion has been made that the one eye might be a lamp tied on the forehead of a gold-miner; so the goat-footed men of the mountain-barrier north of the Argippaeans (Herod., IV, 25) would be mountaineers.

16. cf. 'dog-headed' men of Africa. Schol. ad Aesch., *Prom. Vinct.*, 793. Strabo, I, 43; XV, 520.

17. Herod., IV, 32 (he give the 'Epigoni' a another early source); *Hymn. ad Dionys.*, 241.

18. Steph. Byz., s.v. Ὑπερβόρεοι (from Damastes).

19. Pind., *Pyth.*, X, 30–44. *Olymp.*, III, 14–31.

20. Herod., IV, 13, 32, 33, 35, 36. Mela, II, 1; III, 5, sects. 36–7. Hecataeus of Abdera (not he of Miletus) did much to help the belief at the end of the fourth century B.C. – Schol. ad Apollon. Rhod., II, 675. Ael., *H.A.*, XI, 1. Diodor., II, 47. Pliny, IV, 89–91. We are not concerned with the evolution of the myth into its complete form; for this see O. Schröder, *Archiv. f. Religionswissenschaft*, 1904, VIII, 69. L. R. Farnell, *Cults of the Greek States*, IV, 100. C. Seltman, *Class. Quart.*, July–Oct. 1928, 155 ff., but especially Daebritz in Pauly-Wissowa, s.v. Hyperboreer.

21. Pliny, IV, 95. Elks: Caesar, *B.G.*, VI, 27.

22. Herod., III, 105–6. Megasth., in Strabo, XV, 706. Pliny, XI, 111. Specimen ants kept in Persia: Herod., III, 102. Nearchus saw some skins (of the Red Marmot?). Arr., *Ind.*, 15. Pair of feelers kept at Erythrae: Pliny, XI, 111. Soliman at Constantinople received a present – Busbequius, Ep. IV, 343, ed. Elzevir, but see id., trans, E. S. Forster, 219. Explanations have been sought in the dogs and pick-axes of Tibetan gold-miners. H. Wilson, *Ariana Antiqua*, 135. J. W. McCrindle, *Ancient India as described in Classical Literature*, pp. 3, 51.

23. Herod., III, 106.

24. McCrindle, *Ancient India as described by Ktesias*, 7, 35, 61. Pliny, VII, 23. Steph. Byz., s.v. Σκιάποδες. Philostrat., *Apollon.*, III, 47.

NOTES

25. Strabo, XV, 701. Pliny, VI, 54–5. Mela, III, 7. Ammian. Marcell., XXXIII, 6. Lucian, *Macrob.*, 5. Attacori: Pliny, VI, 55; cf. Ptol., VI, 16, 2, 3, 5; VIII, 24, 7. Lassen, *Ind. Alt.*, I (2nd ed.), 1018, Rohde, 233–4.

26. Herod., IV, 191. Ethiopians: id., III, 17–24; H. Last in *Class. Quart.*, XVII, pp. 35–6, answered by W. R. Halliday, id., XVIII, pp. 53–4. Other strange folk: Mela, III, 9, partly as we saw, p. 101, from oral reports of Eudoxus' voyages.

27. *Pharsalia*, IX, 700–889.

28. Aelian, *de Animalibus*, III, 31; V, 50.

29. Aelius Tubero, ap. Aul. Gellius, VII (VI), 3. For a story of a monster snake which swam out to sea and capsized a trireme, see Aristotle, *Hist. Anim.*, VIII, 27, 6.

30. Diodor., II, 55–60. E. Rohde, *Der Griechische Roman*, 239 ff. (Euhemerus, Iambulus).

31. Diodor., V, 42, 2–7.

32. *Od.*, XI, 11.

33. *In Rufinum*, I, 126 ff.

34. *Bellum Gothicum*, IV, 20. Procopius denied the identity between 'Brittia' and Britain, but without good reason (J. B. Bury, *Klio*, 1906, 79 ff.).

35. *Od.*, IV, 563–9. Although the allusion to Elysium occurs in the prophecy of Proteus, who lived in Egyptian waters, there is no need to suppose that Homer was here reproducing an Egyptian rather than a Greek or Minoan belief.

36. Hesiod, *Works and Days*, 169–72; Pindar, *Olymp.*, II, 68 ff.

37. Hesiod, *Theogony*, 215–16. Schulten, in *Geogr. Zeitschr.*, 1926, 229 ff., suggests that the apples are medlars.

38. *De Facie in Orbe Lunae*, par. 29, pp. 1151–3.

39. C. R. Beazley, *John and Sebastian Cabot*, p. 15 ff.: M. d'Avezac, *Les Îles fantastiques de l'Océan occidental*.

The 'Isle of Antilia' or 'of the Seven Cities', which was marked on Toscanelli's map (c. A.D. 1475) on the latitude of Lisbon, but far out in the Atlantic, lies about midway between the two reputed latitudes of St Brandan's isles. Does this represent a compromise?

40. *Timaeus*, 23–5; *Critias*, 114–18. Recent English commentators assume that Plato's description was nothing more than a parable (Jowett, *Plato* (3rd ed.), Vol. III, pp. 430–3; A. E. Taylor, *A Commentary on Plato's Timaeus*, ad. loc.). Plato's myth would therefore be on the same lines as Theopompus's story of Meropis (fr. 74, ed. Grenfell and Hunt), or Bunyan's *Pilgrim's Progress*, except that Plato, with his keen visualizing instinct, gave his Atlantis a more definite shape and locality. Schulten, in *Geogr. Zeitschr.*, 1926, 229 ff., puts Meropis in the West, without proof. For a fuller discussion, see R. Hennig, *Von rätselhaften Ländern*, pp. 1–37.

41. For a review of the older identifications, see T. H. Martin's edition of the *Timaeus*, Vol. I, pp. 257–333. For the ancient views on Atlantis, see J. Partsch in Pauly-Wissowa, s.v. The chief *locus* is Strabo, II, 102.

NOTES

42. Similarly Defoe transposed Juan Fernandez to the edge of the Caribbean Sea, Dickens removed Betsy Trotwood's Cottage from Broadstairs to Dover Cliff, and H. G. Wells fetched away Mr Shalder's Drapery Emporium from Southsea to Folkestone.

43. Among the more recent writers on the subject, Schulten seeks the harbour of Atlantis on the site of Tartessus (*Tartessos*, pp. 53-5), Borchardt and Hermann at Tell Gullel in southern Tunisia (*Daily Mail*, 15 and 20 March, 1928, 14 April, 1928).

44. R. Hennig, *Das Rätsel der Atlantis*.

45. For these writers see J. G. Dunlop, *History of Prose Fiction* (ed. Wilson), Vol. I, pp. 13 ff. (Diogenes); 16 ff. (Iamblichus); 22 ff. (Heliodorus). For Philostratus on Apollonius, see Rohde, 438 ff.

46. Pliny, II, 248.

INDEX

Abbreviations: (c.) cape; (i.) island; (l.) lake; (m.) mountain; (r.) river.

Names of ancient places, countries, and peoples (but not persons), which are now out of ordinary use, are shown in italics.

Aar, (r.), 148, 151
Abalus (i.), 52–3
Abdera (Greece), 163, 206
Abdera (Spain), 252
Abgarus, 191
Abrocomas, 278
Abu Dhabi, 86
Abu Hamed, 207, 211
Abu Sha'ar, 193
Abu Shahrein, 75
Abu Shehr, 86
Abu Simbel, 206
Abydos (Egypt), 210
Abyssinia, 89–90, 100, 107; exploration of, 208–14, 285, 287, 292
Acesines (r.), 178–9
Achaeans, 25, 38, 236, 250, 251, 253
Acila, 92–3
Acmodae (is.), 258
Acmonia, 161
Acra, 64
Actium, 142
Adam of Bremen, 291
Aden, 76, 80, 89, 95–8, 107, 203, 234, 268, 285
Adriatic, 252, 272, 274
Adulis, 209
Aegean Sea, 23ff., 139, 142, 160–1, 189, 234
Aegeans, 23–32, 37–9, 159. See Minoans
Aegon (r.), 286
Aegyptus (r.), 205
Aelius Gallus, 95, 193
Aemodae (is.), 258
Aeolian Isles, 25, 29
Aeschylus, 79, 121, 150, 162, 263

Afghanistan, 160, 185
Agadir, 64
Agatharchides, 90, 225
Agathyrsi, 135, 138
Agisymba, 220
Agricola, 59–60
Agrippa, 152, 195, 225–6
Agulhas current, 118, 127
Ahaziah, 262
Ainus, 158
Ajanta, 102
Akaba, 73, 89, 192
Akalan, 38–9
Akkad, 23, 76
Akka, 205
Alalia, 70
Alans, 198
Albania (Asia), 186, 190–1, 198
Albania (Europe), 142, 251
Albert Nyanza, 215
Albinovanus, 259
Albion, -es, 44, 255
Alcaeus, 206
Alessio, 142
Alexander (Roman subject), 104, 105, 116
Alexander the Great, 12–15; explorations by sea, 80–7, 92; Europe, 137, 140, 146, 150; Asia, 172–87, 191, 195; Africa, 208, 224, 235–6. See also 240, 244, 272, 279
Alexandretta, 174
Alexandria (ad Aegyptum), 24, 87, 98, 106, 124, 127
Alexandria (near Alexandretta), 174
Alexandria of the Arachosians, 176
Alexandria Arion, 176

295

INDEX

Alexandria (= Bactria), 177
Alexandria of the Caucasus, 178
Alexandria (= Charax), 86, 180
Alexandria Eschate, 177, 185
Alexandria (= Ghazni?), 176
Alexandria (= Merv), 177
Alexandria (= Rhambacia), 180
Algeria, 216, 220–1
Alicante, 35
Allobroges, 150
Alpis (r.), 138, 147, 272
Alps, 146–51, 274
Altai (ms.), 163, 196
Al-'Ula, 88
Aluta (r.), 138
Amalfi, 34
Amanus (m.), 168
Amasis II, 78, 206
Amazon (r.), 271, 272
amber, 25, 43, 44, 52–3, 143–6, 150, 153, 254, 255, 271, 273
Amenemhab, 159
Amenophis I, 159
America, 18; alleged voyages to, 71. See also 76, 117, 153, 231, 243, 290
Amiens, 258
Amisus, 38, 181
Ammon, -ians, 206–7, 210
Amol, 175
Amon Re, 75, 262
Amphipolis, 272
Amritsar, 178
Amu Darya (r.), 176. See Oxus
Amur (r.), 157
Amyntas, 181
Anaku-ki, 23
Anamis, 86
Anatolia, 161, 163
Anaximander, 161, 226
Ancona, 36
Ancyra, 161, 174
Andamans, 104
Andernach, 276
Andhra kings, 99–100, 102–3
Androsthenes, 86

Angola, 269
Angora, 161
Annam, 105
Annius Plocamus, 97, 101
Anson, 248
Antialcidas, 189
Anti-Atlas, 64
Anti-Lebanon, 193
Antilia (i.), 293
Antilles, -ians, 72
Antioch (in Pisidia), 12
Antioch (in Syria), 184, 224
Antioch (= Chodjend), 185
Antiochia Margiana, 185
Antiochus I, 185
Antiochus II, 185, 187
Antiochus III, 93, 188, 225
Antiochus IV, 93, 185, 188
Antipodes, 231
Anti-Taurus, 190
Antoeci, 231
Antoninus Pius, 102, 130, 196, 199
Antony, 192
Antun, 106
Anuradhapura, 103
Anurogrammon, 103
Aornos (m.), 178
Aornos (= Tashkurgan), 176
Apoasis, 88
Apollo, 31, 33, 137, 143
Apollodorus, 189
Apollodotus, 189
Apollonia (Adriatic), 144
Apollonia (Black Sea), 253
Apollonius (traveller), 91
Apollonius Rhodius, 138, 147, 151, 274
Apollonius of Tyana, 244
Apries, 78
Arab, -ia, -ian, 12; explorations and activities by sea before Alexander, 73–80; Alexander, 86–7; Ptolemies, 87–92; Seleucids, 92–3; Romans, 94–101, cf. 101–7, 110, 111, 116–17, 122–3, 131, 214; by land, 157–61,

296

INDEX

Arab, -ia, -ian – contd.
    167, 180; Rome, 192–5. *See also*
    234–5, 240–2, 261, 263
Arabah (c.), 81
Arabis (r.), 80
Arachosia, 176, 180, 188, 224
Aral Sea, 173, 177, 185–6, 199, 277
Arambys (c.), 64
Araxes (r.), 163, 169
Arbela, 161, 175, 181
Archelaus, 195
Archias, 86
Arctinus, 253
Aretas, 192
Argaeus (m.), 184
Argandab, 176, 180
Argaru, 100
Argippaei, 166–7, 292
Argo, -nauts, 37–9, 42, 151, 234, 274
Arguin, 64, 66
Aria, 176
Aristagoras (geographer), 208
Aristagoras (of Miletus), 161–2
Aristeas, 160, 166, 276
Aristo, 88
Aristobulus, 279
Aristocreon, 209
Aristophanes, 80, 244
Aristotle, 17, 47, 122, 133, 173, 231, 259, 269, 270, 275
Arles, 151, 275
Armenia: nature, 158; early exploration, 159, 162–3; Ten Thousand, 167–9, 172; Seleucids, 184, 188; Rome, 189–92, 198–9. *See also* 174, 281
Arrian, 80, 86, 224, 253
Arsacids, 197–8
Artabri, 56
Artaxata, 190
Artaxerxes II, 167–8
Artemidorus, 91
Aryium (c.), 45
Asben, 218–20, 288
Asciburgium, 155, 254, 276

Asia Minor, 22, 36, 39, 42, 157–60, 162–3, 172–3, 184, 188–90, 196, 222, 235, 281
Asoka, 187
Aspasii, 178
Assaceni, 178
Assuan, 115, 202–3
Assurbanipal, 160
Assyria, -ans, 76, 77, 158–60, 269
Astarte, 234
Astola, 82
Astrabad, 175
Astrakhan, 166
Atbara, 89, 202–3, 209–10
Athenaeus, 219
Athens, 112, 122, 162, 207, 231, 243, 261, 291
Atlantis, -ians, 242–3, 269, 294
Atlas (m.), 61, 220
Atlas (r.), 138
Atropatene, 191–2
Attacori, 239
Attila, 17
Attock, 78, 178
Attrek (r.), 186
Aude, 148
Augsburg, 276
Augustus, 58, 60, 94–5, 99, 136, 138, 144–8, 153, 155, 192–3, 224–7, 273, 283
Aujila, 216, 218, 288
Aurelius, M., 106, 199
Aurungabad, 102
Australia, 76, 256
Australis Terra, 230–1, 291
Austria, 140
Auxume, 100
Avienus, 44–6, 133, 224, 275
Axius (r.), 139
Axum, -ites, 100, 105–6
Azerbaijan, 192
Aziris (i.), 33
Azores, 46–7, 70–1
Azov Sea, 15, 55, 228, 258

297

INDEX

Bab el-Mandeb, 75–6, 88–92, 98, 104, 115, 118, 234, 264, 265
Babylon, 23, 29, 39, 77, 86, 89, 158–60, 168, 175, 180, 184–5, 195, 226, 234, 263, 283
Bacare, 99
Bactra, 176–7, 196. See Bactria
Bactria, 160, 162–3, 173, 175–8, 181, 184, 185, 188–9, 191, 195
Badakshan, 160
Baemi, 276
Baeton, 181
Bagamoyo, 214
Bagh, 102
Baghdad, 168
Bagram, 178
Bahardipur, 96, 98
Baharia, 210
Bahmanabad, 98
Bahr el-Abiad, 210
Bahr el-Azrek, 210
Bahr el-Ghazal, 205
Bahrein Isles, 74, 86, 262, 263
Bajour, 178
Bakarawiya, 203
Baker, Sir S., 215
Balcia (i.), 52–3
Balearic Isles, 25, 34–5
Baleocuros, 102
Balisia, 52–3
Balkan Bay (Caspian), 185
Balkan lands, 137, 139–40, 142
Balkan (m.), 142, 272
Balkans (ms., Caspian), 185
Balkh, 160, 173, 176, 195
Baltic, 43, 53, 61; land-connexions, 136, 138, 143–6, 271
Baluchistan, 76, 80
Bambotus (r.), 68
Bampur (r.), 180
Bander-i-Gez, 175
Banks Penins., 256
Banna (i.), 262
Bantam, 284
Bantus, 126, 130

Barbaricon, 96, 98
Barna, 82
Barreta (i.), 255
Barth, H., 288
Barygaza, 92, 96, 99
Basel, 146
Basilia, 52–3
Basilis, 210
Bassadore, 86
Bassein, 105
Batn el-Hagar, 205
Baunonia (i.), 52–3
Beas (r.), 161, 178, 187, 195
Bede, 290
Begerawiyeh, 203
Beheim, 255, 261
Behring Strait, 157
Beilan, 168
Belerium (c.), 49–50, 55, 256
Bengal Bay, 73, 100, 104–5
Benguela current, 119
Bérard, V., 26, 30, 62
Berenice, 94
Beresina (r.), 136
Berezan (i.), 253
Berkeley Fort, 202
Bernese Oberland, 151
Bessus, 175–6
Bezabde, 175
Bhilsa, 189
Biafra Bight, 119
Bilma, 220
Bintenna, 103
Bion, 210
Biscay Bay, 48
Bissagos Bay, 66
Bithynia, 184, 190
Bitlis Chai (r.), 169
Black Sea, 17, 37–42, 55, 134, 136–7, 140, 146, 157–60, 166–7, 172–3, 180–1, 186, 190–1, 199, 224, 234–5, 253
Blair, Capt., 82
Blanco (c.), 66
Bocchus, 125, 126

298

INDEX

Bogus, 125
Bohemia, 137, 143, 145, 147, 153, 250
Bohuslan, 254
Boii, 71
Bojador (c.), 61, 121, 131
Bokhara, 160, 167, 177
Bolan Pass, 180
Bona, 125
Bordeaux, 224
Borkum, 60
Bornholm, 145
Borysthenes (r.), 166
Bosnia, 142
Bosphorus, 37
Boston, 261
Botelho, Diego, 116
Boulogne, 258
Brahmaputra, 272
Brandenburg, 271
Brazil, 271
Brenner Pass, 146
Brigg, 254
Bristol Channel, 50
Britain, British Isles, 12, 43, 44; Himilco, 45-6; Pytheas, 47-52; later knowledge, 55-60. *See also* 242, 257, 258, 293
Britannia, 255
Brittany, 43-8, 254
Brittia, 242, 293
Broach, 94, 96-9, 189
Bruce, 215, 286
Bucarest, 272
Buddha, -ist, 184, 187
Budini, 166
Bug (r. in Poland), 274
Bug (r. in S. Russia), 135, 137
Buhtan Chai (r.), 169
Bujnurd, 175
Bukowina, 274
Buldur (l.), 174
Bulgaria, 140
Burgundian Gate, 148
Burma, -ese, 73, 100, 104-6, 158
Bushire, 86

Bushman Land, 118
Bushmen, 121
Byblus, 22
Byzantium, 31, 42, 140, 172, 253

Cabot, S., 290
Cadamosto, 260
Cadmus, 26-7
Cadusia, -*ians*, 175, 186
Caelius Antipater, 68, 123
Caesar, Julius, 12, 16, 43, 57-8, 153, 192, 222, 258, 274, 288
Caillié, R., 288
Calah, 168
Calama, 81
Callatis, 253
Calinapaxa, 187
Calmucks, 167
Calypso, 29, 38
Camara, 100
Cambay, Gulf, 96, 99
Cambodia, 105
Cambyses, 114, 206-7, 285
camels, 83, 216, 288
Canerum, 63, 67, 121
Canaria, -*ies*, 61, 69, 119, 125, 242, 260
Candace, 211
Cannanore, 103
Cantin (c.), 63, 120
Cantium (c.), 50, 55, 256
Canton, 222
Cape of Good Hope, 118-19, 130-1, 222
Cappadocia, 158, 162, 167, 174, 184
Capua, 144, 155
Carbine (i), 81
Carduchi, 169
Caria, -*an*, 39, 71-2, 78, 172, 174, 253
Carian Fort, 64
Caribbaean, 18, 294
Carmana, 181
Carmania, 81, 83, 93, 176, 188
Carnic Alps, 138
Carnuntum, 144-5

299

INDEX

Carpathians, 135–8, 143–6, 272, 273
Carpini, J. de, 11
Carpis (r.), 138
Carrhae, 191–2
Carteia, 125
Carthage, -inians: in Mediterranean, 35–6; in North Atlantic, 45–8; Britain, 51, 57; West Africa, 62–4, 67–70, 110, 115, 126–8, 130, 206, 219; Spain, 133
Caryanda, 78
Caspian Gates, 175, 181, 194
Caspian Sea, 158, 160, 162; exploration of, 163, 166–7, 175; loss of knowledge, 177–8; Patrocles, 185–6; Pompey, 190–1; Roman re-exploration, 198–9. *See also* 222, 227, 277, 281, 289
Cassiquiare, 272
Cassiterides, 35, 47, 70
Cassites, 234
Cassius Dio, 288
Castile, 133
Cathaeans, 178
Cathay, 11
Cattigara, 105
Caucasus, 37, 42, 160–1; exploration, 163, 166, 173, 176; Pompey, 190; Roman Empire, 198–9. *See also* 234, 236, 263
Cauvery (r.), 100
Celaenae, 167, 174
Celts, 138, 144, 147, 150, 184
Cenis (m.), 147–8
Centrites (r.), 169
Cerasus, 172
Cerne, 64, 66, 122, 260
Ceylon, 76–7, 100, 158; explored, 103, 230; hearsay, 181, 187, 228
Chad (l.), 218, 220, 288
Chaldaeans, 77
Champlain, 12, 186
Chandaka, 99
Chandragupta, 92, 184–7
Charax (Media), 95, 194, 224

Charax (Mesopotamia), 86, 93, 193
Charikar, 176
Charimortus, 88
Chariot of the Gods, 67
Charjui, 177
Charon, 287
Charybdis, 29
Chastana, 102
Chenab (r.), 178–9
Chera, 99, 103, 187
Cherchen, 196
Chersonese (Tauric), 190
Chin dynasty, 100
China, -ese: reached by sea, 99–100, 105–7; Ptolemy, 130, 230, 157–8; Alexander, etc., 176, 184, 187; Graeco-Indians, 189; Roman Empire by land, 195–8, 239. *See also* 220, 228, 239, 261, 263, 267, 282
Chitor, 100
Chitral, 178
Choaspes (r.), 161, 173
Chodjend, 176
Choga (l.), 215
Chola, 100, 103, 187
Chorokh (r.), 172
Chretes (r.), 66
Chryse, 100
Chrysopolis, 172
Cilicia, 23, 174, 188
Cilician Gates, 161, 168, 174
Cimbri, 153
Cimmerians, 38, 44, 254
Cinnamon Country, 88, 123, 287
Cintra, P. de, 260
Circe, 29, 38
Cissian Land, 162
Claudian, 242
Claudius, 96–7, 193–5, 214
Clearchus, 168
Cleomenes I, 161–2
Cleopatra III, 90
Cleopatris, 192
Cloud-Cuckoo-Town, 244
Cnidus, 163

INDEX

Cnossus, 23, 24–5, 28
Cocala, 81, 180
Cochin China, 99, 105–6
Codex Theodosianus, 43
Coimbatore, 103
Col d'Argentière, 148
Col du Clapier, 148
Colaeus, 33–4, 45, 72, 252
Colapis (r.), 272
Colchis, 42, 159, 163, 190, 234
Colchoi, 100
Coloe, 209
Coloe (l.), 214
Columbus, 12, 15, 16, 18, 54, 55–6, 72, 127, 232, 253, 255, 259
Comorin (c.), 73, 100, 103, 289
compass, 17, 21
Congo (r.), 130–1, 272
Constance (l.), 146
Constantine I, 130
Constantinople, 107, 292
Cook, Capt., 12, 230, 256
Cook, F., 54
Cophas, 82
Corbilo, 151–2, 258
Corbulo, 198
Corcyra, 142
Cornelius Balbus, 219
Cornelius Nepos, 272
Cornwall, -ish, 43–9, 57, 151
Corobius, 32
Corrientes (c.), 131
Corsica, 34, 35, 252
Cortés, 19
Corunna, 47, 51, 56–7, 70
Cosmas, 92
Cossacks, 166
Cossaeans, 180
Cotyora, 172
Covilhão, P. de, 107
Cranganore, 98, 99
Crassus, M. (sen.), 142, 191
Crassus, M. (jun.), 142
Crassus, P., 56–7
Craterus, 180, 280, 292

Crates, 123, 226
Crete-, ans, 22–8, 32, 250
Crimea, 38–9, 42
Crocola, 80
Crommyussa (i.), 34
Cronian Sea, 51
Cronus (i.), 242
Crusaders, 26
Ctesias, 80, 238, 239, 245
Ctesiphon, 194
Cuba, 259
Cuervo (i.), 71, 261
Culpa (r.), 143, 272
Cumae, 33, 36, 70
Cunaxa, 168
Curula, 104
Cush, 203
Cutch, 96, 99, 179, 188
Cybele, 12
Cybistra, 161
Cyclops, 29
Cyiza, 82
Cyprus, 23–5, 29, 162
Cyrano de Bergerac, 245
Cyrene, -aica, 31, 33, 218
Cyrus the Elder, 160
Cyrus the Younger, 168, 278
Cysa, 82
Cyzicus, 123, 253

D'Ailly, 259
Daira (i.), 262
Dakka, 208
Dalion, 209
Dalmatia, 36–7, 142, 272
Damascus, 174
Damastes, 166, 271
Damghan, 175
Damirice, 282
Damocritus, 287
Danaus, 22
Danube: Hesiod, 39; exploration, 134–46, 152–3, 155, 222, 227. See also 272, 276
Danzig, 143

301

INDEX

Dardanelles, 37, 139
Dardistan, 161
Dar-es-Salaam, 214
Darial Pass, 191, 198
Darius I, 39, 78–9; Indian waters, 116, 120, 227, 228; Europe, 139; Asia, 160–2, 179
Darius III, 172–5
Dasht (r.), 180
Daxata, 196
Deccan, 187
Deimachus, 187
Del Tegghia, 261
Delgado (c.), 101, 130
Delgado, Porto, 222
Delos, 137, 143, 189
Delphi, 31–3
Delta (Nile), 32
Demetrius, 188
Democedes, 162
Democritus, 162–3, 206
Denia, 35
Denmark, 146, 155
Derar, 209
Derbe, 12
Derbent, 177
Dhofar, 75
di Recco, 261
Diamasa, 282
Diarbekr, 190
Diaz, 131
Dicaearchia, 124
Dicaearchus, 47, 183, 225
Dieppe, 290
Dilmun (i.), 262
Diodorus, 48–50, 69–70, 151, 225, 240–1
Diodotus, 188
Diogenes (author), 244
Diogenes (explorer), 101, 214
Diognetus, 181
Dionysius Periegetes, 195, 199
Dionysodorus, 244
Dionysus, 235, 245
Dioscorus, 101, 130

Dioscurias, 160, 186
Dioscuridu Nesos, 92
Diridotis, 86
Dnieper, 134–7, 166, 271
Dniester, 135, 190
Domitian, 195
Don, 53, 134–5, 157, 160, 166, 173, 176, 190, 198–9, 236, 270. *See* Tanais
Donetz, 135
Dorians, 25–6, 159
Dosarene, 100
Douro (r.), 134
Draa (r.), 64, 66, 124
Drake, 116, 230
Drangiana, 176, 188
Drapsaca, 176
Drave (r.), 138
Drusus, 60, 146, 153
Dvipa Sukhadara, 76, 92, 240, 264. *See* Socotra
Dyrrachium, 144

Eannes, Gil, 259
Ebro (r.), 134, 270
Ecbatana, 159, 175, **180–1**
Egnatia Via, 142
Egra, 193
Egypt, -ian, 12; in Mediterranean, 22ff., 38; in Red Sea, Somali, Arabia, 74ff.; Scylax, 78–9; Persians, 78–80; Ptolemies, 87 ff.; Romans, 93 ff., 106; circumnavigation of Africa, 110 ff.; Eudoxus, 123 ff.; in Asia, 159–60; Nile, 202 ff. *See also* 226, 233–6, 239, 243, 269, 286, 293
El Hasr, 76
El Haura, 192
El Katr, 89, 93
Elam, -ite, 76, 158, 234, 263, 278
Elbe (r.), 44, 53, 55, 60, 145–6, 153, 155, 276
Elburz (m.), 157, 159, 175
Eldorados, 13
elephant-hunts, 88, 264

302

INDEX

Elephantine, 207–8, 286
Elisar, 193
Elysian field, 242, 293
Emporiae, 151
Ems (r.), 60, 153
Ephesus, 163, 188
Ephorus, 122, 136, 138, 147, 269, 286
Er, 72
Eratosthenes, 16, 49, 123, 192, 225, 227, 264, 280, 281
Erech, 234
Ergamenes, 209
Eridanus (r.), 36, 44, 150, 254
Eridu, 76
Erythrae, 292
Erythraean Sea, 74–80, 86, 95, 98, 107, 121, 130, 235, 240
Erythras, 74
Erzeroum, 169
Essex, 58
Ethiopia, -ian, 62, 64, 76, 79, 90, 111, 123, 125, 129, 203, 205–6, 220, 234–6, 239–40, 287
Etruria, -scan, 36, 70, 144, 146
Eucratides, 188
Eudoxus, 12, 90, 91, 101, 123–31, 248, 261, 269, 293
Euergetes. *See* Ptolemy III, VII
Euhemerus, 240
Euphemus, 72
Euphrates (r.), 22, 38, 76, 79, 86, 92, 159–63, 168, 174, 188–95, 224, 235, 278
Euripides, 80, 274
Euthydemus, 188
Euthymenes, 62, 259
Euxine Sea, 186
Evans, Sir A., 23
Ezekiel, 77

Fahraj, 180
Faisabad, 177
Faranj, 176
Fellujah, 168
Ferghana, 176, 189

Fernandez, 260
Finns, proto-, 135
Finsteraarhorn (m.), 275
Fish-eaters, 83, 89, 206
Flaccus, Septimius, 220
Flavian dynasty, 101
Fola Rapids, 213
Foreland, S., 55
Forth estuary, 60
Fortunate Islands, 69, 226, 245, 264
France, French, 49, 55, 147–8, 150–2
Frankincense Country, 90
Frisia, 53–4, 60, 258
Fuerteventura (i.), 69, 125
Funchal, 70
Furah, 176

Gabun (r.), 63
Gades, 28, 34–5, 45–8, 53, 55–6, 70, 110, 123–9, 133, 224, 243, 270
Gades (i.), 255
Gaetulians, 220
Gaius Caesar, 129
Galloway, 49
Gambia (r.), 66
Gandhara, 102
Ganges: reached by sea, 100, 104; Alexander, 179, 235–6, 280; Megasthenes, 187; Graeco-Indians, 189
Ganjam, 104
Garama, 219
Garamantes, 219–20, 288
Garbatus (m.), 213
Garonne (r.), 148, 151–2
Gascony, 54
Gaul, -s, 52, 58, 124, 128, 147, 152
Gaza, 163, 174
Gedrosia, 81, 176, 179
Geneva, 275
Geneva (l.), 151
Genèvre (m.), 148, 274
Genoa, 128
Ger (r.), 220
Germanicus, 286

INDEX

Germany, 145–6, 153, 155, 222, 275, 276, 289
Gerrha, -aeans, 86, 92–3
Gessi Pasha, 215
Ghadames, 288
Ghazni, 176
Gibraltar, 25, 33, 45–6, 63, 71, 119, 231, 236, 255, 261
Gilgamesh, 29, 76, 234, 242
Giotto, 243
Gir (r.), 221, 289
Gironde (r.), 58, 257
Glausae, 178
Gnbti, 75
Goa, 116
Gobi, 157
Gondophares, 98
Gordium, 174
Goree, 66
Gorillas, 67
Gosselin, 115
Gothland (i.), 273
Goths, 53
Gourund, 81
Graham Land, 256
Granicus (r.), 174
Guadalquivir (r.), 25, 133
Guanches, 64, 69
Guardafui (c.), 73–4, 88–91, 97, 100, 118, 123–4, 129, 230, 289
Guinea, 67, 119, 121, 269, 270, 290
Gujarat, 188
Gulashkird, 180
Gunek Su (r.), 169
Gurdaspur, 178
Gurgan (r.), 175
Guriev, 166
Gwadar, 180
Gwattar, 83
Gymnias, 172
Gyptis, 35

Habb (r.), 81
Hades, 242, 245
Hadramut, 75, 87, 91, 95, 192–3

Hadria, 36
Hadrian, 102–4, 136, 195–6, 199, 224
Haemus, 272
Haiderabad, 179
Hajr, 193
Hakra, 179
Halil (r.), 180
Halys (r.), 161
Hamadan, 175, 194
Han dynasty, 184, 196
Hannibal, 147–8
Hanno, 63–4, 66–8, 114, 116, 119, 121, 222, 259, 260, 269
Happy Isles, 70
Haran, 192
Harkhuf, 203
Harmozia, 86
Harmozica, 199
Harpasus (r.), 172
Hatshepsut, 75, 262
Havilah, 76, 262
Hebrews, 78. *See* Jews
Hebrides, 60
Hebrus (r.), 139
Hebudes, 258
Hecataeus (of Abdera), 292
Hecataeus (of Miletus), 36, 79, 122, 133, 139, 162, 205–6, 208, 225–6, 239, 259, 272, 292
Hecatompylus, 175
Hedjaz, 88
Heligoland, 53, 55, 273
Heliocles, 188
Heliodorus (author), 244
Heliodorus (explorer), 189
Hellanicus, 255, 271
Hellespont, 174
Helmund (r.), 176, 180
Helvetia, 275
Henry of Portugal, 48, 61, 114, 131
Heracleides (author), 120
Heracleides (explorer), 180
Heracles, 235, 245, 274
Herat, 176, 181, 194, 196
Hercynian Forest, 138, 146–7, 276

INDEX

Hercynian Rock, 151
Herne (i.), 66, 122
Herodotus, 17, 26, 36, 55, 79, 111–12, 114–22, 131–8, 143–4, 161–6, 173, 206–8, 218–21, 225–8, 239, 245, 254, 255, 259, 288
Hertfordshire, 58
Heruli, 242
Hesiod, 39, 111, 160, 205, 242, 254
Hesperides, 235, 242
Hibernians, 44
Hiberus (r.), 270
Hieron, 87
Hierosykaminos, 211
Himalayas, 157; hearsay, 173; Alexander, 178–9, 181; Seleucids, 186–8; Rome, 195, 225–6, 236
Himera, 47
Himilco, 45–7, 63, 70, 115, 255
Himyarites, 75, 91, 193
Hindoos, 12, 71, 102, 158
Hindu Kush, 157; hearsay, 173; Alexander, 176, 178, 236; Graeco-Indians, 188; Rome, 196
Hingol (r.), 81, 180
Hippalus, 96, 104, 114, 227, 255
Hipparchus, 16, 53, 123, 225–6
Hippo Regius, 125
Hiram, 250
Hisn Ghorab, 97
Hissar, 177
Histri, 272
Hittites, 25, 77
Hoggar (m.), 219, 288
Homer, 17, 21, 25–30, 44, 62, 78, 134, 234, 242, 293. *See also Iliad, Odyssey*
Horace, 273
Hosidius Geta, 288
Hottentots, 126
Hudson, 127
Hughli (r.), 187
Hungary, 137, 140
Huns, 189, 198–9
Huron (l.), 186
Huzha, 175

Hwang-ho (r.), 196
Hydaspes (r.), 178, 181
Hydraces, 82
Hydraotis, (r.), 178
Hyères, 35
Hyksos, 235
Hyperboreans, 136, 227, 237
Hypernotians, 237
Hyphasis (r.), 178
Hypsicrates, 190
Hyrcani, -ia, 175, 186
Hyrcanian Sea, 173
Hyspasines, 189

Iambulus, 240–1, 245
Iaxartes, 160, 173, 176–7, 181, 184–5, 199, 236
Iberia (Spain), 150
Iberia (Caucasus), 198–9
Ibrim, 211
Iceland, 51
Ichnussa, 34
Iconium, 12
Ictis, 49, 256
Idrisi, 260
Igharghar (r.), 221, 289
Iliad, 37–8
Iliazzu, 193
Ilisaros, 193
Illyria, -ians, 25, 139
Imaus (m.), 178, 226
India, -ians, 71; early voyages to and from, 73 ff.; Scylax, 78–9; Nearchus, 80 ff.; Eudoxus, 90–1, 123–5, 130; late Ptolemies, 92; Seleucids, 92–3, 184 ff.; Roman Empire, the sea-passage, 94 ff.; Persians, 160 ff.; Alexander, 177 ff.; Megasthenes, 187–8; Graeco-Indians, 188–9; Roman Empire, land connexions, 194 ff. *See also* 231–41, 259, 282, 284, 292
Indus (r.), 55, 77; Scylax, 78–9; Nearchus, 80 ff.; Seleucus, 92, 185; Hippalus, etc., 95–6, 98; Persians,

305

INDEX

Indus – contd.
   160–3, 172–3; Alexander, 177–80; Graeco-Indians, 189; Maes, 196
Inn (r.), 146
Insala, 288
Ionia,-ians, 34, 78–9, 137, 147, 161–2, 174, 252
Ipsus, 161
Iran, -ians, 80, 92, 158, 160, 181, 184–5, 187–8, 194–5, 222, 227, 234
Ireland, 43–4, 49, 54, 59–60, 242, 254, 258
Iron Gate, 137, 142
Iron Gate Pass, 274
Ischia, 34
Isesi, 262
Isidore, 95, 194, 224
Isis, 12
Iskam Keui, 167
Isonzo (r.), 138
Issedones, 160, 166
Issus, 160, 168, 172–4, 278
Ister, 39, 272
Istria, 137, 253
Itineraries, 195, 224, 226
Iurcae, 166
Iviza (i.), 34

Jaddi (c.), 82
Jaigarh, 96, 98
Jalalpur, 178
Japan, 252, 284
Jask (c.), 81, 83
Jason, 37–8, 234, 236, 253
Jataka, 77
Java, 105
Jebel Barkhal, 205
Jebel Moya, 205
Jebel Tak, 194
Jedi, 220–1
Jehoshaphat, 262
Jerma, 219–20, 288
Jews, 12, 101, 184, 192
Jezireh, 168

Jhelum (r.), 178–9
João II, 131
Joliba (r.), 272
Jonah Pass, 168
Jonah Pillar, 174
Joppa, 78
Jordan, 193
Juan Fernandez, 294
Juba, 69, 91, 125, 129, 192, 210, 215, 220, 225, 264, 265, 287
Juby (c.), 64, 66, 69
Judaea, 191
Jumna, 185, 187, 189
Jura, 150, 151, 153
Justinian, 197, 199
Jutland, 43, 53, 61, 143, 146, 255
Juverna, 59

Kabul (r.), 78, 161, 173, 176, 188–9
Kabul (town), 178, 181
Kafiristan, 161
Kagera, 215
Kakulima (m.), 67
Kalahari, 119
Kalami (r.), 81–2
Kalhat, 92
Kampot, 105
Kan Ying, 105
Kandahar, 176, 180–1, 194–6, 224
Kaoshan (pass), 178
Kaptara-Ki, 23
Kara Kum, 185
Karachi, 80, 158
Kara-Kapu, 174
Karikal, 104
Karri, 178
Kars, 172
Karwar Point, 102
Kashaf Rud (r.), 175
Kashgar, 176, 196–7, 239
Kashmir, 161, 178
Kaufmann, 177
Kaviripaddinam, 100, 103
Keftiu, 24, 27

INDEX

Kent, 58, 257
Kenya (m.), 208, 214–15
Kerman, 83, 176
Kertch, 160, 190
Khabur (r.), 168
Kharga (oasis), 207, 210
Khian, 235
Khiva, 167, 177
Khiva Bay, 185–6
Khodja Su (r.), 174
Khor Reiri, 95
Khotan, 189, 196
Khyber Pass, 178
Kieff, 135–6
Kilif, 176–7
Kilimanjaro (m.), 208, 215
Kioga (l.), 215
Kirghiz, 173
Kishm (i.), 86
Kizil Irmak (r.), 161
Kizil Kum, 167
Kizil Uzain (r.), 185
Knoblecher, 215
Koblenz, 276
Kohaito, 209
Kolkai, 100
Kols, 158
Konkan, 102
Kopet Dagh, 167
Korosko, 207
Kosseir, 75, 79, 118, 268
Kottayam, 99
Krapf, 215
Kuban, 166–7
Kuchri (c.), 81
Kuen Lun, 157
Kufara, 218
Kuma (r.), 166
Kunar (r.), 178
Kunduz, 176
Kunish, 168
Kur (r.), 172, 186, 190–1
Kurdistan, 158, 168–9
Kurds, 169
Kuren (r.), 175

Kuria Muria (is.), 75, 264
Kusha, 196
Kushans, 195

La Tène, 153
Laccadives (is.), 99
Laestrygones, 44
Lagash, 76
Laghman, 178
Lahn, 153
Laing, Major, 288
Landai, 178
Land's End, 49, 55, 57, 257
Lanzarote, 125
Laodicea the Burnt, 161
Larissa, 140
Las Bela, 179
Lathyrus. *See* Ptolemy VIII
Lebanon, 22, 193
Leman (l.), 148, 151, 275
Lemnos, 37
Leon, 88; Roman subject, 106
Leonnatus, 81
Libau, 53
Libya, -yan, 32, 33, 64, 111, 112, 120, 126, 129, 202, 207, 208, 210, 219, 243, 259
Libyphoenicians, 63
Lichas, 88
Ligurian, 275
Lions, Gulf of, 28, 34, 35, 147
Lipari, 29
Lippe, 153, 155
Lisbon, 45, 116
Lissus, 142
Lithuania, 136
Livingstone, 12
Livy, 148
Lixus (r.), *Lixite*, 64, 124, 259
Lizas, 259
Lobo, 286
Loh-yang, 196
Loire, 58, 148, 151, 152, 258
Lop Nor, 197
Lualuba, 272

307

INDEX

Lucan, 240
Lucian, 97, 196, 244-5
Lucullus, 190, 191
Lun, 106
Luna, 276
Luristan, 180, 188
Lycaonia, 167
Lydia, 160, 161, 162, 172, 174, 235
Lyon, Captain, 260
Lyon, 148

Maagrammon, 103
Macauley (i.), 67
Macedon, -ia, -ians, 140, 142, 147, 150-1, 272. *See* Philip, Alexander
Machin, Robert, 261
Madagascar, 76, 129-30, 251, 261,
Madeira, 69, 70, 125, 242, 261
Madras, 100
Maeander, 167
Maelstrom, 61
Maenace, 34, 252
Maeonia, -ians, 160, 276
Maeotis, 199; as Aral, 279
Maes, 196, 197
Magan, 76, 262
Magellan, 116, 230, 248
Mago, 219
Magus, 120
Mahaban, 178, 280
Maharrakah, 211
Mahavamsa, 77
Mahlava, 179
Main, 153, 155
Mainland, 51
Makran, 76; Nearchus and Alexander, 81, 176, 181
Makwardam, 213
Malabar, 76; exploration of, 98, 101, 104; pirates of, 97, 99, 266
Malacca, 104, 105
Malaga, 34, 252
Malan, *Malana*, 81
Malay, 73; exploration of, 100, 104, 105, 187, 249

Malayans, Malays, 261, 267
Maldives (is.), 103, 267
Malindi, 116, 131
Malli, 179
Mallorca, 34
Malocello, 260
Maloja Pass, 146
Manetho, 210
Mangalore, 102
maps, 161-2, 195, 211-12, 214, 226, 227, 284, 289-90
Maracanda, 160
Marchand, Capt., 13
Marco Polo, 11, 12, 19, 256, 260, 266, 267
Marcus Agrippa, 152, 195, 225, 226
Marcus Aurelius, 106, 199
Mardi, 175
Mare-Milkers, 237
Mari(a)ba, 283
Mar'ib, 95, 193, 283
Marinus, 214, 225, 227
Maritza, 139, 140
Marmora, Sea of, 42, 252
Maros, 138
Marosh, 138
Marseille, 34, 148
Marsiaba, 193
Masalia, 100
Masira, 90, 264
Masius (m.), 191
Massagetae, 167
Massilia, -ian, 34-5, 44, 47-8, 61-2, 124, 251, 258
Massiliote, 224
Massowa, 209
Maste, 213
Masulipatam, 100, 104
Maternus, Julius, 220, 221
Matienoi, 162
Mauretania, 126, 129. *See* Morocco
Maurusia, 125. *See* Morocco
Maurya, 179, 184, 185, 187, 188
Mazaca, 161
Mazenderan, 175

INDEX

Media, 159, 168, 184, 188, 194, 278
Mediterranean: conditions, 17, 21–2; exploration, 22 ff., 222, 224, 225, 228
Μεγαρε(ῖ)ς, 253
Megasthenes, 12, 92, 174, 187, 238, 282
Mehedia, 63
Mehemet Ali, 215
Mela, 59, 71, 125, 198, 225, 228, 273
Melissa, 259
Melitta, 64
Melukhkha, 76
Melussa, 34
Memel, 53
Memnon, 172
Memphis, 286
Menander, 189
Menelaus, 205
Menon, 168
Menorca, 34
Menuthias, 101, 129
Merawi, 205
Mercator, 230
Mernere, 230
Meroe, exploration of, 203, 205, 206, 208, 209–10, 212, 213–14. *See also* 244, 285, 287
Meropis, 293
Merv, 160, 177, 185, 192, 194, 195, 196
Mervdasht, 175
Meshed, 175
Mesopotamia, -ian, 22–3, 74, 75–6, 163, 184, 190, 191, 193–4, 195, 262
Messina, Straits of, 25, 29, 33
Metellus Celer, 71, 198
Metrodorus, 190
Mexico, 71
Midacritus, 45, 61–2
Miletus, Milesian, 42, 161, 162, 225, 226, 253
Minab, 86
Minaeans, 193
Minnagara, 98; another, 100

Minoans, 23–6, 27–8, 29, 30–1, 32, 39, 159, 249, 251
Minos, 23, 24–5, 26–7
Miran, 196, 197
Mississippi, 272
Mithradates, 189, 190
Modura, 102–3
Moeris (l.), 210
Mogador, 64
Mogan, 190
Moldau, 146
Moldavia, 137
Mong, 178
monsoons, 73, 76, 77, 80, 91; discovery of, 96. *See also* 114, 118, 129, 131, 227
Mt Cenis, 147
Mt Genèvre, 148, 274
Monte Rosa, 147
Morang, 187
Morava, 140
Moravian Gate, 143, 144
Morocco, Moroccan, 61, 63, 68–9, 119, 121, 125, 216, 269
Mosarna, 82
Moscha, 262
Moslem(s), 12, 111
Mosul, 175
Mountain(s) of the Moon, 214, 215, 230
Mozambique, 118, 119, 127, 130
Muchiri, 102. See *Muziris*
Mucklow, 81
Mula Pass, 180
Multan, 179
Mungo Park, 260
Murad Su (r.), 169, 190
Murghab, 175, 176, 194
Murray (r.), 272
Murzuk, 218, 219, 220, 288
Muscat, 76
Mush, 169
Mussendam. *See* Ras Mussendam
Muziris, 97, 99, 102
Mycenae, 38–9

309

INDEX

Myos Hormos, 94, 98, 193
Myriandrus, 168, 174

Nabataeans, 91, 192, 193
Nahapana, 99, 100
Nahr Belik, 191
Nambanos, 99
Nanus, 35
Napata, 205, 206, 211, 212
Naples, 33, 34, 36
Napoleon Bay, 215
Narbo, 152, 258
Narbonne, 148
Nardjiki, 190
Nasamones, 218–19, 221
Nasik, 102
Nassau, 105
Nastesen, 285
Naucratis, 32, 78, 206
Naxos, 33
Nazretabad, 180
Nearchus, 15, 55, 79; his voyage, 80 ff. See also 92–3, 179, 222, 279, 289, 290
Nebuchadrezzar, 77, 160, 236
Necho, 78; sends Phoenicians round Africa, 111–12, 114, 115, 116, 118, 119, 120, 121, 126, 130, 227, 228
Nedunj-Cheliyan, 102
Negrais, 104
Negrana, 193
Nejd, 193
Nejran, 193
Nelcynda, 99
Nepos, 198, 269
Nerbudda, 96, 99
Nero, 98, 129, 144, 195, 211–12, 213, 231
Nervii, 275
Neuchâtel (l.), 153
Neuri, 135
New Zealand, 230, 256, 261
N'Gher, 289
Nias, 105
Nicaea, 178

Nice, 35
Nicobars, 104
Niemen, 289
Niger, 119, 215, 216, 218, 272
Nigeria, 219, 222
Nigir, 221, 289
Nile, 22, 24, 31–2, 62, 78, 79, 88, 89, 101, 111, 115, 118, 121, 122, 179; explorations of, 202 ff. See also 216, 220–1, 222, 236, 259, 272, 285, 286
Nile, Blue, 202, 203, 205, 210, 213, 215
Nile, White, 202, 203, 205, 210, 213–14, 215
Nineveh, 161, 168
Ninguaria, 69
Ninus, 234
Nisibis, 190
Niy, 159
Normans, 26
Norsemen, 18, 253, 291
North Sea, 60
Northmen, 117
Norway, 51–2, 61, 257
Nosala, 81
Νοῦβαι, 210. See Nubia, -ian
Noviomagus, 155
Nubia, -ian, 202, 203, 205, 206, 207, 210, 285
Numenius, 93
Numidia, 125, 220
Nun (c.), 63, 128
Nymwegen, 155
Nysa, 235
Nyses, 286

Oasis, 207
objects, 11 ff
'Ocean', 111, 131, 167, 173, 185, 186, 198, 199, 208, 227, 240, 242
Ochus, 186
Octavian, 142. See Augustus
Oder, 143, 155
Odoric, 11

INDEX

Odysseus, 29, 38, 233, 234, 236, 242, 251, 254
Odyssey, 26, 27, 29–30, 37, 38, 234, 236. *See* Homer
Oeonae (is.), 238
Oestrymnis, 44, 254
Ohind, 178
Oland, 145
Olbia, 34, 135, 136, 137, 271
Olti, 172
Oman, 76, 93; Gulf of, 101
Omana, 93
Onesicritus, 92, 263
Ontario (l.), 186
Ophelas, 260
Ophir, 76, 78, 180, 262
Ophiussa, 45
Opian, 176
Orang (i.), 66
Orca (c.), 50, 256
Orcades, 258
Orissa, 100
Oritae, 81, 180
Orkneys, 60
Ormuz, 86; (i.) 86
Orodes, 192
Ortegal (c.), 45, 48, 54
Ottorocorrae, 239
Oxus, (r.), 160, 163, 173; Alexander, 176, 177; Patrocles, 186, 281; Pompey, 191; Roman Empire, 199
Ozene, 100

Pa-anch, 75, 240
Paethana, 100
Paithan, 100, 102
Palestine, 157, 158, 187, 194. *See* Syria
Palibothra, 179, 187, 241
Palmas (c.), 67, 119, 260
Palmyra, 193
Palmyrenes, 101
Palura, 104
Pamir, 157, 158, 196
Panara, 241
Panchaea, 241, 264

Pandya, 99–100, 103, 187
Pangani, 287
Panjshir, 176
Panticapaeum, 160
Paraetacene (Hissar), 177
παράπλους, 289
Parmenio, 174, 175, 279
Parnasus, 173. *See* Hindu Kush, Himalayas
Paropanisus, 178. *See* Hindu Kush
Parry, 290
Parthia, -ian, 89, 92, 95, 99, 101, 106, 181, 184, 188, 189, 191–2, 194, 195, 196, 198, 283. *See also* Persia, -ian
Pasargadae, 175
Pasitigris, 86, 175, 180
Pasni, 82, 180
Patala, 179
Pataliputra, 179
Patna, 179, 187, 189, 281
Patrocles, 185–6, 281
Patta Kesar, 177
Paul, 12, 15
Paurara, 178
Pausanias, 71, 136, 283
Peake, Major, 202
Pcitholaus, 88
Peking, 197
Pemba, 101, 129–30
Peninsular War, 134
Pentland Firth, 50
Pepi, 203
Perdiccas, 178
Pergamum, 184
Perioeci, 231
Periplus, περίπλους, 44, 91, 224, 225, 226
Periplus, The, 96, 222, 266
Persepolis, 175, 180
Persia, -ian, 73, 78 ff., 89, 93, 95, 107, 112, 120, 121, 139, 158, 159, 160 ff., 184, 205–8, 238, 261, 278, 292. *See also* Parthia, -ian
Persian Gates, 279
Persian Gulf, 23; early voyages, 55,

INDEX

Persian Gulf – contd.
73, 74, 76, 77; Persians, 78, 80; Nearchus, 86; Androsthenes, etc., 86; Ptolemies and Seleucids, 89, 91, 92–3; Rome, 95–6, 101, 106, 107. *See also* 179, 180, 181, 187, 193, 195, 196, 263

Persis, 93, 188, 241
Perth, 59
Pessinus, 161
Petra, 191, 192, 193
Petronius, 211
Peucelaotis, 178
Peutinger Table, 227
Phaeacia, -ian, 29, 236
Pharasmanes, 177
Phasis (r.), 17, 160, 169, 186, 191, 234–5
Phasis (town), 160, 199, 253
Pherecydes, 252
Philadelphus. *See* Ptolemy II
Philae, 209
Philemon, 144, 258, 273
Philip of Macedon, 137, 140, 146, 173, 174
Philip V, 140, 142
Philippeville, 125
Philo, 87, 88
Philonides, 181
Philostratus, 244
Philoxenus, 175
Phocaea, -aean, 34, 35, 36, 42, 45, 133
Phoenician, -ians, 13, 15, 23, 24; in the Mediterranean, 26–9, 30–6; in the Atlantic and along W. Africa, 46, 56–7, 61 ff.; in America (?), 71; in eastern waters, 74–5, 76–7; circumnavigation of Africa, 110, 111–19, 121, 123, 126, 130, 131; in Asia, 158, 159, 160, 172, 178; in Africa, 219. *See also* 222, 227, 228, 249, 250, 251, 254, 259, 261, 276
Φοινικοῦσσαι, etc., 252
Phounoi, 189
Phrada, 176, 181

Phrygia, -ian, 160, 162, 174, 276
Phrynoi, 189
Physcon. *See* Ptolemy VII
Pir-sar, 178, 280
Pillar of Jonah, 174
Pillars of Heracles, 112, 120, 127, 129, 255. *See* Gibraltar
Pindar, 235, 237, 242
Pisidia, -ian, 167, 174
Pithecussa, 34
Pityussa, 34
Pius, 199. *See* Antoninus Pius
Pizarro, 13, 19
Platea (i.), 32, 33
Plato, 72, 242, 243, 293
Pliny, 45, 46, 53, 59, 61, 68, 71, 91, 193, 198, 211, 212, 213, 225, 228, 229, 247, 256, 257, 258, 260
Plutarch, 54, 70, 242
Po (r.), 36, 147, 150, 151, 274
Poduce, 100
Poland, 135, 136
Pole Star, 48, 115
Polybius, 48, 49, 52, 53, 54, 63, 68, 112, 114, 123, 148, 189, 225, 227, 260, 273, 275
Polynesians, 261
Pompey, 148, 190, 191
Pondicherry, 100
Pontus, 184, 188, 189, 283
Porakad, 99
Portugal, Portuguese, 12, 13, 30, 131, 132, 215, 252, 258, 275
Porus, 178
Poseidonius, 90, 100, 112, 123, 126, 127, 128, 225, 258, 275
Posnania, 145, 271, 273
Potters' Mart, 167
Premnis, 211, 212
Primis, 212
Pripet, 136
Priscus, 17
Proconnesus, 160
Procopius, 242, 293
Prometheus, 236

312

INDEX

Prophthasia, 176
Protis, 35
Prussia, 274
Psammetichus I, 78, 111, 205, 208
Psammetichus II, 206; Psammis 114,
Psammis. *See* Psammetichus II.
Psebo (l.), 89, 209, 210, 214
Pselchis, 211, 212
Ptolemais of the Hunts, 209
Ptolemies, 87 ff., 123–4, 208–11
Ptolemy, as Geographer, 15, 16, 51, 60, 61, 68, 103, 105, 107, 130, 131, 136, 139, 145, 155, 194, 196, 213, 214, 215, 221, 225, 226, 227, 228, 229, 230, 231, 239, 247, 273, 276, 287
Ptolemy I, 87–8, 279
Ptolemy II Philadelphus, 87–9, 187, 208, 209, 272
Ptolemy III, 205
Ptolemy IV, 88, 209
Ptolemy V, 209
Ptolemy VI, 211
Ptolemy VII Euergetes II Physcon, 90, 123, 128
Ptolemy VIII Lathyrus, 124, 128
Ptolemy XI Auletes, 92
Pudukottai, 103, 266
Puenta de Yfach, 34
Punic. *See* Carthage, Carthaginian
Punjab, 79, 98, 178, 185, 189
Punt, Puoni, 75–6, 203, 233, 262
Pura, 180
Purali, 80
Puranas, 77
Purple Is., 69
Puru, 178
Puteoli, 124
Pygmies, 64, 205
Pylae, 168
Pyramus, 168
Pyrene, 138
Pyrenees, 17, 133
Pythagoras, 88, 206, 277
Pythangelus, 88

Pytheas, 12, 13; explorations, 47–56. *See also* 57–8, 60–1, 68, 117, 153, 227, 256, 257, 273

Quadi, 276
Quinctilius Varus, 155
Quorra, 272

Raab, 143
Rachias, 97
Rajapur, 96
Ramses II, 235
Ramses III, 205
Ramu, 105
Rangoon, 105
Ras Benas, 88
Ras el-Had, 92
Ras Fartak, 96
Ras Mussendam, 83, 86, 89, 91, 93
Ras Shamal Bunder, 82
Ravi, 178, 179
Rebmann, 215
di Recco, 261
Red Tower Pass, 274
Red Sea, 73 ff., 111–12, 115, 118, 119, 122, 129, 202, 203, 209, 210, 211. *See also Erythraean Sea*
Redesiya, 98
Redskins, 71–2
Regulus, 240
Resht, 185
results, 222 ff.
Rha (r.), 199
Rhadman, 193
Rhagae, 175, 281
Rhambacia, 180
Rhammanitae, 193
Rhapta, 214, 287
Rhine, 44, 52; Rome, 58, 60, 139, 143, 146, 147, 148, 152–3, 155. *See also* 222, 254, 257, 276
Rhipaean Mountains, 61, 135, 136, 227, 230, 237, 271
Rhoda, 151
Rhodes, 137, 226

313

INDEX

Rhodesia, -ian, 76, 118, 262
Rhodope, 140
Rhodopha, 187
Rhône, 34, 147, 148, 150, 274, 275; Perte du, 275
Rig Veda, 77
Riga, 145
Rio de Ouro, 64
Rio Roosevelt, 271
Rion, 160, 169, 191
Ripon Falls, 215
Roca (c.), 45
Rome, Romans, 12, 13, 14, 17, 19, 26, 28; in the Mediterranean, 36–7; in the Atlantic and Britain, 55 ff., 68, 69; in eastern waters, 94 ff.; circumnavigation of Africa, 117, 123, 125, 128–31; in Europe, 132 ff., 136, 138–9, 140, 142–8, 151, 152–3, 155, 156; in Asia, 157, 174, 188, 189 ff.; in Africa, 205, 211 ff., 219 ff. *See also* 225–7, 238, 239, 240, 258, 273, 274
Ross, 290
Royal Road, Mauryan, 161
Royal Road, Persian, 187
Rufu, 287
Rügen, 53
Ruhr (r.), 155, 254
Rumania, 140
Rusicada, 125
Russia, -ian: earliest visits, via Black Sea, 38, 39; exploration of southwest, 134–7; round the Caspian, 163, 166. *See also* 222, 228, 271
Ruwenzora, 208, 215, 287

Saale, 146, 153
Saba, 79
Sabaeans, 87, 89, 95, 105, 193, 264, 283
Sabos, 193
Sacae, 161, 194. *See* Sakas
Sada, 104
Sahara, 61, 64, 68, 110, 121; exploration of, 216 ff. *See also* 222, 228, 248, 270, 287, 288, 289

Sahuri, 75
St Bernard, Great, 147, 148, 150
St Bernard, Little, 147, 148, 150
St Brandan's Is., 293
St Helena Bay, 118
St Michael's Mount, 47, 256 (Ictis)
St Thomas, 98, 107
St Vincent, Cape, 46, 255
Saka, 100, 102, 188, 194
Sala, 129
Sallee, 129
Salween (r.), 106
Samarkand, 160, 176, 196
Sambre, 275
Samos, Samian, 32, 34, 121
San Martin, 13
Sandanes, 43
Sandoway, 104, 105
Sandrocottus, 179. *See* Chandragupta
Sangala, 178
Sangarius (r.), 174
Santhals, 158
Saône, 148
Saracens, 131
Sardanapalus, 160
Sardinia, 25, 28, 34, 35
Sardis, 161, 163, 167, 174
Sargasso Sea, 46
Sargon of Akkad, 23
Sargon II, 159
Sarikol, 196
Sarmatians, 136
Sarus, 168
Sassanids, 198
Sataspes, 119–21, 127
Satyrus, 88
Sauromatae, 166
Sauu, 75
Save (r.), 138, 143, 144, 272
Scadinavia, 155
Scanderoon, Gulf of, 160, 168
Scandinavia, -ian, 43, 51, 61, 155, 228, 242
Schubin, 271
Sciapodes, 239

314

INDEX

Scillies, 47, 255
Scipio Aemilianus, 68, 152, 258
Scotland, 50, 52, 60, 228, 258
Scott, Capt., 19, 247
Scutari, 172
Scylax, 39, 78–9, 116, 161, 227, 228, 277, 290
'Scylax', 78–9, 120, 122, 138, 224, 252, 260
Scymnus (pseudo-), 136, 147
Scythia, -ian, 52, 53; = S.W. Russia, exploration, 134–7, 139, 144, 173; east of the Don, exploration, 161, 166, 167, 173, 177, 192, 199. *See also* 235, 271, 292
Scyths, Royal, 167
Sebni, 203
Seine, 58, 148, 151, 256
Seistan, 176, 180
Seleucia, 93, 184, 185, 194
Seleucids, 89, 92–3, 184 ff.
Seleucus I, 184, 185, 186–7, 281
Sembritae, 208
Semiramis, 234
Semite, 158
Seneca, 195, 213, 231
Senegal, 121, 269. *See* Senegal (r.)
Senegal (r.), 61, 62, 66, 68, 269, 289
Senegambia, 62, 270
Senia, 142
Senmout, 233
Sennacherib, 77
Sennar, 208
Senwosri I, 203
Senwosri III, 235
Septimius Flaccus, 220
Septimius Severus, 12, 226
Sera, 196
Serbia, 142
Seres, 189, 192, 194, 195, 196, 239, 267, 282. *See* China, -ese
Seret, 137
Sertorius, 70
Sesostris, 235, 262
Seti I, 226, 235

Sevo, Mount, 61
Shackleton, 247
Shady-feet, 239
Shahderi, 178
Shahkot, 178
Shahrud, 175
Shantung, 76
Sherbro Sound, 67, 120
Shetland(s), 51, 60, 61
ships, 14–15, 247; Egyptian, 22, 23–4, 117, 202–3; N. European, 124; of Gades, 123; Indian, 254; Minoan, 23–5; Phoenician, 27–8
Siam, -ese, 105, 158
Siberia, 157, 197
Sibiru, 105
Sicily, 25, 26, 28, 33, 34, 35, 228
Side, 174
Sidon, 27
Sierra Leone, 67–8, 120, 222, 260
Sierra Morena, 133
Sigynni, 275
Sikkim, 187
Silesia, -ian, 144, 145
Simmias, 88, 264
Simonides, 209
Simplon, 275
Simus, 35
Sinae, 100, 106
Sinai, 262
Sind, 78
Sindbad, 233, 251
Singanfu, 196, 222
Sinope, 38–9, 136, 160, 172, 253
Siriptolemaeos, 102
Siscia, 142
Siszeg, 142
Siwah, 206, 207, 208
Skager (c.), 60–1
Slavs, proto-, 135
Smolensk, 135
Snefru, 22, 203
Sobat (r.), 202, 208, 209, 213
Socotra (i.), 75, 76, 89, 92, 101, 234, 240, 242, 264

315

INDEX

Sofala, 76, 107, 130, 131, 262, 269
Sogdiana, 176, 177, 188; town, 177
Solinus, 229
Soloeis: as Cantin (c.), 63; as Spartel (c.), 120
Solomon, 76
Solon, 12, 206
Somali, -land, 62, 75-6, 79-80, 87, 88, 89, 91, 98, 100, 123, 129, 205, 210, 216, 262, 285, 292
Somme, 152
Son (r.), 282
Sonmiani Bay, 95
Sopatma, 100
Sophagasenus, 188
Sophon, 98, 266
Σουίωνες, 274
South Foreland, 55
Southern Horn (E. Africa), 91
Southern Horn (W. Africa), 67
Spain, Spanish, 23, 25, 26, 28, 34, 35, 36, 38, 45, 46, 47, 48, 51, 54, 57, 58, 70, 72, 90, 111, 122, 123, 124, 125, 127, 128, 129, 130, 132-4 (explored), 151, 224, 231, 250, 251, 256, 258, 270
Spanish Indies, 261
Sparta, 161
Spartel (c.), 120
Speke, 186
Spina, 36
Splügen Pass, 146
Sri Pulumayi, 102
Stadiasmus Maris Magni, 224, 248
Stanley, 19, 215, 230
Statius, 195
Statius Sebosus, 260
Stein, Sir Aurel, 197
Stewart (j.), 256
Stone Tower, 196
Strabo, 26, 48, 51, 54, 56, 57, 58, 59, 90, 94, 123-4, 125, 126, 127, 129, 136, 139, 174, 186, 189, 191, 193, 194, 205, 210, 211, 225, 227, 231, 249, 272, 273, 274, 276, 285, *passim* in the notes
Struma, 139, 140, 272
Strymon, 139. *See* Struma
Styria, -ian, 144, 273
Subhagasena, 188
Sudan, -ese, 62, 202, 203, 216, 219, 220, 222
Suetonius Paulinus, 220-1, 288
Suevi, 198
Suez, Gulf of, 86-7, 90, 157, 161, 192, 222, 269
Suez (town), 78, 79, 88, 89, 98, 118
Suez, Peninsula, 78
Suiones, 145, 155
Sukhum Kaleh, 160
Sumatra, 73, 76, 104, 105, 260
Sumer, -*ian*, 74, 76, 77
Superga, 147
Sur, 92
Suram, 191
Susa, 161, 162, 175, 180, 181, 283
Susiana, 185
Sutlej, 179, 187
Swat, 178
Sweden, Swedish, 61, 145
Sweserenre Khian, 235
Swiss, Switzerland, 147, 151, 155
Syene, 202, 203, 207-8, 210, 211
Syllaeus, 192
Syr Darya (r.), 160. *See* Iaxartes
Syracusan, 36, 47, 48
Syria, -ian, 22, 24, 101, 102, 106, 118, 159, 160, 162, 184, 185, 191, 192, 193, 194, 197
Syrtis Major, 32

Tabaristan, 175
Tabis, 281
Tacitus, 59, 145, 155, 276
Tagara, 100
Tagus, 133
Tahiti, 12
Taloi (ms.), 180
Tamil, 97, 99-100, 102-3, 187, 282

INDEX

Tamluk, 100
Tanais (r.), 52, 53, 134, 157, 186, 236, 278
Tanais (town), 186, 253
Tanjier, 288
Taprobane, 97
Tapuri, 175
Tarentum, 147, 151, 258
Tarnak, 176
Tarshish, 26, 28. *See* Tartessus
Tarsus, 168, 174
Tartessus, -ian, 25, 34, 35, 44, 45, 46, 47, 63, 110, 133, 250, 255, 292, 294
Tartessus (r.), 270
Tashkurgan (Aornos), 176
Tashkurgan (in Sarikol), 196
Tasman, 230
Ta-ts'in, 106
Taurus (m.), 161, 163, 168, 173, 181, 184, 188, 189–90, 278, 281
Taxila, 178, 189
Taxiles, 178
Tearchus, 235
Teima, 159
Tejend, 194
Teleboas, 169
Telephanes, 162
Tell Gullel, 294
Temala, 104
Teneriffe, 69
Tensift, 64
Ten Thousand, 17, 139, 167–9, 172
Ter, 100
Terra Australis, 230–1, 291
Teutones, 153
Texel, 53
Thade, 104
Thales, 208
Thanet, 256
Thapsacus, 168, 174, 281
Thar, 164, 177
Thasos, 137, 142
Thatung, 105
Thebes (Boeotia), 26–7

Thebes (Egypt), 78, 286. *See also* 206, 207
Theches (m.), 172
Theinni, 106
Theiss, 138
Theophanes, 190, 283
Theophilus, 101
Theophrastus, 77
Theopompus, 136, 138, 140
Thera, 32
Thessaly, -ian, 37, 38, 140, 142
Thinae (city), 106
Thinae (people), 100
Thomas, St, 98, 107, 256
Thothmes III, 24, 159
Thrace, -ians, 42, 139, 235, 237
Thucydides, 26, 33
Thule, 51 ff., 54, 61, 231, 244, 256
Thuringia, 147
Thymiaterium, 63
Thyssagetae, 166
Tian Shan (ms.), 157, 176, 196
Tiastanes, 102
Tiberius, 60, 94, 95, 138–9, 145, 146, 153, 198, 211
Tibesti, 218–19
Tibet, -an, 99, 187, 238, 292
Tibisis (r.), 138
Tierra del Fuego, 230
Tiglath Pileser I, 158
Tiglath Pileser III, 159
Tigranes, 190
Tigranocerta, 190, 283
Tigre, 203
Tigris (r.), 77, 93, 161, 163, 174–5, 180, 184, 190, 195
Timaeus, 48–9, 138, 250, 252, 256, 257, 260, 272, 274, 275
Timagetus = Timosthenes, 272
Timbuctu, 218, 221, 288
Timosthenes, 272
Tin Islands, 45, 47, 56, 255
Tin Land, 49
Tirhaqa, 236

317

INDEX

Tissaphernes, 168
Tomerus, 81
Tongking, 105, 116, 222
Torres, 256
Toscanelli, 293
Trajan, 101, 130, 145, 193, 195, 199, 274
Transylvania, -ian, 137, 140, 144, 145, 147, 274
Trapezus, 160, 172
Trebenishte, 272
Trepizond, 172
Trichinopoly, 100
Trieste, 143
Tripoli, 29, 220; Bay of, 216, 218
Trog(l)odytes, 64, 88, 92, 219
Trojan War, 27-8, 37-8, 39, 50
Troy, 27, 38, 39, 159
Trutulum, 59
Tsana (l.), 209, 213, 214
Tsanpo, 272
Tuareg, 220, 289
Tunisia, 216, 219, 240, 294
Turcomans, 166, 177
Turin, 147
Turkestan, 39, 160, 177, 189
Turks, 199
Tyana, 167
Tylos (i.), 86
Tyre, -ian, 27, 34, 112, 163, 174, 250
Tyrolese, 146, 147
Tyrrenia, 243. See Etruria

Ujjain, 100, 102
Umbrian, 138, 147
Ur, 74, 76-7
Uraiyur, 100
Urals, 38, 166, 237
Uralsk, 167, 173
Uruk, 234
Ushant (i), 49, 55, 256
Ust Urt, 167
Utica, 28, 56
Utopias, 234, 236 ff.

Uttarakuru, 239
Uxian (Uxii), 175
Uzboi, 185

Vaes, 261
Vakhsh, 177
Van (l.), 190, 195
Vancouver, 256
Vardar, 139, 140
Vasco da Gama, 107, 126, 131, 247
Veddahs, 158
Veneti, 43
Venetia, -ian, 25, 131
Venice, 143, 144
Verde (c.), 61, 64, 66, 67, 121, 131
Vespasian, 199
Vetterfelde, 271
Victoria Nyanza, 186, 214, 215, 222
Vienna, 144
Viking, 43
Vilivayacura, 102
Vindusara, 187
Vistula, 53, 143
Vitosh (m.), 140
Vivaldo, Vivaldi brothers, 128, 260
Volcae, 150
Volga (r.), 137, 166, 173, 186, 198, 277
Volusenus, 57

Wadi Draa, 124. See Draa
Wadi Halfa, 205
Wallachia, 144
Walmer shingle, 58
Wawat, 233
Wellesley, 152
Weser, 153
West Horn, 66
Wight, Isle of, 256
William of Rubrouck, 11, 270, 271
Wisby, 273
Wolseley, Sir G., 13
Wonderlands, 236 ff.
Wood, Sir Evelyn, 13

318

INDEX

Xanthus, 162
Xenocles, 181
Xenophon, 17, 139, 168
Xerxes, 120, 121, 162

Yam, 203
Yang-tsze-Kiang (r.), 196
Yarkand, 196
Yavana, 102
Yemen, 75, 79, 87, 89, 193
Yolof, 260
Yueh-chi, 188
Yusafzai, 178

Za Hakale, 100
Zab, Greater (r.), 161, 168, 175
Zab, Lesser (r.), 161, 168
Zabae, 105
Zadracarta, 175
Zagros, 278
Zaire, 272
Zanzibar, 101, 129–30, 215
Zarcho, 261
Zariaspa, 177, 279
Zebirget (i.), 87
Zengg, 142
Zeno brothers, 290
Zenzibar, 256, 291
Zeugma, 161, 191, 195, 224
Zimbabwe, 130, 262
Zoscales, 100
Zuyder, 60